AF389426

The Role of Extracellular Matrix in Cancer Development and Progression

The Role of Extracellular Matrix in Cancer Development and Progression

Editors

George Tzanakakis
Dragana Nikitovic

MDPI • Basel • Beijing • Wuhan • Barcelona • Belgrade • Manchester • Tokyo • Cluj • Tianjin

Editors
George Tzanakakis
The University of Crete
Greece

Dragana Nikitovic
The University of Crete
Greece

Editorial Office
MDPI
St. Alban-Anlage 66
4052 Basel, Switzerland

This is a reprint of articles from the Special Issue published online in the open access journal *Biomolecules* (ISSN 2218-273X) (available at: https://www.mdpi.com/journal/biomolecules/special_issues/Extracellular_Matrix_in_Cancer).

For citation purposes, cite each article independently as indicated on the article page online and as indicated below:

LastName, A.A.; LastName, B.B.; LastName, C.C. Article Title. *Journal Name* **Year**, *Volume Number*, Page Range.

ISBN 978-3-0365-3405-3 (Hbk)
ISBN 978-3-0365-3406-0 (PDF)

Contents

About the Editors

George Tzanakakis is a professor of Anatomy—Histology at the Medical School of the University of Crete (UOC), Heraklion, Greece. Dr. Tzanakakis obtained his medical degree from the University of Athens in 1980 and the titles of specialist in pathological anatomy in 1985 (Greece) and clinical cytology in 1995 (Sweden). He graduated with a Doctor of Medical Sciences degree from the Medical Department of the University of Patras in 1989. His research focuses on tumor biology and its implications in the diagnosis and therapy of cancer, work that has resulted in approximately 200 original publications and book chapters. During his long academic career, Dr. Tzanakakis contributed significantly to the teaching, research, administration, and promotion of the UOC.

Dragana Nikitovic is currently an associate professor and Head of the Laboratory of Histology-Embryology at the Medical School of the University of Crete (UOC). She earned a Doctor of Medical Sciences degree from the UOC, Greece and has carried out postgraduate research at the Queen Mary and Westfield College, London, UK, and the Karolinska Institute, Stockholm, Sweden. Her scientific interests include the effects of proteoglycans (PGs) and glycosaminoglycans (GAGs) in disease progression, including cancer and inflammation, and their mechanisms of action. She serves as an editor for several journals and is an active member of scientific societies. She has authored more than 100 original publications and book chapters.

 *biomolecules*

Editorial

Preface of the Special Issue on the Role of Extracellular Matrix in Development and Cancer Progression

George Tzanakakis * and Dragana Nikitovic *

Laboratory of Histology-Embryology, School of Medicine, University of Crete, 71003 Heraklion, Greece
* Correspondence: tzanakak@uoc.gr (G.T.); nikitovic@uoc.gr (D.N.); Tel.: +30-2810394557 (D.N.)

Citation: Tzanakakis, G.; Nikitovic, D. Preface of the Special Issue on the Role of Extracellular Matrix in Development and Cancer Progression. *Biomolecules* 2022, 12, 362. https://doi.org/10.3390/biom12030362

Received: 21 February 2022
Accepted: 23 February 2022
Published: 25 February 2022

Publisher's Note: MDPI stays neutral with regard to jurisdictional claims in published maps and institutional affiliations.

The consecutive steps of tumor growth, local invasion, intravasation, extravasation, invasion of anatomically distant sites, and immunosuppression are obligatorily perpetrated through specific interactions of the tumor cells with their microenvironment. During cancer progression, significant changes can be observed in the properties of extracellular matrix (ECM) components, which deregulate the behavior of stromal cells, promote tumor-associated angiogenesis and inflammation, and lead to the generation of a tumorigenic microenvironment. Thus, mediators originating from the ECM have a vital effect on all cellular functions implicated in cancer development and progression. ECM components, including fibrillar proteins, proteoglycans (PGs), and glycosaminoglycans (GAGs), modulate the bioavailability of active mediators, control the stiffness of the stroma immediately correlated to cancer cell invasion, and regulate the metastatic processes and angiogenesis. Various ECM-derived components can also modulate the immune response and affect the response to therapy and thus need to be taken into account when designing efficient anticancer treatment. Indeed, the ECM effectors regulate processes correlated to chemoresistance, including autophagy and apoptosis.

The Special Issue of *Biomolecules* entitled "Role of Extracellular Matrix in Development and Cancer Progression" focuses on recent findings in the structural and functional characterization of ECM components and how they relate to the processes involved in cancer pathogenesis and response to therapy. The Special Issue focusing on the crosstalk between the ECM and cellular processes features original research and review articles. These articles address several relevant topics, including HA roles in cancer, MMPs regulation of carcinoma progression, and proteoglycans mediation of cancer growth, invasion, angiogenesis, and response to therapy. Furthermore, the role of ECM effectors as diagnostic tools and targets is elaborated.

1. Hyaluronan in Breast Cancer

HA, the unique nonsulfated GAG, and its CD44 and RHAMM receptors have been strongly implicated in cancer progression. HA, as elaborated by Tolg et al., also has a major role in tissue injury by sequentially promoting and then suppressing inflammation and fibrosis, a duality that is featured and regulated in wound repair [1]. Tolg et al. thus focus on the hijacking of the essential response-to-injury functions of HA by tumor cells to promote their invasion and avoidance of immune detection. This is followed by a discussion of how HA metabolism is deregulated in malignant progression and how targeting HA might be used to better manage breast cancer progression.

While HA is the only GAG not normally substituted with sulfate groups, several studies suggest that sulfated hyaluronan (sHA) exhibits promising antitumor results. Koutsakis et al. show that sHA fragments attenuate breast cancer cell proliferation, migration, and invasion while increasing adhesion on collagen type I in a manner related to their estrogen receptor (ER) status [2]. Furthermore, sHA modulates the expression of epithelial-to-mesenchymal transition (EMT) markers and downregulate matrix remodeling enzymes such as the matrix metalloproteinases (MMPs). Since sHA exhibits a stronger effect on the breast cancer cell properties than the nonsulfated counterpart, a deeper understanding of the mechanism of its action could contribute to the development of novel therapeutic strategies.

2. MMPs Regulate Carcinoma Progression

MMPs are endopeptidases characterized by a broad range of substrate specificities and are important in ECM remodeling. However, as MMPs also have a high affinity for membrane receptors, ligands, and signaling molecules, these enzymes can be defined as cell signal regulators. In the skin, the expression of MMPs is increased in response to various stimuli, including ultraviolet (UV) radiation, one of the main factors involved in the development of basal cell carcinoma (BCC). Tampa et al. discuss the role of MMPs in the pathogenesis and evolution of BCC, as molecules involved in tumor aggressiveness and risk of recurrence, to offer an updated perspective on this field [3].

During tumor progression, elastin fragments, including a nonapeptide, AG-9, are released in the tumor microenvironment. AG-9 affects tongue squamous cell carcinoma invasive properties. Bretaudeau et al. demonstrated that AG-9 stimulates tongue squamous carcinoma cell invasion, increasing MMP-2 secretion and MT1-MMP expression [4]. The green-tea-derived polyphenol, (−)-epigallocatechin-3-gallate (EGCG), abolished AG-9-induced invasion, MMP-2 secretion, and MT1-MMP expression. The feasibility of utilizing matrix-derived signaling axes to develop novel anticancer therapeutics is underlined here.

3. Proteoglycans Regulate Cancer Growth, Invasion, Angiogenesis, and Response to Therapy

Syndecans, a family of transmembrane heparan sulfate proteoglycans (HSPGs), are involved in key biological processes, such as cell proliferation, adhesion, and migration, supporting homeostasis. Still, their expression/activities are often deregulated in cancer. In addition to their roles as transmembrane PGs, syndecans' extracellular domain can be "shedded" from the cell surface by the action of MMPS, converting them into soluble molecules capable of binding distant cell and ECM molecules. Sousa Onyeisi et al. discuss the input of the syndecan-4 member in the pathogenesis of various cancer types as its expression is commonly aberrant [5]. These authors elaborate on the point that anticancer drugs modulate syndecan-4 expression. Therefore the "take-home" line is that the syndecan-4 emerges as an important target for cancer therapy and diagnosis.

Bertriou et al. focus on the role of syndecan-1 and -2 in the progression of pancreatic ductal adenocarcinoma (PDAC), a fatal disease with a poor prognosis [6]. Furthermore, in this review, the authors explore the potential of syndecans as therapeutic targets for this devastating disease.

Lumican, a small leucine-rich (SLRP) PGs member, is a secreted PG. Notably, this PG is involved in cellular processes associated with tumorigeneses, such as EMT, cellular proliferation, migration, invasion, and adhesion. Furthermore, lumican is expressed in various cancer tissues and is reported to have a positive or negative correlation with tumor progression. Giatagana et al. discuss the effects of lumican on cancer cell growth, invasion, motility, and metastasis, together with the repercussions on autophagy and apoptosis [7]. Finally, in light of the available data, the authors propose novel roles for lumican as a cancer prognosis marker, chemoresistance regulator, and cancer therapy target.

4. ECM Effectors as Diagnostic Tools and Target

In addition to cell signaling functions, the ECM molecules determine the mechanical properties of the tumor and stroma tissues. Indeed, the stroma's mechanical properties, e.g., stiffness, are directly correlated to the pathogenesis of cancer. Ahmad et al. note that the PDAC is characterized by a dense, fibrotic stroma composed of ECM proteins that poses a significant physical barrier that is immunosuppressive and obstructs penetration of cytotoxic chemotherapy agents into the tumor microenvironment (TME) [8]. These authors discuss the significant contribution of fibrosis to the pathogenesis of pancreatic cancer, with a focus on the crosstalk between immune cells and pancreatic stellate cells that contribute to

ECM deposition. Furthermore, the therapeutic strategies that target the stroma and hinder immune cell promotion of fibrogenesis, which led to mixed results, are summarized.

Extracellular vesicles (EVs), comprising exosomes, microvesicles, and apoptotic bodies, are released by all cells into the ECM and body fluids. They play important roles in intercellular communication and matrix remodeling in various pathological conditions. Jamadi et al. characterized the tumor heterogeneity and extracellular vesicle diversity in pleural effusion exosomes as diagnostic or prognostic markers for malignant pleural mesothelioma (MPM) [9]. The corresponding ratios of mesothelin, galectin-1, osteopontin, and VEGF were higher in MPM effusions exosomes than those in the benign group. Therefore, these authors suggest that relevant diagnostic markers can be recovered from exosomes.

5. Conclusions

The articles featured within this Special Issue freshly highlight the multifaceted roles of ECM molecules in cancer pathogenesis. In addition to their roles as signaling mediators, ECM molecules define the mechanical aspects of the formed tumor tissue directly correlated to the disease progression. We thank the authors for their timely contributions and hope that the "Role of Extracellular Matrix in Development and Cancer Progression" Special Issue will form an incentive for further focused research.

Funding: D.N. was partially funded by the Research Committee of University of Crete (ELKE), grant number (KA:10028).

Institutional Review Board Statement: Not applicable.

Informed Consent Statement: Not applicable.

Data Availability Statement: Not applicable.

Acknowledgments: This article is part of the Innogly Cost action initiative.

Conflicts of Interest: The authors declare no conflict of interest.

References

1. Tolg, C.; Messam, B.J.-A.; McCarthy, J.B.; Nelson, A.C.; Turley, E.A. Hyaluronan Functions in Wound Repair That Are Captured to Fuel Breast Cancer Progression. *Biomolecules* **2021**, *11*, 1551. [CrossRef] [PubMed]
2. Koutsakis, C.; Tavianatou, A.-G.; Kokoretsis, D.; Baroutas, G.; Karamanos, N.K. Sulfated Hyaluronan Modulates the Functional Properties and Matrix Effectors Expression of Breast Cancer Cells with Different Estrogen Receptor Status. *Biomolecules* **2021**, *11*, 1916. [CrossRef]
3. Tampa, M.; Georgescu, S.R.; Mitran, M.I.; Mitran, C.I.; Matei, C.; Caruntu, A.; Scheau, C.; Nicolae, I.; Matei, A.; Caruntu, C.; et al. Current Perspectives on the Role of Matrix Metalloproteinases in the Pathogenesis of Basal Cell Carcinoma. *Biomolecules* **2021**, *11*, 903. [CrossRef]
4. Bretaudeau, C.; Baud, S.; Dupont-Deshorgue, A.; Cousin, R.; Brassart, B.; Brassart-Pasco, S. AG-9, an Elastin-Derived Peptide, Increases In Vitro Oral Tongue Carcinoma Cell Invasion, through an Increase in MMP-2 Secretion and MT1-MMP Expression, in a RPSA-Dependent Manner. *Biomolecules* **2021**, *11*, 39. [CrossRef]
5. Onyeisi, J.O.S.; Lopes, C.C.; Götte, M. Syndecan-4 as a Pathogenesis Factor and Therapeutic Target in Cancer. *Biomolecules* **2021**, *11*, 503. [CrossRef] [PubMed]
6. Betriu, N.; Bertran-Mas, J.; Andreeva, A.; Semino, C.E. Syndecans and Pancreatic Ductal Adenocarcinoma. *Biomolecules* **2021**, *11*, 349. [CrossRef] [PubMed]
7. Giatagana, E.-M.; Berdiaki, A.; Tsatsakis, A.; Tzanakakis, G.N.; Nikitovic, D. Lumican in Carcinogenesis—Revisited. *Biomolecules* **2021**, *11*, 1319. [CrossRef] [PubMed]
8. Ahmad, R.S.; Eubank, T.D.; Lukomski, S.; Boone, B.A. Immune Cell Modulation of the Extracellular Matrix Contributes to the Pathogenesis of Pancreatic Cancer. *Biomolecules* **2021**, *11*, 901. [CrossRef] [PubMed]
9. Javadi, J.; Görgens, A.; Vanky, H.; Gupta, D.; Hjerpe, A.; EL-Andaloussi, S.; Hagey, D.; Dobra, K. Diagnostic and Prognostic Utility of the Extracellular Vesicles Subpopulations Present in Pleural Effusion. *Biomolecules* **2021**, *11*, 1606. [CrossRef] [PubMed]

biomolecules

Review

Hyaluronan Functions in Wound Repair That Are Captured to Fuel Breast Cancer Progression

Cornelia Tolg [1], Britney Jodi-Ann Messam [2], James Benjamin McCarthy [3], Andrew Cook Nelson [4] and Eva Ann Turley [5,*]

[1] London Regional Cancer Program, Lawson Health Research Institute, London, ON N6A 5W9, Canada; Cornelia.Toelg@lhsc.on.ca
[2] Department Biochemistry, Schulich School of Medicine and Dentistry, Western University, London, ON N6A 5C1, Canada; bmessam@uwo.ca
[3] Department of Laboratory Medicine and Pathology, University of Minnesota, Minneapolis, MN 55455, USA; mccar001@umn.edu
[4] Masonic Cancer Center, University of Minnesota, Minneapolis, MN 55455, USA; nels2055@umn.edu
[5] London Regional Cancer Program, Lawson Health Research Institute, Department Oncology, Biochemistry and Surgery, Schulich School of Medicine and Dentistry, Western University, London, ON N6A 5C1, Canada
* Correspondence: Eva.Turley@lhsc.on.ca

Abstract: Signaling from an actively remodeling extracellular matrix (ECM) has emerged as a critical factor in regulating both the repair of tissue injuries and the progression of diseases such as metastatic cancer. Hyaluronan (HA) is a major component of the ECM that normally functions in tissue injury to sequentially promote then suppress inflammation and fibrosis, a duality in which is featured, and regulated in, wound repair. These essential response-to-injury functions of HA in the microenvironment are hijacked by tumor cells for invasion and avoidance of immune detection. In this review, we first discuss the numerous size-dependent functions of HA and emphasize the multifunctional nature of two of its receptors (CD44 and RHAMM) in regulating the signaling duality of HA in excisional wound healing. This is followed by a discussion of how HA metabolism is de-regulated in malignant progression and how targeting HA might be used to better manage breast cancer progression.

Keywords: hyaluronan; RHAMM; CD44; wound repair; breast cancer

Citation: Tolg, C.; Messam, B.J.-A.; McCarthy, J.B.; Nelson, A.C.; Turley, E.A. Hyaluronan Functions in Wound Repair That Are Captured to Fuel Breast Cancer Progression. *Biomolecules* **2021**, *11*, 1551. https://doi.org/10.3390/biom11111551

Academic Editors: Dragana Nikitovic and George Tzanakakis

Received: 15 September 2021
Accepted: 14 October 2021
Published: 20 October 2021

Publisher's Note: MDPI stays neutral with regard to jurisdictional claims in published maps and institutional affiliations.

1. Background

Decades ago, Dvorak concluded that cancer is spawned in chronic non-resolving wounds, implicating a role for the status of the host microenvironment in cancer initiation, and focusing attention on identifying the processes of wound repair that are coopted by mutant cells to initiate tumors and support their progression to metastasis [1]. Subsequent studies have identified the processes of inflammation and fibrosis, which are critical to efficient wound repair, as key microenvironmental factors that promote cancer progression [2–4]. Hyaluronan (HA) has emerged as an important extracellular matrix (ECM) regulator of inflammation and fibrosis in the setting of excisional wound repair [5–11], and this polymer is also implicated in breast cancer progression [5,8,12–15]. Here, we review the well-regulated inflammatory and fibrogenic functions of HA polymers and two HA receptors—CD44 and RHAMM (HMMR)—in cutaneous wound repair and consider evidence that these functions also contribute to the progression of breast cancer.

2. The Hyaluronome

The collection of genes that controls the synthesis, metabolism, and signaling properties of the tissue polysaccharide, HA, are collectively called the hyaluronome, and include HA synthases responsible for the production of HA, HA receptors, which bind HA and activate cellular signaling cascades; and hyaluronidases, which break the native HA polymer into fragments that differ from the native polymer in their signaling functions [5].

2.1. Hyaluronan

HA is a simple linear polysaccharide consisting of repeated saccharides (N–acetylglucosamine and B–glucuronic acid disaccharide units form the HA polymer), which was historically considered to be an 'inert' structural component. At that time, effects on cell behaviour and tissue homeostasis were postulated to result from the physicochemical characteristics of HA that provide tissue hydration, expansion and elasticity [16–18]. Although these physicochemical characteristics of HA are impressive and critical to the homeostasis of organs such as skin [19,20], the demonstration that HA activates kinase cascades in cultured fibroblasts [21] and binds to specific cell receptors such as CD44 [22] and RHAMM (HMMR) [23] provided initial evidence of its signaling properties. HA has since been shown to regulate MAP kinase, PI3 kinase, Hippo, and multiple growth factor signaling networks [5,24]. The complex functional information provided by this simple linear polymer is due in large part to metabolism related changes in both its expression level and in its molecular weight. For example, the native newly synthesized HA polymer (defined here as high molecular weight HA, HMW–HA, >500–700 kDa) blunts cell proliferation and migration and is anti–inflammatory as shown by studies demonstrating its functions to suppress an M1 and enhance M2 polarization of macrophages [14,25]. These properties of HMW–HA are considered important for maintaining tissue architecture and homeostasis particularly in skin. In contrast, smaller HA polymers created by enzymatic and/or chemical degradation of HMW–HA (e.g., low molecular weight HA, LMW–HA, 10–250 kDa; and oligosaccharides, O–HA, <10 kDa) function as 'danger alerts' (DAMPs [26,27]) that are released by cell death/stress and are strongly immunogenic. These tissue damage–induced HA oligomers provide pro–inflammatory (e.g., support M1 macrophage polarization), proliferation and migration signals [5,28,29] (Figure 1), and are critical for initiating a response to injury. It is intriguing that DAMPs released in response to tumor cell death, are also implicated in breast cancer progression [30], providing another example of the similarities between wounds and breast tumors.

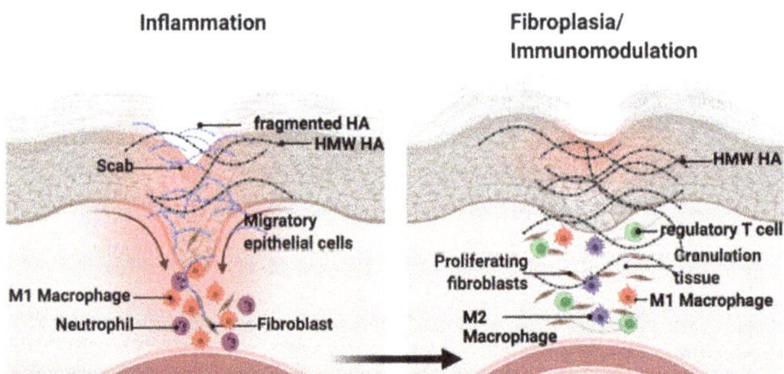

Figure 1. Hyaluronan contributes to inflammation and resolution of excisional wounds. Native (high molecular weight, HMW–HA) hyaluronan production is increased upon and throughout tissue injury. Low molecular weight (LMW–HA) and oligosaccharide (O–HA) fragments are rapidly generated from HMW–HA hyaluronidases and ROS/NOS, resulting in a mixture of HMW and fragmented HA polymers. LMW– and O–HA promote macrophage and fibroblast influx into the wound that initiate inflammation while HMW–HA restrains the extent of inflammation. At later stages of wound repair, HMW–HA predominates and supports macrophage polarization into the immunosuppressive M2 phenotype. Created with BioRender.com (accessed on 12 September 2021).

2.2. Hyaluronan Synthases

In mammals, HA is synthesized by one or more of three genetically distinct cell membrane isoenzymes, hyaluronan synthases 1–3 (HAS1–3). These synthases differ in their enzymatic properties, regulation by stimuli, and contribution to normal and pathological processes. For example, whereas HAS1 and 2 synthesize HMW–HA (avg. 2×10^3 kDa), HAS3 synthesizes shorter HA polymers (avg. 2×10^2 kDa). HAS2 is expressed during early embryonic development and its genomic deletion leads to embryonic lethality resulting from cardiac defects [31]. In contrast, *Has1$^{-/-}$*, *Has3$^{-/-}$*, and *Has1:Has3$^{-/-}$* mice are viable [32–34]. HAS2 is expressed in most tissues including skin and mammary gland ductal epithelial and stromal cells [35], and elevated HAS2 expression has been linked to promoting breast cancer progression [13,36]. HAS1 and 3 are also expressed in the epidermis and dermis of the skin [37] although keratinocytes primarily express HAS3 [13,37,38] while dermal fibroblasts primarily express HAS1 [39,40]. All three HA synthases are upregulated during cutaneous wound repair and in some cancers [34,39] but, primarily, HAS2 is upregulated in breast cancer [41,42].

2.3. Hyaluronidases

HMW–HA produced by HAS1–3 is degraded into heterogeneously sized fragments by both hyaluronidases (HYALs) and cell metabolism by–products such as reactive oxygen and nitrogen species (ROS/NOS). HYALs are hyalurono–glucosidases that cleave the beta–(1,4) linkage between N–acetylglucosamine and glucuronate [43]. The human genome encodes *HYAL1–HYAL5* and one pseudogene (*HYAL6*). Out of these, HYAL1 (present in lysosomes) is mainly responsible for HA degradation into oligo–, di–, and monosaccharides while HYAL2 (localized to the cell surface via a GPI linker) degrades HA into fragments of about 20 kDa. These HYAL2 created fragments are either released into the microenvironment or internalized by HA receptors such as CD44 to be further degraded by HYAL1 [44–46]. Two additional proteins with hyaluronidase activity have more recently been discovered that process HA into intermediate–sized LMW fragments: the transmembrane protein TMEM2 [47] and KIAA1199 (CEMIP, HYBID) [48]. TMEM2 is a widely expressed membrane protein that can digest native HA into 5kDa fragments [47]. Pro–inflammatory cytokines that are released early in cutaneous repair such as IL–6, regulate KIAA1199, which is expressed by activated skin fibroblasts [49] and macrophages [50–52]. IL–6 and the resulting HA fragments [28,53] promote leukocyte infiltration into the wound. However, IL–6 also plays a key modulatory role in the switch from a pro–inflammatory to immunosuppressive microenvironment required for wound resolution [54]. The increased expression of IL–6 is also linked to breast cancer progression [48], where it performs both immunomodulatory functions similar to those in wound repair [55]. This cytokine has also been shown to modulate tumor cell plasticity, which impacts breast cancer progression and chemo–resistance [55]. Therefore, deeper analyses of the links between CEMIP and IL–6 are potentially important for understanding the commonalities between wound repair and cancer metastasis, as well as providing potential therapeutic targets to control both of these processes.

2.4. Hyaluronan Receptors, CD44 and RHAMM

To date, characterized HA receptors include CD44 [56], RHAMM (HMMR) [23], LYVE1 [57], TLR2/4 [29,58], STAB2/HARE [59], and LAYN [49]. CD44, LYVE1, and STAB2/HARE bind to HA via link modules [56,60]. In contrast, the HA binding domain of RHAMM has been localized to alpha helical clusters of positively charged residues [61,62]. TLR2 and 4 contain similar clusters of positively charged amino acids as RHAMM. Although these receptors are required for responses to HA fragments there is controversy as to whether or not they directly bind to this polysaccharide [60]. LAYN has been shown to bind directly to HA but contains neither link module nor clusters of positively charged amino acids, and the sequences responsible for this interaction have presently not been reported [5,28]. In this review, we focus upon the biology of CD44 and RHAMM because— unlike LYVE1, STAB2/HARE, and LAYN—these receptors have been studied in detail in the context of cutaneous wound repair and breast cancer [8] and because they are well characterized to directly bind to HA [60]. Readers interested in the biology of these HA receptors are directed to additional reviews [15,59,63–67].

Although CD44 and RHAMM clearly differ in key biochemical and structural properties, some commonalities include a likely evolution from heparin–binding ancestors, complex functions due to isoform expression generated by alternative mRNA splicing and coordinated intracellular and extracellular functions. CD44 is a non–kinase cell surface HA receptor that contributes to the proliferation, migration/invasion, adhesion, polarity, plasticity, and differentiation of many cell types, including resident skin cells [68,69]. CD44 is constitutively and widely expressed in tissues such as skin, and binds to HMW, LMW, and O–HA via a link module, which is distinct from the HA binding sequences of RHAMM [61,70]. The signaling properties of CD44 result not only from its association with HA polymers but also its interactions with other cell surface and extracellular proteins (e.g., growth factor receptors, osteopontin, metalloproteinases, and collagens) [71]. The small, intracellular domain (ICD) of CD44 additionally binds to intracellular adaptor and cytoskeletal proteins [72]. The binding of HA polymers to CD44 promotes homotypic CD44 clustering that can activate or impede oncogenic signaling cascades depending upon the HA polymer size and its partnering with other proteins. As an example, HMW–HA stimulates tumor–suppressive Hippo signaling by clustering CD44, which recruits polarity–regulating kinase (PAR1b) to the intracellular domain of CD44 and leads to activation of Hippo signaling [73]. In contrast, LMW–HA inhibits this Hippo signaling by disrupting HMW–HA/CD44 clustering. However, the interaction of CD44 with HMW–HA [74] and RHAMM can also result in expression of genes such as MMP9 that are utilized for both cutaneous wound repair and breast cancer progression [71].

In contrast to CD44, RHAMM expression is low and primarily intracellular in most homeostatic tissues but expression and extracellular export increases with pathologic stress, injury, and neoplastic transformation [8]. Under injury conditions, small amounts of intracellular RHAMM are released from cells, which bind to LMW and O–HA via alpha–helical clusters of positively charged amino acids [8,61,75]. The three–dimensional organization of these clusters is similar to those located in the alpha–helical glycosaminoglycan binding sites of lectins (e.g., GRO cytokines) [76,77]. RHAMM: HA complexes associate with integral HA receptors such as CD44 and TLR4 to activate signaling cascades (Figure 2), initiating an early response–to–injury through the NRLP3 inflammasome and other signaling cascades [14,28,29]. The functional and physical association of RHAMM with CD44 is influenced by the presentation of HA polymers in both soluble and ECM–immobilized form [78]. These interactions regulate cell motility and gene expression. The intracellular functions of RHAMM are complex and multifunctional, and include regulation of microtubule stability, mitotic spindle dynamics, intracellular signaling complexes, and gene transcription (Figure 2). Collective study of CD44 and RHAMM signaling predict that binding preferences for HA polymer sizes (which regulate receptor clustering), as well as mode of HA presentation, are two mechanisms for how cells detect and differentially signal in response to HMW–, LMW–, and O–HA polymers.

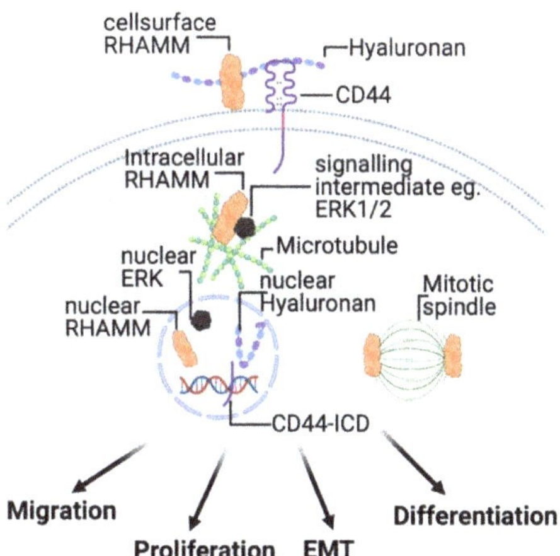

Figure 2. HA receptors CD44 and RHAMM regulate signaling that control skin cell migration, proliferation, plasticity, and differentiation during response–to–injury. CD44 is an integral membrane protein that coordinates signaling through growth factor receptors (e.g., EGFR) and cell surface RHAMM. The intracellular domain (ICD) of CD44, which can be released under injury conditions, forms part of transcriptional complexes that regulate expression of injury response genes. RHAMM also occurs in multiple intracellular compartments including the microtubule and actin cytoskeleton and, like CD44, is a component of transcriptional complexes regulating expression of extracellular matrix proteins that are required for wound repair. Created with BioRender.com (accessed on 12 September 2021).

3. Functions of Hyaluronan: Size Matters in Cutaneous Repair and Breast Cancer

Are the functions of the hyaluronome in excisional wound repair replicated in breast cancer progression? While skin wound healing is a tightly regulated and orderly physiological response to injury, breast cancer is not. For example, cutaneous repair can be reproducibly simplified into three sequential stages: inflammation, fibroplasia, and the final immunomodulation/tissue remodeling required for wound resolution [79] (Figure 3). In contrast, during breast cancer initiation and progression, inflammation and fibroplasia are simultaneous and chronic with an evolution towards immunomodulation/remodeling that culminates in progression to metastasis (Figure 3). HA affects all three of the wound repair stages, particularly targeting immune and fibroblast functions (Figure 1). In general, HMW–HA is anti–inflammatory, anti–fibrotic, and pro–regenerative while HA fragments are pro–inflammatory, support fibroplasia, which results in scar formation in excisional wound repair [8,80] and alter the immune landscape of cancer microenvironments [15]. Nevertheless, some sizes of LMW–HA can be useful therapeutically since they can promote the rapid closure of wounds and reduce infection [15].

3.1. HA and Cutaneous Injury

In injured tissues, HMW–HA synthesis is closely coupled to the generation of LMW–HA fragments that initiate a robust inflammatory and fibrogenic response resulting in the rapid wound closure and control of opportunistic pathogens [81]. However, the reliance on inflammation as the initial response–to–injury results in a dermal scar that compromises skin elasticity and strength [25,80,82]. In contrast to adult tissue repair, embryonic wound repair, which occurs in a sterile environment, and proceeds in the absence of extensive HMW–HA fragmentation or immune cell influx, is regenerative, healing without a scar. Elevated expression of HAS1,2 by skin cells is responsible for the increased production of HMW–HA, which occurs throughout the repair stages [34,83]. ROS/NOS, in combination with released hyaluronidases, rapidly fragment a portion of the newly synthesized HMW–HA into a highly heterogeneous pool of LMW and O–HA polymers [81]. Thus, a mixture of HMW–, LMW–, and O–HA collectively contributes to the repair and resolution of excisional wounds.

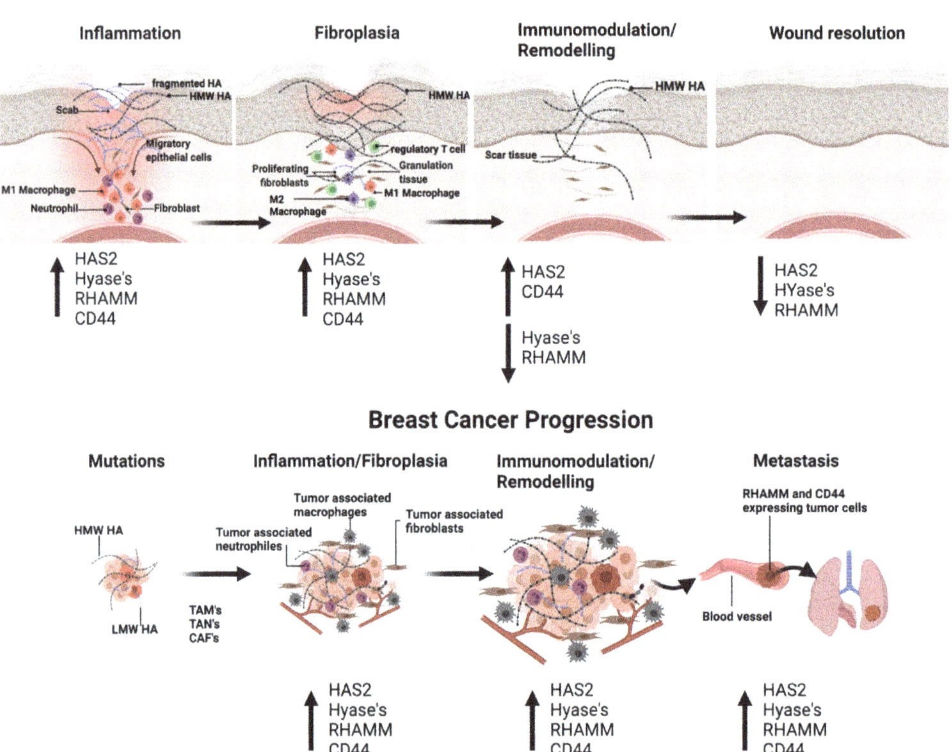

Figure 3. HA polymers, CD44 and RHAMM, regulate wound repair and promote breast cancer progression. During early stages of wound repair and during breast tumor progression, expression of HAS2, Hyal's, CD44 and RHAMM is upregulated, providing optimal conditions for infiltration of immune cells and cell proliferation. During later stages of wound repair, Hyal and RHAMM expression is reduced, resulting in a prevalence of HMW HA and signaling via CD44. At the final stage of wound repair, HA production and HA receptor expression return to levels seen in unwounded skin. In contrast, HA synthesis, fragmentation and HA receptor expression remain high throughout breast cancer progression. Created with BioRender.com (accessed on 12 September 2021).

The properties of HMW–HA perform multiple functions during excisional wound repair. HMW–HA provides a source for the generation of LMW– and O–HA polymer sizes, activates specific immunogenic signaling pathways and regulates fibrogenesis. As a source for generating HA fragments, HMW–HA contributes to inflammation. However, as a native polymer, it restrains HA fragment–induced inflammation by inhibiting MAP kinase, NFkB and other pathways, which blunt the expression of pro–inflammatory cytokines— such as TNFA, IL1B, IL–6, and CCL2 [84]—thereby suppressing the M1 pro–inflammatory polarization of macrophages [84,85]. HMW–HA also inhibits wound fibroblast expression of pro–inflammatory cytokines, e.g., IL–6 and other chemokines [86]. HMW–HA further contributes to dampening inflammation by promoting the polarization of M1 macrophages into an immunosuppressive M2 macrophage with the concomitant expression of cytokines such as TGFB1, IL10, IL11, and ARG1 [84]. These immunosuppressive cytokines are required for wound resolution [80,84,85,87]. For example, the HMW–HA/IL–10 axis affects adaptive immune response by modulating CD4+ effector T cells and promoting T regulatory cell function to reduce both innate immune activity and wound scarring [25,80]. In addition, HMW–HA reduces innate immune cell and fibroblast migration [88,89] and proliferation [90,91], which collectively control the extent of wound fibroplasia/fibrosis [92,93]. In contrast to these effects on immune cells and fibroblasts, HMW–HA (2290 kDa) stimulates keratinocyte migration and wound re–epithelialization [94]. These collective properties of HMW–HA have been utilized clinically to reduce inflammation and fibrosis tipping wound repair to a more fetal–like regenerative repair. Thus, injection of HMW–HA into keloids inhibits fibroblast proliferation [92], and reduces the fibrogenic properties of keloid fibroblasts [95] while topical application of HMW–HA [88] or forced overexpression of HAS1 to excisional skin wounds speeds repair [96] and reduces scarring [83].

In opposition to HMW–HA, LMW and O–HA fragments drive inflammation and fibroplasia during the early stages of excisional repair to enhance the speed of wound closure. Indeed, topical application of HYAL2 to full–thickness wounds speeds their closure [97]. SDS–PAGE analyses of wound and tumor HA reveal a continuous gradient of polymer sizes that is a complex mixture of biological cues to responding cells [5,81]. Studies have shown that HA fragments can have a precise size–dependent effect on excisional wound repair. For example, 40 kDa LMW–HA inhibits while 6mer O–HA ($\approx$1 kDa) [98] and 250 kDa LMW–HA [99] promote wound closure in vivo, and selectively regulate expression of pro–inflammatory and immunosuppressive cytokines [100] as well as production of chemokines that attract fibroblasts into excisional wounds [101]. Although some HA polymer sizes have distinct functions during repair, others exhibit functional duality. For example, 500 kDa HA exerts both pro– and anti–inflammatory effects on macrophages [84]. This mixture of distinct and overlapping functions likely provides an exquisitely subtle control of inflammation and fibrosis. It is noteworthy that acute application of HYAL2 to full thickness skin wounds speeds wound closure [97]; continuous application of large amounts of O–HA to excisional wounds prevents wound repair [96], indicating that tight control of fragmentation is necessary for normal wound repair.

3.2. HA and Breast Cancer

Currently, there are no clear genetic abnormalities associated with the critical transition from DCIS to invasive cancer; however, there is emerging evidence linking this progression to tumor–induced changes in the microenvironment [3]. In particular, evidence supports a role for the immune/inflammation [102–104] and fibrogenic functions [12,105,106] of a wound–like host microenvironment in providing conditions to support early breast cancer cell spread and progression to a metastatic state. It is important to note that both host and tumor cells contribute to a cancer microenvironment, and that the evolution of a tumor–supporting microenvironment is chaotic in comparison to the defined stages of wound repair. Furthermore, tumor cells are highly plastic making both their contributions and responses to the microenvironment dynamic. These properties and the heterogeneity of breast cancers complicate efforts to dissect the roles of ECM components in breast cancer progression. Nevertheless, a change in HA metabolism has repeatedly emerged as one of the microenvironmental factors linked to breast cancer progression (Figure 3). For example, recent meta–analysis of published data from breast cancer patient tumors shows that increased HA accumulation in the tumor stroma [107] and LMW–HA in tumor patient plasma [108] are biomarkers for poor outcome. Experimental evidence shows that HA synthesis contributes to tumor supporting microenvironment [109], blocking HA production by knockdown of *HAS2* [110] and the use of inhibitors such as 4–Methylumbelliferone [111] inhibits tumorigenesis and metastasis of breast cancer cell lines. The concept that HA fragments fuel breast cancer progression is also supported by evidence that elevated *HYAL* expression (in particular CEMIP and TMEM2) is linked to breast tumor initiation [13]. However, the contributions of HMW–HA and LMW–HA to the wound–like inflammatory and fibrogenic properties of the breast cancer microenvironment support rather than resolve disease progression.

Like wounds, tumors contain a heterogeneous mixture of HMW–, LMW–, and O–HA polymers, which affect the function of both tumor and host cells [5]. However, unlike the coordinated synthesis and transient degradation of HMW–HA evident during wound repair [81], HA synthesis and HA fragmentation in tumors are deregulated, uncoupled and remain elevated during tumor progression with consequences to both tumor and host cells that support tumor progression rather than its resolution. For example, the beneficial functions of HMW–HA that facilitate wound resolution are co–opted by breast tumor cells to suppress immune detection and reduce exposure to therapy. Similar to its functions in wounds, HMW–HA promotes an immunosuppressive M2 macrophage polarization, particularly in the context of breast cancer [112,113]. While this function is essential for wound resolution, it contributes to signaling that supports immune evasion and progression of breast tumors [114]. For example, HMW–HA stimulates in–trafficking and primes tumor–associated macrophages to produce pro–angiogenic cytokines, which stimulates neoangiogenesis that contributes to disease progression [33,112]. HA also targets cancer–associated fibroblasts to promote their migration towards tumor spheroids, where their close proximity supports a paracrine tumor cell growth and migration [15,115,116]. The viscous properties of HMW–HA, which concentrate essential growth and other signaling factors near migrating cells to facilitate wound closure, also impedes therapeutic responses in cancer by reducing drug perfusion of tumors [117,118]. Two potentially tumor–suppressive effects of HMW–HA are its ability to arrest tumor cell proliferation [119,120] and increase breast tumor cell apoptosis [120]. However, the anti–proliferation function of HMW–HA is a two–edge sword since limiting tumor cell proliferation may actually attenuate the efficacy of cytotoxic chemotherapy that best targets proliferating cells.

Tumor cells, such as wound cells, detect and differentially respond to various sizes of HA fragments. The continual de–regulated synthesis of HMW–HA provides a constant source of LMW–HA and O–HA, which sustains host inflammation and fibrosis [13,14,121], and directly promotes breast tumor cell invasion and successful colonization of distant tissues [5,122], an event that does not happen during wound repair. LMW– and O–HA also promote expression of pro–inflammatory cytokines such as CCL2, which attract pro–tumorigenic circulating monocytes and stromal cells into the tumor microenvironment [33,123,124], and ECM regulators that support pro–tumor immunogenic and fibrogenic functions [125]. LMW–HA notably promotes invasion and migration of breast tumor cells, which is particularly observed in triple negative breast cancer [126,127]. Furthermore, triple negative breast cancer cell subpopulations that bind high levels of LMW–HA are more invasive and metastatic than tumor cells that bind only low levels [128]. Consistent with these experimental results, high levels of LMW–HA in the serum of breast cancer patients correlates with increased incidence of lymph node metastasis [108].

These collective observations predict that the wound–like functions of HMW–HA, LMW–HA, and O–HA are oncogenic in the context of breast cancer but are chronically sustained, which culminates in disease progression rather than resolution. However, additional analyses of which HMW–HA, LMW–HA, and O–HA polymers exert immune and fibrogenic functions [110,129], the cell types that are targeted by these polymers and their functional consequence to cancer cells is needed.

4. Roles of HA Receptors in De–Coding HA Polymer Size

4.1. Cutaneous Wound Repair

To date, a mechanistic understanding of how immune and mesenchymal cells detect and respond to differences in HA polymer size during physiological and disease processes is not well understood [59]. CD44 is constitutively expressed in skin cells and performs multiple functions during tissue injury, which can either promote or resolve inflammation. This multifunctional property is likely context dependent, since CD44 binds to multiple sizes of HA polymers. Total or basal keratinocyte–targeted (K14) loss of CD44 does not detectably affect uninjured skin architecture [130,131] but embryonic deletion of CD44 mildly increases the inflammatory phase of excisional repair. Thus, neutrophils, M1 and

M2 macrophages, and CD3+ T cells are slightly but significantly enhanced, and this increase is accompanied by elevated IL1B and IL4 expression. In contrast to its mild immunogenic effects, genomic loss of CD44 substantially alters the temporal profile and wound distribution of SMA[+] and FAP[+] fibroblasts subsets resulting in increased fibro–proliferation and scar formation relative to wildtype wounds. These results predict that CD44 signaling suppresses fibroplasia and may contribute to the anti–fibrotic impact of HMW–HA.

In contrast to CD44, RHAMM is not constitutively expressed in skin but is upregulated with excisional injury [132,133] and preferentially binds to LMW and O–HA [98,134,135]. Genomic *Rhamm*–loss and RHAMM function–blocking reagents robustly reduce inflammation and fibrosis [29,75,81,98,133,136]. Specifically, *Rhamm*–loss, RHAMM mimetic peptides, which bind to and sequester LMW– and O–HA to limit access of these polymers to RHAMM [75], and RHAMM blocking antibodies alter fibroblast heterogeneity, reduce wound macrophage number/cytokine expression, blunt fibroplasia, and promote expression of dermal markers such as Tenascin–C for regenerative repair [137,138]. The immunogenic and fibrogenic effects of RHAMM match closely with those of HA fragments. For example, RHAMM expression is required for dermal fibroblast migration and wound macrophage influx–promoting effects of 6mer O–HA [98].

4.2. Breast Cancer Progression

Several studies show that the functions of HA during tumorigenesis are associated with the expression and display of HA receptors on tumor and host cells. For example, the invasive/metastatic triple negative breast cancer cell subsets that bind to high levels of HA display high levels of CD44 and RHAMM [128]. CD44 is widely used as a marker for breast tumor–initiating cells [139,140] and experimental analyses show that CD44 contributes to the pro–tumorigenic behaviour of breast cancer cells by stimulating cell proliferation, migration, invasion, and plasticity [141,142]. These effects of CD44 expression are linked to activation of pro–tumorigenic signaling pathways via partnership with growth factor receptors and RHAMM. The oncogenic functions of CD44 are complex and are affected by posttranslational modification and alternative splicing of this transmembrane protein as well as its epigenetic regulation of gene expression. For example, CD44 mediates uptake of iron–bound hyaluronan that supports the iron–dependent demethylation of histones and upregulation of cell plasticity genes [143]. Intriguingly, CD44 can suppress or support tumorigenicity in a context–dependent manner. Xenograft studies of human cell lines show a role for CD44:HA interactions in promoting breast cancer progression [120] while conversely, lung metastasis is enhanced rather than suppressed in a CD44$^{-/-}$ mouse model of mammary gland susceptibility [144]. These experimental differences suggest that CD44 can be oncogenic or tumor suppressive depending upon the host immune microenvironment. Despite these known oncogenic and tumor–suppressing functions of CD44, this HA receptor is being explored as a therapeutic target, imaging agent and tumor marker in breast and other cancers [145,146]. Successful use of its clinical potential, particularly for therapeutic targeting, will likely require a greater mechanistic understanding for the biological and molecular contexts of the tumor–supporting vs. tumor–suppressing properties of CD44.

High RHAMM expression in tumor cell subsets is a marker for increased peripheral metastasis and poor outcome [147]. The wound–like functions of RHAMM that contribute to breast cancer malignancy include increased cell migration, invasion, and proliferation [8]. Other functions which may or may not be linked to its HA binding properties include effects on cellular polarity, plasticity, genomic stability, chemo–resistance, and de–regulation of oncogenic driver pathways. The tumorigenic consequence of de–regulated RHAMM expression is influenced by the molecular subtype. For example, whereas RHAMM expression is increased in most breast cancers compared to adjacent normal breast tissue, luminal A subtype breast cancers displays a relatively low RHAMM expression compared to other breast cancer types [148] and *RHAMM* knockdown in cell lines derived from this subtype increases rather than decreases migration and metastasis. However, blocking

RHAMM signaling in triple negative breast cancer blunts invasion and metastasis and ablation of the HA binding capability of RHAMM destroys its transforming potential [149]. These results and the restricted expression of RHAMM in normal tissues, which contrasts with constitutive and widespread CD44 expression, predict that RHAMM is an attractive potential cancer therapy target for the breast cancer subtypes that use this HA receptor to promote invasion and metastasis.

5. Conclusions

In summary, the dynamic changes of HA concentration and fragment size distribution in the remodelling microenvironments of wounds and breast tumors provide cells with important contextual information, that promotes but also limits specific immune and fibroblast functions. This contextual information is "interpreted" by a dynamic expression of HA receptors in particular CD44 and RHAMM, which couple signaling pathways that control cellular migration, invasion, proliferation, and immune regulation required for both efficient wound repair and metastatic spread of tumors. Despite this functional complexity, the medical and cosmetic use of the HA polymer is a growing industry, and in particular experimental studies predict that targeting HA synthesis, hyaluronidases, and HA receptors has enormous therapeutic potential for improving wound repair and management of cancer. For example, application of HMW–HA and its modified derivatives improve cutaneous wound repair [25,150]. HMW–HA is also being developed in experimental models and clinical trials to target CD44 for both imaging and delivery of therapeutics to cancer stem cells [145,151]. Conversely, blocking HA synthesis with 4MU reduces tumor spread [152–155], increases exposure of tumor cell HER2 for PET imaging of tumors [156] and sensitizes tumor cells to trastuzumab [157]. Modifying the remodeling tumor microenvironment using stabilized hyaluronidases to remove HMW–, LMW–, and O–HA is a novel method for improving delivery of therapeutic drugs to multiple cancers including breast [158–161]. Targeting HA receptors has also met with success in moderating fibrotic wound repair and managing cancer. For example, RHAMM peptide mimetics that bind to LMW–HA and O–HA reduce fibroplasia in bleomycin–induced lung and skin injury, and promote a regenerative repair in excisional wounds [29,75]. Finally, CD44–HA interactions are actively investigated for their therapeutic potential in particular as a target for HA–based drug formulations [162]. As well, CD44 monoclonal antibodies are being assessed in pre–clinical and clinical trials for both imaging and treating cancers and cancer stem cells that overexpress CD44 [145,146,163,164]. As knowledge of the hyaluronome's multifunctionality deepens, the number of medical uses, particularly in the realm of wound repair and cancer, will undoubtedly increase.

Author Contributions: Conceptualization, C.T., J.B.M., B.J.M., A.C.N. and E.A.T.; Writing—original draft preparation, C.T., J.B.M., B.J.M., A.C.N. and E.A.T.; Writing—review and editing, C.T., J.B.M., B.J.M., A.C.N. and E.A.T. All authors have read and agreed to the published version of the manuscript.

Funding: Andrew C. Nelson: American Cancer Society Clinical Scientist Development Scholar program [132574-CSDG-18-139-01-CSM].

Conflicts of Interest: The authors declare no conflict of interest.

References

1.   Dvorak, H.F. Tumors: Wounds That Do Not Heal–A Historical Perspective with a Focus on the Fundamental Roles of Increased Vascular Permeability and Clotting. *Semin. Thromb. Hemost.* **2019**, *45*, 576–592. [CrossRef]
2.   Bissell, M.J.; Hines, W.C. Why don't we get more cancer? A proposed role of the microenvironment in restraining cancer progression. *Nat. Med.* **2011**, *17*, 320–329. [CrossRef]
3.   Allen, M.D.; Jones, L.J. The role of inflammation in progression of breast cancer: Friend or foe? *Int. J. Oncol.* **2015**, *47*, 797–805. [CrossRef]
4.   Nakamura, N. A hypothesis: Radiation carcinogenesis may result from tissue injuries and subsequent recovery processes which can act as tumor promoters and lead to an earlier onset of cancer. *Br. J. Radiol.* **2020**, *93*, 20190843. [CrossRef]
5.   Liu, M.; Tolg, C.; Turley, E. Dissecting the Dual Nature of Hyaluronan in the Tumor Microenvironment. *Front. Immunol.* **2019**, *10*, 947. [CrossRef]

6. Abatangelo, G.; Vindigni, V.; Avruscio, G.; Pandis, L.; Brun, P. Hyaluronic Acid: Redefining Its Role. *Cells* **2020**, *9*, 1743. [CrossRef] [PubMed]

7. Muto, J.; Sayama, K.; Gallo, R.L.; Kimata, K. Emerging evidence for the essential role of hyaluronan in cutaneous biology. *J. Derm. Sci.* **2019**, *94*, 190–195. [CrossRef] [PubMed]

8. Tolg, C.; McCarthy, J.B.; Yazdani, A.; Turley, E.A. Hyaluronan and RHAMM in wound repair and the "cancerization" of stromal tissues. *Biomed. Res. Int.* **2014**, *2014*, 103923. [CrossRef] [PubMed]

9. Turley, E.A.; Wood, D.K.; McCarthy, J.B. Carcinoma Cell Hyaluronan as a "Portable" Cancerized Prometastatic Microenvironment. *Cancer Res.* **2016**, *76*, 2507–2512. [CrossRef] [PubMed]

10. Kavasi, R.M.; Berdiaki, A.; Spyridaki, I.; Corsini, E.; Tsatsakis, A.; Tzanakakis, G.; Nikitovic, D. HA metabolism in skin homeostasis and inflammatory disease. *Food Chem. Toxicol.* **2017**, *101*, 128–138. [CrossRef] [PubMed]

11. Litwiniuk, M.; Krejner, A.; Speyrer, M.S.; Gauto, A.R.; Grzela, T. Hyaluronic Acid in Inflammation and Tissue Regeneration. *Wounds* **2016**, *28*, 78–88.

12. Zhao, Y.; Zheng, X.; Zheng, Y.; Chen, Y.; Fei, W.; Wang, F.; Zheng, C. Extracellular Matrix: Emerging Roles and Potential Therapeutic Targets for Breast Cancer. *Front. Oncol.* **2021**, *11*, 650453. [CrossRef]

13. Velesiotis, C.; Vasileiou, S.; Vynios, D.H. A guide to hyaluronan and related enzymes in breast cancer: Biological significance and diagnostic value. *FEBS J.* **2019**, *286*, 3057–3074. [CrossRef] [PubMed]

14. Schwertfeger, K.L.; Cowman, M.K.; Telmer, P.G.; Turley, E.A.; McCarthy, J.B. Hyaluronan, Inflammation, and Breast Cancer Progression. *Front. Immunol.* **2015**, *6*, 236. [CrossRef] [PubMed]

15. Tavianatou, A.G.; Caon, I.; Franchi, M.; Piperigkou, Z.; Galesso, D.; Karamanos, N.K. Hyaluronan: Molecular size–dependent signaling and biological functions in inflammation and cancer. *FEBS J.* **2019**, *286*, 2883–2908. [CrossRef] [PubMed]

16. Balazs, E.A. Hyaluronan as an ophthalmic viscoelastic device. *Curr. Pharm. Biotechnol.* **2008**, *9*, 236–238. [CrossRef] [PubMed]

17. Balazs, E.A.; Högberg, B.; Laurent, T.C. The Biological Activity of Hyaluron Sulfuric Acid. *Acta Physiol. Scand.* **1951**, *23*, 168–178. [CrossRef] [PubMed]

18. Snetkov, P.; Zakharova, K.; Morozkina, S.; Olekhnovich, R.; Uspenskaya, M. Hyaluronic Acid: The Influence of Molecular Weight on Structural, Physical, Physico–Chemical, and Degradable Properties of Biopolymer. *Polymers* **2020**, *12*, 1800. [CrossRef]

19. Valachova, K.; Soltes, L. Hyaluronan as a Prominent Biomolecule with Numerous Applications in Medicine. *Int. J. Mol. Sci.* **2021**, *22*, 7077. [CrossRef]

20. Ruiz Martinez, M.A.; Peralta Galisteo, S.; Castan, H.; Morales Hernandez, M.E. Role of proteoglycans on skin ageing: A review. *Int. J. Cosmet. Sci.* **2020**, *42*, 529–535. [CrossRef]

21. Turley, E.A. Hyaluronic acid stimulates protein kinase activity in intact cells and in an isolated protein complex. *J. Biol. Chem.* **1989**, *264*, 8951–8955. [CrossRef]

22. Aruffo, A.; Stamenkovic, I.; Melnick, M.; Underhill, C.B.; Seed, B. CD44 is the principal cell surface receptor for hyaluronate. *Cell* **1990**, *61*, 1303–1313. [CrossRef]

23. Hardwick, C.; Hoare, K.; Owens, R.; Hohn, H.P.; Hook, M.; Moore, D.; Cripps, V.; Austen, L.; Nance, D.M.; Turley, E.A. Molecular cloning of a novel hyaluronan receptor that mediates tumor cell motility. *J. Cell Biol.* **1992**, *117*, 1343–1350. [CrossRef] [PubMed]

24. Turley, E.A.; Noble, P.W.; Bourguignon, L.Y. Signaling properties of hyaluronan receptors. *J. Biol. Chem.* **2002**, *277*, 4589–4592. [CrossRef]

25. Kaul, A.; Short, W.D.; Keswani, S.G.; Wang, X. Immunologic Roles of Hyaluronan in Dermal Wound Healing. *Biomolecules* **2021**, *11*, 1234. [CrossRef] [PubMed]

26. Li, N.; Geng, C.; Hou, S.; Fan, H.; Gong, Y. Damage–Associated Molecular Patterns and Their Signaling Pathways in Primary Blast Lung Injury: New Research Progress and Future Directions. *Int. J. Mol. Sci.* **2020**, *21*, 6303. [CrossRef] [PubMed]

27. Roedig, H.; Damiescu, R.; Zeng–Brouwers, J.; Kutija, I.; Trebicka, J.; Wygrecka, M.; Schaefer, L. Danger matrix molecules orchestrate CD14/CD44 signaling in cancer development. *Semin. Cancer Biol.* **2020**, *62*, 31–47. [CrossRef] [PubMed]

28. Garantziotis, S.; Savani, R.C. Hyaluronan biology: A complex balancing act of structure, function, location and context. *Matrix Biol.* **2019**, *78*, 1–10. [CrossRef] [PubMed]

29. Savani, R.C. Modulators of inflammation in Bronchopulmonary Dysplasia. *Semin. Perinatol.* **2018**, *42*, 459–470. [CrossRef]

30. Zheng, H.; Siddharth, S.; Parida, S.; Wu, X.; Sharma, D. Tumor Microenvironment: Key Players in Triple Negative Breast Cancer Immunomodulation. *Cancers* **2021**, *13*, 3357. [CrossRef]

31. Camenisch, T.D.; Spicer, A.P.; Brehm–Gibson, T.; Biesterfeldt, J.; Augustine, M.L.; Calabro, A., Jr.; Kubalak, S.; Klewer, S.E.; McDonald, J.A. Disruption of hyaluronan synthase–2 abrogates normal cardiac morphogenesis and hyaluronan–mediated transformation of epithelium to mesenchyme. *J. Clin. Investig.* **2000**, *106*, 349–360. [CrossRef] [PubMed]

32. Bai, K.J.; Spicer, A.P.; Mascarenhas, M.M.; Yu, L.; Ochoa, C.D.; Garg, H.G.; Quinn, D.A. The role of hyaluronan synthase 3 in ventilator–induced lung injury. *Am. J. Respir. Crit. Care Med.* **2005**, *172*, 92–98. [CrossRef]

33. Kobayashi, N.; Miyoshi, S.; Mikami, T.; Koyama, H.; Kitazawa, M.; Takeoka, M.; Sano, K.; Amano, J.; Isogai, Z.; Niida, S.; et al. Hyaluronan deficiency in tumor stroma impairs macrophage trafficking and tumor neovascularization. *Cancer Res.* **2010**, *70*, 7073–7083. [CrossRef]

34. Mack, J.A.; Feldman, R.J.; Itano, N.; Kimata, K.; Lauer, M.; Hascall, V.C.; Maytin, E.V. Enhanced inflammation and accelerated wound closure following tetraphorbol ester application or full–thickness wounding in mice lacking hyaluronan synthases Has1 and Has3. *J. Investig. Dermatol.* **2012**, *132*, 198–207. [CrossRef]

35. Tolg, C.; Yuan, H.; Flynn, S.M.; Basu, K.; Ma, J.; Tse, K.C.K.; Kowalska, B.; Vulkanesku, D.; Cowman, M.K.; McCarthy, J.B.; et al. Hyaluronan modulates growth factor induced mammary gland branching in a size dependent manner. *Matrix Biol.* **2017**, *63*, 117–132. [CrossRef]

36. Heldin, P.; Basu, K.; Olofsson, B.; Porsch, H.; Kozlova, I.; Kahata, K. Deregulation of hyaluronan synthesis, degradation and binding promotes breast cancer. *J. Biochem.* **2013**, *154*, 395–408. [CrossRef]

37. Sayo, T.; Sugiyama, Y.; Takahashi, Y.; Ozawa, N.; Sakai, S.; Ishikawa, O.; Tamura, M.; Inoue, S. Hyaluronan synthase 3 regulates hyaluronan synthesis in cultured human keratinocytes. *J. Investig. Dermatol.* **2002**, *118*, 43–48. [CrossRef]

38. Sugiyama, Y.; Shimada, A.; Sayo, T.; Sakai, S.; Inoue, S. Putative hyaluronan synthase mRNA are expressed in mouse skin and TGF–beta upregulates their expression in cultured human skin cells. *J. Investig. Dermatol.* **1998**, *110*, 116–121. [CrossRef]

39. Tammi, R.; Pasonen–Seppanen, S.; Kolehmainen, E.; Tammi, M. Hyaluronan synthase induction and hyaluronan accumulation in mouse epidermis following skin injury. *J. Investig. Dermatol.* **2005**, *124*, 898–905. [CrossRef] [PubMed]

40. Yamada, Y.; Itano, N.; Narimatsu, H.; Kudo, T.; Morozumi, K.; Hirohashi, S.; Ochiai, A.; Ueda, M.; Kimata, K. Elevated transcript level of hyaluronan synthase1 gene correlates with poor prognosis of human colon cancer. *Clin. Exp. Metastasis* **2004**, *21*, 57–63. [CrossRef] [PubMed]

41. Heldin, P.; Lin, C.Y.; Kolliopoulos, C.; Chen, Y.H.; Skandalis, S.S. Regulation of hyaluronan biosynthesis and clinical impact of excessive hyaluronan production. *Matrix Biol.* **2019**, *78*, 100–117. [CrossRef]

42. Heldin, P.; Basu, K.; Kozlova, I.; Porsch, H. HAS2 and CD44 in breast tumorigenesis. *Adv. Cancer Res.* **2014**, *123*, 211–229.

43. Sindelar, M.; Jilkova, J.; Kubala, L.; Velebny, V.; Turkova, K. Hyaluronidases and hyaluronate lyases: From humans to bacterio- phages. *Colloids Surf. B Biointerfaces* **2021**, *208*, 112095. [CrossRef]

44. Triggs–Raine, B.; Natowicz, M.R. Biology of hyaluronan: Insights from genetic disorders of hyaluronan metabolism. *World J. Biol. Chem.* **2015**, *6*, 110–120. [CrossRef] [PubMed]

45. Pibuel, M.A.; Poodts, D.; Diaz, M.; Hajos, S.E.; Lompardia, S.L. The scrambled story between hyaluronan and glioblastoma. *J. Biol. Chem.* **2021**, *296*, 100549. [CrossRef] [PubMed]

46. Piperigkou, Z.; Kyriakopoulou, K.; Koutsakis, C.; Mastronikolis, S.; Karamanos, N.K. Key Matrix Remodeling Enzymes: Functions and Targeting in Cancer. *Cancers* **2021**, *13*, 1441. [CrossRef]

47. Irie, F.; Tobisawa, Y.; Murao, A.; Yamamoto, H.; Ohyama, C.; Yamaguchi, Y. The cell surface hyaluronidase TMEM2 regulates cell adhesion and migration via degradation of hyaluronan at focal adhesion sites. *J. Biol. Chem.* **2021**, *296*, 100481. [CrossRef] [PubMed]

48. Liu, J.; Yan, W.; Han, P.; Tian, D. The emerging role of KIAA1199 in cancer development and therapy. *Biomed. Pharmacother.* **2021**, *138*, 111507. [CrossRef] [PubMed]

49. Bono, P.; Rubin, K.; Higgins, J.M.; Hynes, R.O. Layilin, a novel integral membrane protein, is a hyaluronan receptor. *Mol. Biol. Cell* **2001**, *12*, 891–900. [CrossRef]

50. Soroosh, A.; Albeiroti, S.; West, G.A.; Willard, B.; Fiocchi, C.; de la Motte, C.A. Crohn's Disease Fibroblasts Overproduce the Novel Protein KIAA1199 to Create Proinflammatory Hyaluronan Fragments. *Cell. Mol. Gastroenterol. Hepatol.* **2016**, *2*, 358–368.e4. [CrossRef]

51. Yoshida, H.; Nagaoka, A.; Kusaka–Kikushima, A.; Tobiishi, M.; Kawabata, K.; Sayo, T.; Sakai, S.; Sugiyama, Y.; Enomoto, H.; Okada, Y.; et al. KIAA1199, a deafness gene of unknown function, is a new hyaluronan binding protein involved in hyaluronan depolymerization. *Proc. Natl. Acad. Sci. USA* **2013**, *110*, 5612–5617. [CrossRef]

52. Yoshida, H.; Okada, Y. Role of HYBID (Hyaluronan Binding Protein Involved in Hyaluronan Depolymerization), Alias KIAA1199/CEMIP, in Hyaluronan Degradation in Normal and Photoaged Skin. *Int. J. Mol. Sci.* **2019**, *20*, 5804. [CrossRef]

53. Johnson, P.; Arif, A.A.; Lee–Sayer, S.S.M.; Dong, Y. Hyaluronan and Its Interactions with Immune Cells in the Healthy and Inflamed Lung. *Front. Immunol.* **2018**, *9*, 2787. [CrossRef] [PubMed]

54. Johnson, B.Z.; Stevenson, A.W.; Prele, C.M.; Fear, M.W.; Wood, F.M. The Role of IL–6 in Skin Fibrosis and Cutaneous Wound Healing. *Biomedicines* **2020**, *8*, 101. [CrossRef] [PubMed]

55. Abaurrea, A.; Araujo, A.M.; Caffarel, M.M. The Role of the IL–6 Cytokine Family in Epithelial–Mesenchymal Plasticity in Cancer Progression. *Int. J. Mol. Sci.* **2021**, *22*, 8334. [CrossRef] [PubMed]

56. Toole, B.P. The CD147–HYALURONAN Axis in Cancer. *Anat. Rec.* **2020**, *303*, 1573–1583. [CrossRef]

57. Jackson, D.G. Hyaluronan in the lymphatics: The key role of the hyaluronan receptor LYVE-1 in leucocyte trafficking. *Matrix Biol.* **2019**, *78*, 219–235. [CrossRef]

58. Garantziotis, S.; Matalon, S. Sugarcoating Lung Injury: A Novel Role for High–Molecular–Weight Hyaluronan in Pneumonia. *Am. J. Respir. Crit. Care Med.* **2019**, *200*, 1197–1198. [CrossRef]

59. Weigel, P.H. Planning, evaluating and vetting receptor signaling studies to assess hyaluronan size–dependence and specificity. *Glycobiology* **2017**, *27*, 796–799. [CrossRef]

60. Weigel, P.H.; Baggenstoss, B.A. What is special about 200 kDa hyaluronan that activates hyaluronan receptor signaling? *Glycobiol- ogy* **2017**, *27*, 868–877. [CrossRef]

61. Ziebell, M.R.; Prestwich, G.D. Interactions of peptide mimics of hyaluronic acid with the receptor for hyaluronan mediated motility (RHAMM). *J. Comput. Aided. Mol. Des.* **2004**, *18*, 597–614. [CrossRef]

62. Yang, B.; Yang, B.L.; Savani, R.C.; Turley, E.A. Identification of a common hyaluronan binding motif in the hyaluronan binding proteins RHAMM, CD44 and link protein. *EMBO J.* **1994**, *13*, 286–296. [CrossRef]

63. Johnson, L.A.; Jackson, D.G. Hyaluronan and Its Receptors: Key Mediators of Immune Cell Entry and Trafficking in the Lymphatic. *System Cells* **2021**, *10*, 2061.

64. Tammi, M.I.; Oikari, S.; Pasonen–Seppanen, S.; Rilla, K.; Auvinen, P.; Tammi, R.H. Activated hyaluronan metabolism in the tumor matrix—Causes and consequences. *Matrix Biol.* **2019**, *78*, 147–164. [CrossRef]

65. Altman, R.; Bedi, A.; Manjoo, A.; Niazi, F.; Shaw, P.; Mease, P. Anti–Inflammatory Effects of Intra–Articular Hyaluronic Acid: A Systematic Review. *Cartilage* **2019**, *10*, 4352. [CrossRef] [PubMed]

66. Heldin, P.; Kolliopoulos, C.; Lin, C.Y.; Heldin, C.H. Involvement of hyaluronan and CD44 in cancer and viral infections. *Cell Signal.* **2020**, *65*, 109427. [CrossRef] [PubMed]

67. Marozzi, M.; Parnigoni, A.; Negri, A.; Viola, M.; Vigetti, D.; Passi, A.; Karousou, E.; Rizzi, F. Inflammation, Extracellular Matrix Remodeling, and Proteostasis in Tumor Microenvironment. *Int. J. Mol. Sci.* **2021**, *22*, 8102. [CrossRef]

68. Anderegg, U.; Simon, J.C.; Averbeck, M. More than just a filler—The role of hyaluronan for skin homeostasis. *Exp. Dermatol.* **2014**, *23*, 295–303. [CrossRef]

69. Kleiser, S.; Nystrom, A. Interplay between Cell–Surface Receptors and Extracellular Matrix in Skin. *Biomolecules* **2020**, *10*, 1170. [CrossRef] [PubMed]

70. Banerji, S.; Wright, A.J.; Noble, M.; Mahoney, D.J.; Campbell, I.D.; Day, A.J.; Jackson, D.G. Structures of the CD44–hyaluronan complex provide insight into a fundamental carbohydrate–protein interaction. *Nat. Struct. Mol. Biol.* **2007**, *14*, 234–239. [CrossRef]

71. Senbanjo, L.T.; AlJohani, H.; AlQranei, M.; Majumdar, S.; Ma, T.; Chellaiah, M.A. Identification of sequence–specific interactions of the CD44–intracellular domain with RUNX2 in the transcription of matrix metalloprotease–9 in human prostate cancer cells. *Cancer Drug. Resist.* **2020**, *3*, 586–602. [CrossRef]

72. Al–Othman, N.; Alhendi, A.; Ihbaisha, M.; Barahmeh, M.; Alqaraleh, M.; Al–Momany, B.Z. Role of CD44 in breast cancer. *Breast Dis.* **2020**, *39*, 1–13. [CrossRef]

73. Ooki, T.; Murata–Kamiya, N.; Takahashi–Kanemitsu, A.; Wu, W.; Hatakeyama, M. High–Molecular–Weight Hyaluronan Is a Hippo Pathway Ligand Directing Cell Density–Dependent Growth Inhibition via PAR1b. *Dev. Cell* **2019**, *49*, 590–604.e9. [CrossRef]

74. Jou, I.M.; Wu, T.T.; Hsu, C.C.; Yang, C.C.; Huang, J.S.; Tu, Y.K.; Lee, J.S.; Su, F.C.; Kuo, Y.L. High molecular weight form of hyaluronic acid reduces neuroinflammatory response in injured sciatic nerve via the intracellular domain of CD44. *J. Biomed. Mater. Res. B Appl. Biomater.* **2021**, *109*, 673–680. [CrossRef]

75. Hauser–Kawaguchi, A.; Luyt, L.G.; Turley, E. Design of peptide mimetics to block pro–inflammatory functions of HA fragments. *Matrix Biol.* **2019**, *78*, 346–356. [CrossRef]

76. Gulati, K.; Jamsandekar, M.; Poluri, K.M. Mechanistic insights into molecular evolution of species–specific differential glycosaminoglycan binding surfaces in growth–related oncogene chemokines. *R. Soc. Open Sci.* **2017**, *4*, 171059. [CrossRef]

77. Boittier, E.D.; Gandhi, N.S.; Ferro, V.; Coombe, D.R. Cross–Species Analysis of Glycosaminoglycan Binding Proteins Reveals Some Animal Models Are "More Equal" than Others. *Molecules* **2019**, *24*, 924. [CrossRef]

78. Carvalho, A.M.; Soares da Costa, D.; Paulo, P.M.R.; Reis, R.L.; Pashkuleva, I. Co–localization and crosstalk between CD44 and RHAMM depend on hyaluronan presentation. *Acta Biomater.* **2021**, *119*, 114–124. [CrossRef] [PubMed]

79. Proksch, E.; Brandner, J.M.; Jensen, J.M. The skin: An indispensable barrier. *Exp. Dermatol.* **2008**, *17*, 1063–1072. [CrossRef] [PubMed]

80. Singampalli, K.L.; Balaji, S.; Wang, X.; Parikh, U.M.; Kaul, A.; Gilley, J.; Birla, R.K.; Bollyky, P.L.; Keswani, S.G. The Role of an IL–10/Hyaluronan Axis in Dermal Wound Healing. *Front. Cell Dev. Biol.* **2020**, *8*, 636. [CrossRef] [PubMed]

81. Tolg, C.; Hamilton, S.R.; Zalinska, E.; McCulloch, L.; Amin, R.; Akentieva, N.; Winnik, F.; Savani, R.; Bagli, D.J.; Luyt, L.G.; et al. A RHAMM mimetic peptide blocks hyaluronan signaling and reduces inflammation and fibrogenesis in excisional skin wounds. *Am. J. Pathol.* **2012**, *181*, 1250–1270. [CrossRef]

82. Condorelli, A.G.; El Hachem, M.; Zambruno, G.; Nystrom, A.; Candi, E.; Castiglia, D. Notch–ing up knowledge on molecular mechanisms of skin fibrosis: Focus on the multifaceted Notch signalling pathway. *J. Biomed. Sci.* **2021**, *28*, 36. [CrossRef]

83. Caskey, R.C.; Allukian, M.; Lind, R.C.; Herdrich, B.J.; Xu, J.; Radu, A.; Mitchell, M.E.; Liechty, K.W. Lentiviral–mediated over-expression of hyaluronan synthase–1 (HAS–1) decreases the cellular inflammatory response and results in regenerative wound repair. *Cell Tissue Res.* **2013**, *351*, 117–125. [CrossRef]

84. Lee, B.M.; Park, S.J.; Noh, I.; Kim, C.H. The effects of the molecular weights of hyaluronic acid on the immune responses. *Biomater. Res.* **2021**, *25*, 27. [CrossRef]

85. Shi, Q.; Zhao, L.; Xu, C.; Zhang, L.; Zhao, H. High Molecular Weight Hyaluronan Suppresses Macrophage M1 Polarization and Enhances IL–10 Production in PM2.5–Induced Lung Inflammation. *Molecules* **2019**, *24*, 1766. [CrossRef] [PubMed]

86. Vistejnova, L.; Safrankova, B.; Nesporova, K.; Slavkovsky, R.; Hermannova, M.; Hosek, P.; Velebny, V.; Kubala, L. Low molecular weight hyaluronan mediated CD44 dependent induction of IL–6 and chemokines in human dermal fibroblasts potentiates innate immune response. *Cytokine* **2014**, *70*, 97–103. [CrossRef] [PubMed]

87. Kim, H.; Cha, J.; Jang, M.; Kim, P. Hyaluronic acid–based extracellular matrix triggers spontaneous M2–like polarity of monocyte/macrophage. *Biomater. Sci.* **2019**, *7*, 2264–7771. [CrossRef] [PubMed]

88. Meyer–Siegler, K.L.; Leifheit, E.C.; Vera, P.L. Inhibition of macrophage migration inhibitory factor decreases proliferation and cytokine expression in bladder cancer cells. *BMC Cancer* **2004**, *4*, 34. [CrossRef]

89. Tamoto, K.; Nochi, H.; Tada, M.; Shimada, S.; Mori, Y.; Kataoka, S.; Suzuki, Y.; Nakamura, T. High–molecular–weight hyaluronic acids inhibit chemotaxis and phagocytosis but not lysosomal enzyme release induced by receptor–mediated stimulations in guinea pig phagocytes. *Microbiol. Immunol.* **1994**, *38*, 73–80. [CrossRef]

90. Schimizzi, A.L.; Massie, J.B.; Murphy, M.; Perry, A.; Kim, C.W.; Garfin, S.R.; Akeson, W.H. High–molecular–weight hyaluronan inhibits macrophage proliferation and cytokine release in the early wound of a preclinical postlaminectomy rat model. *Spine J.* **2006**, *6*, 550–556. [CrossRef]

91. Sheehan, K.M.; DeLott, L.B.; West, R.A.; Bonnema, J.D.; DeHeer, D.H. Hyaluronic acid of high molecular weight inhibits proliferation and induces cell death in U937 macrophage cells. *Life Sci.* **2004**, *75*, 3087–3102. [CrossRef]

92. Huang, L.; Gu, H.; Burd, A. A reappraisal of the biological effects of hyaluronan on human dermal fibroblast. *J. Biomed. Mater. Res. A* **2009**, *90*, 1177–1785. [CrossRef]

93. Evanko, S.P.; Potter–Perigo, S.; Petty, L.J.; Workman, G.A.; Wight, T.N. Hyaluronan Controls the Deposition of Fibronectin and Collagen and Modulates TGF–beta1 Induction of Lung Myofibroblasts. *Matrix Biol.* **2015**, *42*, 74–92. [CrossRef]

94. Kawano, Y.; Patrulea, V.; Sublet, E.; Borchard, G.; Iyoda, T.; Kageyama, R.; Morita, A.; Seino, S.; Yoshida, H.; Jordan, O.; et al. Wound Healing Promotion by Hyaluronic Acid: Effect of Molecular Weight on Gene Expression and In Vivo Wound Closure. *Pharmaceuticals* **2021**, *14*, 301. [CrossRef]

95. Hoffmann, A.; Hoing, J.L.; Newman, M.; Simman, R. Role of Hyaluronic Acid Treatment in the Prevention of Keloid Scarring. *J. Am. Coll. Clin. Wound Spec.* **2012**, *4*, 23–31. [CrossRef] [PubMed]

96. D'Agostino, A.; Stellavato, A.; Corsuto, L.; Diana, P.; Filosa, R.; La Gatta, A.; De Rosa, M.; Schiraldi, C. Is molecular size a discriminating factor in hyaluronan interaction with human cells? *Carbohydr. Polym.* **2017**, *157*, 21–30. [CrossRef]

97. Fronza, M.; Caetano, G.F.; Leite, M.N.; Bitencourt, C.S.; Paula–Silva, F.W.; Andrade, T.A.; Frade, M.A.; Merfort, I.; Faccioli, L.H. Hyaluronidase modulates inflammatory response and accelerates the cutaneous wound healing. *PLoS ONE* **2014**, *9*, e112297. [CrossRef] [PubMed]

98. Tolg, C.; Telmer, P.; Turley, E. Specific sizes of hyaluronan oligosaccharides stimulate fibroblast migration and excisional wound repair. *PLoS ONE* **2014**, *9*, e88479. [CrossRef] [PubMed]

99. Damodarasamy, M.; Johnson, R.S.; Bentov, I.; MacCoss, M.J.; Vernon, R.B.; Reed, M.J. Hyaluronan enhances wound repair and increases collagen III in aged dermal wounds. *Wound Repair Regen.* **2014**, *22*, 521–526. [CrossRef] [PubMed]

100. Radrezza, S.; Baron, G.; Nukala, S.B.; Depta, G.; Aldini, G.; Carini, M.; D'Amato, A. Advanced quantitative proteomics to evaluate molecular effects of low–molecular–weight hyaluronic acid in human dermal fibroblasts. *J. Pharm. Biomed. Anal.* **2020**, *185*, 113199. [CrossRef] [PubMed]

101. Nagy, N.; Kuipers, H.F.; Marshall, P.L.; Wang, E.; Kaber, G.; Bollyky, P.L. Hyaluronan in immune dysregulation and autoimmune diseases. *Matrix Biol.* **2019**, *78*, 292–313. [CrossRef]

102. Jin, J.; Li, Y.; Zhao, Q.; Chen, Y.; Fu, S.; Wu, J. Coordinated regulation of immune contexture: Crosstalk between STAT3 and immune cells during breast cancer progression. *Cell Commun. Signal.* **2021**, *19*, 50. [CrossRef]

103. Jenkins, S.; Wesolowski, R.; Gatti–Mays, M.E. Improving Breast Cancer Responses to Immunotherapy–a Search for the Achilles Heel of the Tumor Microenvironment. *Curr. Oncol. Rep.* **2021**, *23*, 55. [CrossRef] [PubMed]

104. Deligne, C.; Midwood, K.S. Macrophages and Extracellular Matrix in Breast Cancer: Partners in Crime or Protective Allies? *Front. Oncol.* **2021**, *11*, 620773. [CrossRef] [PubMed]

105. Joshi, R.S.; Kanugula, S.S.; Sudhir, S.; Pereira, M.P.; Jain, S.; Aghi, M.K. The Role of Cancer–Associated Fibroblasts in Tumor Progression. *Cancers* **2021**, *13*, 1399. [CrossRef] [PubMed]

106. Deepak, K.G.K.; Vempati, R.; Nagaraju, G.P.; Dasari, V.R.; Nagini, S.; Rao, D.N.; Malla, R.R. Tumor microenvironment: Challenges and opportunities in targeting metastasis of triple negative breast cancer. *Pharmacol. Res.* **2020**, *153*, 104683. [CrossRef] [PubMed]

107. Wu, W.; Chen, L.; Wang, Y.; Jin, J.; Xie, X.; Zhang, J. Hyaluronic acid predicts poor prognosis in breast cancer patients: A protocol for systematic review and meta analysis. *Medicine* **2020**, *99*, e20438. [CrossRef]

108. Wu, M.; Cao, M.; He, Y.; Liu, Y.; Yang, C.; Du, Y.; Wang, W.; Gao, F. A novel role of low molecular weight hyaluronan in breast cancer metastasis. *FASEB J.* **2015**, *29*, 1290–2198. [CrossRef]

109. Bohrer, L.R.; Chuntova, P.; Bade, L.K.; Beadnell, T.C.; Leon, R.P.; Brady, N.J.; Ryu, Y.; Goldberg, J.E.; Schmechel, S.C.; Koopmeiners, J.S. Activation of the FGFR–STAT3 pathway in breast cancer cells induces a hyaluronan–rich microenvironment that licenses tumor formation. *Cancer Res.* **2014**, *74*, 374–386. [CrossRef] [PubMed]

110. Whiteside, T.L. Regulatory T cell subsets in human cancer: Are they regulating for or against tumor progression? *Cancer Immunol. Immunother.* **2014**, *63*, 67–72. [CrossRef]

111. Urakawa, H.; Nishida, Y.; Wasa, J.; Arai, E.; Zhuo, L.; Kimata, K.; Kozawa, E.; Futamura, N.; Ishiguro, N. Inhibition of hyaluronan synthesis in breast cancer cells by 4–methylumbelliferone suppresses tumorigenicity in vitro and metastatic lesions of bone in vivo. *Int. J. Cancer* **2012**, *130*, 454–466. [CrossRef]

112. Spinelli, F.M.; Vitale, D.L.; Icardi, A.; Caon, I.; Brandone, A.; Giannoni, P.; Saturno, V.; Passi, A.; García, M.; Sevic, I.; et al. Hyaluronan preconditioning of monocytes/macrophages affects their angiogenic behavior and regulation of TSG–6 expression in a tumor type–specific manner. *FEBS J.* **2019**, *286*, 3433–3449. [CrossRef]

113. Spinelli, F.M.; Vitale, D.L.; Sevic, I.; Alaniz, L. Hyaluronan in the Tumor Microenvironment. *Adv. Exp. Med. Biol.* **2020**, *1245*, 67–83.

114. Witschen, P.M.; Chaffee, T.S.; Brady, N.J.; Huggins, D.N.; Knutson, T.P.; LaRue, R.S.; Munro, S.A.; Tiegs, L.; McCarthy, J.B.; Nelson, A.C.; et al. Tumor Cell Associated Hyaluronan–CD44 Signaling Promotes Pro–Tumor Inflammation in Breast Cancer. *Cancers* **2020**, *12*, 1325. [CrossRef] [PubMed]

115. McCarthy, J.B.; El–Ashry, D.; Turley, E.A. Hyaluronan, Cancer–Associated Fibroblasts and the Tumor Microenvironment in Malignant Progression. *Front. Cell Dev. Biol.* **2018**, *6*, 48. [CrossRef] [PubMed]

116. Zhao, C.; Thompson, B.J.; Chen, K.; Gao, F.; Blouw, B.; Marella, M.; Zimmerman, S.; Kimbler, T.; Garrovillo, S.; Bahn, J.; et al. The growth of a xenograft breast cancer tumor model with engineered hyaluronan–accumulating stroma is dependent on hyaluronan and independent of CD44. *Oncotarget* **2019**, *10*, 6561–6576. [CrossRef] [PubMed]

117. Lompardia, S.; Diaz, M.; Pibuel, M.; Papademetrio, D.; Poodts, D.; Mihalez, C.; Álvarez, É.; Hajos, S. Hyaluronan abrogates imatinib–induced senescence in chronic myeloid leukemia cell lines. *Sci. Rep.* **2019**, *9*, 10930. [CrossRef] [PubMed]

118. Tansi, F.L.; Frobel, F.; Maduabuchi, W.O.; Steiniger, F.; Westermann, M.; Quaas, R.; Teichgräber, U.K.; Hilger, I. Effect of Matrix–Modulating Enzymes on The Cellular Uptake of Magnetic Nanoparticles and on Magnetic Hyperthermia Treatment of Pancreatic Cancer Models In Vivo. *Nanomaterials* **2021**, *11*, 438. [CrossRef] [PubMed]

119. Chen, X.; Du, Y.; Liu, Y.; He, Y.; Zhang, G.; Yang, C.; Gao, F. Hyaluronan arrests human breast cancer cell growth by prolonging the G0/G1 phase of the cell cycle. *Acta Biochim. Biophys. Sin.* **2018**, *50*, 1181–1189. [CrossRef] [PubMed]

120. Zhao, Y.; Qiao, S.; Hou, X.; Tian, H.; Deng, S.; Ye, K.; Nie, Y.; Chen, X.; Yan, H.; Tian, W. Bioengineered tumor microenvironments with naked mole rats high–molecular–weight hyaluronan induces apoptosis in breast cancer cells. *Oncogene* **2019**, *38*, 4297–4309. [CrossRef] [PubMed]

121. Zhao, Y.F.; Qiao, S.P.; Shi, S.L.; Yao, L.F.; Hou, X.L.; Li, C.F.; Lin, F.H.; Guo, K.; Acharya, A.; Chen, X.B.; et al. Modulating Three–Dimensional Microenvironment with Hyaluronan of Different Molecular Weights Alters Breast Cancer Cell Invasion Behavior. *ACS Appl. Mater. Interfaces* **2017**, *9*, 9327–9338. [CrossRef] [PubMed]

122. Bourguignon, L.Y.; Wong, G.; Earle, C.A.; Xia, W. Interaction of low molecular weight hyaluronan with CD44 and toll–like receptors promotes the actin filament–associated protein 110–actin binding and MyD88–NFkappaB signaling leading to proin-flammatory cytokine/chemokine production and breast tumor invasion. *Cytoskeleton* **2011**, *68*, 671–693.

123. Koyama, H.; Hibi, T.; Isogai, Z.; Yoneda, M.; Fujimori, M.; Amano, J.; Kawakubo, M.; Kannagi, R.; Kimata, K.; Taniguchi, S.; et al. Hyperproduction of hyaluronan in neu–induced mammary tumor accelerates angiogenesis through stromal cell recruitment: Possible involvement of versican/PG–M. *Am. J. Pathol.* **2007**, *170*, 1086–1099. [CrossRef] [PubMed]

124. Koyama, H.; Kobayashi, N.; Harada, M.; Takeoka, M.; Kawai, Y.; Sano, K.; Fujimori, M.; Amano, J.; Ohhashi, T.; Kannagi, R.; et al. Significance of tumor–associated stroma in promotion of intratumoral lymphangiogenesis: Pivotal role of a hyaluronan–rich tumor microenvironment. *Am. J. Pathol.* **2008**, *172*, 179–193. [CrossRef] [PubMed]

125. Tavianatou, A.G.; Piperigkou, Z.; Barbera, C.; Beninatto, R.; Masola, V.; Caon, I.; Onisto, M.; Franchi, M.; Galesso, D.; Karamanos, N.K. Molecular size–dependent specificity of hyaluronan on functional properties, morphology and matrix composition of mammary cancer cells. *Matrix Biol. Plus* **2019**, *3*, 100008. [CrossRef] [PubMed]

126. Lien, H.C.; Lee, Y.H.; Jeng, Y.M.; Lin, C.H.; Lu, Y.S.; Yao, Y.T. Differential expression of hyaluronan synthase 2 in breast carcinoma and its biological significance. *Histopathology* **2014**, *65*, 328–339. [CrossRef]

127. Hamilton, S.R.; Fard, S.F.; Paiwand, F.F.; Tolg, C.; Veiseh, M.; Wang, C.; McCarthy, J.B.; Bissell, M.J.; Koropatnick, J.; Turley, E.A. The hyaluronan receptors CD44 and Rhamm (CD168) form complexes with ERK1,2 that sustain high basal motility in breast cancer cells. *J. Biol. Chem.* **2007**, *282*, 16667–16680. [CrossRef]

128. Veiseh, M.; Kwon, D.H.; Borowsky, A.D.; Tolg, C.; Leong, H.S.; Lewis, J.D.; Turley, E.A.; Bissell, M.J. Cellular heterogeneity profiling by hyaluronan probes reveals an invasive but slow–growing breast tumor subset. *Proc. Natl. Acad. Sci. USA* **2014**, *111*, 1731–1739. [CrossRef]

129. Glasner, A.; Plitas, G. Tumor resident regulatory T cells. *Semin. Immunol.* **2021**, 101476, in press. [CrossRef]

130. Govindaraju, P.; Todd, L.; Shetye, S.; Monslow, J.; Pure, E. CD44–dependent inflammation, fibrogenesis, and collagenolysis regulates extracellular matrix remodeling and tensile strength during cutaneous wound healing. *Matrix Biol.* **2019**, *75*, 314–330. [CrossRef]

131. Shatirishvili, M.; Burk, A.S.; Franz, C.M.; Pace, G.; Kastilan, T.; Breuhahn, K.; Hinterseer, E.; Dierich, A.; Bakiri, L.; Wagner, E.F.; et al. Epidermal–specific deletion of CD44 reveals a function in keratinocytes in response to mechanical stress. *Cell Death Dis.* **2016**, *7*, e2461. [CrossRef] [PubMed]

132. Lovvorn, H.N., III; Cass, D.L.; Sylvester, K.G.; Yang, E.Y.; Crombleholme, T.M.; Adzick, N.S.; Savani, R.C. Hyaluronan receptor expression increases in fetal excisional skin wounds and correlates with fibroplasia. *J. Pediatr. Surg.* **1998**, *33*, 1062–1069. [CrossRef]

133. Tolg, C.; Hamilton, S.R.; Nakrieko, K.A.; Kooshesh, F.; Walton, P.; McCarthy, J.B.; Bissell, M.J.; Turley, E.A. Rhamm–/– fibroblasts are defective in CD44–mediated ERK1,2 motogenic signaling, leading to defective skin wound repair. *J. Cell Biol.* **2006**, *175*, 1017–1028. [CrossRef] [PubMed]

134. Kouvidi, K.; Berdiaki, A.; Nikitovic, D.; Katonis, P.; Afratis, N.; Hascall, V.C.; Karamanos, N.K.; Tzanakakis, G.N. Role of receptor for hyaluronic acid–mediated motility (RHAMM) in low molecular weight hyaluronan (LMWHA)–mediated fibrosarcoma cell adhesion. *J. Biol. Chem.* **2011**, *286*, 38509–38520. [CrossRef]

135. Wu, K.; Kim, S.; Liu, V.M.; Sabino, A.; Minkhorst, K.; Yazdani, A.; Turley, E.A. Function–blocking RHAMM peptides attenuate fibrosis and promote anti–fibrotic adipokines in a bleomycin–induced murine model of systemic sclerosis. *J. Investig. Dermatol.* **2020**, *141*, 1482–1492. [CrossRef]

136. Gao, F.; Yang, C.X.; Mo, W.; Liu, Y.W.; He, Y.Q. Hyaluronan oligosaccharides are potential stimulators to angiogenesis via RHAMM mediated signal pathway in wound healing. *Clin. Investig. Med.* **2008**, *31*, 106–116. [CrossRef]
137. Seifert, A.W.; Monaghan, J.R.; Voss, S.R.; Maden, M. Skin regeneration in adult axolotls: A blueprint for scar–free healing in vertebrates. *PLoS ONE* **2012**, *7*, e32875.
138. Imanaka–Yoshida, K.; Tawara, I.; Yoshida, T. Tenascin–C in cardiac disease: A sophisticated controller of inflammation, repair, and fibrosis. *Am. J. Physiol. Cell Physiol.* **2020**, *319*, 781–796. [CrossRef]
139. Walsh, H.R.; Cruickshank, B.M.; Brown, J.M.; Marcato, P. The Flick of a Switch: Conferring Survival Advantage to Breast Cancer Stem Cells Through Metabolic Plasticity. *Front. Oncol.* **2019**, *9*, 753. [CrossRef]
140. Saeg, F.; Anbalagan, M. Breast cancer stem cells and the challenges of eradication: A review of novel therapies. *Stem Cell Investig.* **2018**, *5*, 39. [CrossRef]
141. Louderbough, J.M.; Schroeder, J.A. Understanding the dual nature of CD44 in breast cancer progression. *Mol. Cancer Res.* **2011**, *9*, 1573–1586. [CrossRef]
142. Huang, P.; Chen, A.; He, W.; Li, Z.; Zhang, G.; Liu, Z.; Liu, G.; Liu, X.; He, S.; Xiao, G.; et al. BMP–2 induces EMT and breast cancer stemness through Rb and CD44. *Cell Death Discov.* **2017**, *3*, 17039. [CrossRef]
143. Muller, S.; Sindikubwabo, F.; Caneque, T.; Lafon, A.; Versini, A.; Lombard, B.; Loew, D.; Wu, T.D.; Ginestier, C.; Charafe–Jauffret, E.; et al. CD44 regulates epigenetic plasticity by mediating iron endocytosis. *Nat. Chem.* **2020**, *12*, 92–938. [CrossRef]
144. Lopez, J.I.; Camenisch, T.D.; Stevens, M.V.; Sands, B.J.; McDonald, J.; Schroeder, J.A. CD44 attenuates metastatic invasion during breast cancer progression. *Cancer Res.* **2005**, *65*, 6755–6763. [CrossRef]
145. Xu, H.; Niu, M.; Yuan, X.; Wu, K.; Liu, A. CD44 as a tumor biomarker and therapeutic target. *Exp. Hematol. Oncol.* **2020**, *9*, 36. [CrossRef]
146. Ma, L.; Dong, L.; Chang, P. CD44v6 engages in colorectal cancer progression. *Cell Death Dis.* **2019**, *10*, 30. [CrossRef] [PubMed]
147. Wang, C.; Thor, A.D.; Moore, D.H., II; Zhao, Y.; Kerschmann, R.; Stern, R.; Watson, P.H.; Turley, E.A. The overexpression of RHAMM, a hyaluronan–binding protein that regulates ras signaling, correlates with overexpression of mitogen–activated protein kinase and is a significant parameter in breast cancer progression. *Clin. Cancer Res.* **1998**, *4*, 567–576. [PubMed]
148. Wang, J.; Li, D.; Shen, W.; Sun, W.; Gao, R.; Jiang, P.; Wang, L.; Liu, Y.; Chen, Y.; Zhou, W.; et al. RHAMM inhibits cell migration via the AKT/GSK3beta/Snail axis in luminal A subtype breast cancer. *Anat. Rec.* **2020**, *303*, 2344–2356. [CrossRef]
149. Hall, C.L.; Yang, B.; Yang, X.; Zhang, S.; Turley, M.; Samuel, S.; Lange, L.A.; Wang, C.; Curpen, G.D.; Savani, R.C.; et al. Overexpression of the hyaluronan receptor RHAMM is transforming and is also required for H–ras transformation. *Cell* **1995**, *82*, 19–26. [CrossRef]
150. Kotla, N.G.; Bonam, S.R.; Rasala, S.; Wankar, J.; Bohara, R.A.; Bayry, J.; Rochev, Y.; Pandit, A. Recent advances and prospects of hyaluronan as a multifunctional therapeutic system. *J. Control. Release* **2021**, *336*, 598–620. [CrossRef]
151. Salari, N.; Mansouri, K.; Valipour, E.; Abam, F.; Jaymand, M.; Rasoulpoor, S.; Dokaneheifard, S.; Mohammadi, M. Hyaluronic acid–based drug nanocarriers as a novel drug delivery system for cancer chemotherapy: A systematic review. *Daru* **2021**, 1–9. [CrossRef]
152. Kultti, A.; Pasonen–Seppänen, S.; Jauhiainen, M.; Rilla, K.J.; Kärnä, R.; Pyöriä, E.; Tammi, R.H.; Tammi, M.I. 4–Methylumbelliferone inhibits hyaluronan synthesis by depletion of cellular UDP glucuronic acid and downregulation of hyaluronan synthase 2 and 3. *Exp. Cell Res.* **2009**, *315*, 1914–1923. [CrossRef] [PubMed]
153. Yates, T.J.; Lopez, L.E.; Lokeshwar, S.D.; Ortiz, N.; Kallifatidis, G.; Jordan, A.; Hoye, K.; Altman, N.; Lokeshwar, V.B. Dietary Supplement 4–Methylumbelliferone: An Effective Chemopreventive and Therapeutic Agent for Prostate Cancer. *J. Natl. Cancer Inst.* **2015**, *107*, djv085. [CrossRef] [PubMed]
154. Kudo, D.; Suto, A.; Hakamada, K. The Development of a Novel Therapeutic Strategy to Target Hyaluronan in the Extracellular Matrix of Pancreatic Ductal Adenocarcinoma. *Int. J. Mol. Sci.* **2017**, *18*, 600. [CrossRef] [PubMed]
155. Nagy, N.; Kuipers, H.F.; Frymoyer, A.R.; Ishak, H.D.; Bollyky, J.B.; Wight, T.N.; Bollyky, P.L. 4–methylumbelliferone treatment and hyaluronan inhibition as a therapeutic strategy in inflammation, autoimmunity, and cancer. *Front. Immunol.* **2015**, *6*, 123. [CrossRef] [PubMed]
156. Pereira, P.M.R.; Ragupathi, A.; Shmuel, S.; Mandleywala, K.; Viola, N.T.; Lewis, J.S. HER2–Targeted PET Imaging and Therapy of Hyaluronan–Masked HER2–Overexpressing Breast Cancer. *Mol. Pharm.* **2020**, *17*, 327–337. [CrossRef]
157. Palyi–Krekk, Z.; Barok, M.; Isola, J.; Tammi, M.; Szollosi, J.; Nagy, P. Hyaluronan–induced masking of ErbB2 and CD44–enhanced trastuzumab internalisation in trastuzumab resistant breast cancer. *Eur. J. Cancer* **2007**, *43*, 2423–2433. [CrossRef]
158. Wong, K.M.; Horton, K.J.; Coveler, A.L.; Hingorani, S.R.; Harris, W.P. Targeting the Tumor Stroma: The Biology and Clinical Development of Pegylated Recombinant Human Hyaluronidase (PEGPH20). *Curr. Oncol. Rep.* **2017**, *19*, 47. [CrossRef]
159. Adel, N. Current treatment landscape and emerging therapies for pancreatic cancer. *Am. J. Manag. Care* **2019**, *25*, 3–10.
160. Verdaguer, H.; Arroyo, A.; Macarulla, T. New Horizons in the Treatment of Metastatic Pancreatic Cancer: A Review of the Key Biology Features and the Most Recent Advances to Treat Metastatic Pancreatic Cancer. *Target Oncol.* **2018**, *13*, 691–704. [CrossRef]
161. Brundel, D.H.; Feeney, O.M.; Nowell, C.J.; Suys, E.J.; Gracia, G.; Kaminskas, L.M.; McIntosh, M.M.; Kang, D.W.; Porter, C.J. Depolymerization of hyaluronan using PEGylated human recombinant hyaluronidase promotes nanoparticle tumor penetration. *Nanomedicine* **2021**, *16*, 275–292. [CrossRef] [PubMed]

162. Chaudhry, G.E.; Akim, A.; Naveed Zafar, M.; Safdar, N.; Sung, Y.Y.; Muhammad, T.S.T. Understanding Hyaluronan Receptor (CD44) Interaction, HA–CD44 Activated Potential Targets in Cancer Therapeutics. *Adv. Pharm. Bull.* **2021**, *11*, 426–438. [CrossRef] [PubMed]
163. Diebolder, P.; Mpoy, C.; Scott, J.; Huynh, T.T.; Fields, R.; Spitzer, D.; Bandara, N.; Rogers, B.E. Preclinical Evaluation of an Engineered Single–Chain Fragment Variable–Fragment Crystallizable Targeting Human CD44. *J. Nucl. Med.* **2021**, *62*, 137–143. [CrossRef]
164. Menke–van der Houven van Oordt, C.W.; Gomez–Roca, C.; van Herpen, C.; Coveler, A.L.; Mahalingam, D.; Verheul, H.M.; van der Graaf, W.T.; Christen, R.; Rüttinger, D.; Weigand, S.; et al. First–in–human phase I clinical trial of RG7356, an anti–CD44 humanized antibody, in patients with advanced, CD44–expressing solid tumors. *Oncotarget* **2016**, *7*, 80046–80058. [CrossRef] [PubMed]

 *biomolecules*

Article

Sulfated Hyaluronan Modulates the Functional Properties and Matrix Effectors Expression of Breast Cancer Cells with Different Estrogen Receptor Status

Christos Koutsakis [1], Anastasia-Gerasimoula Tavianatou [1], Dimitris Kokoretsis [1], Georgios Baroutas [1] and Nikos K. Karamanos [1,2,*]

[1] Biochemistry, Biochemical Analysis and Matrix Pathobiology Research Group, Laboratory of Biochemistry, Department of Chemistry, University of Patras, 265 04 Patras, Greece; ckoutsakis@upatras.gr (C.K.); natasha.tavianatou@hotmail.gr (A.-G.T.); kokoretsisdimitris@gmail.com (D.K.); baroutasg@gmail.com (G.B.)

[2] Foundation for Research and Technology-Hellas (FORTH)/Institute of Chemical Engineering Sciences (ICE-HT), 265 04 Patras, Greece

* Correspondence: n.k.karamanos@upatras.gr; Tel.: +30-261-099-7915

Abstract: Hyaluronan (HA) is an extracellular matrix glycosaminoglycan (GAG) that plays a pivotal role in breast cancer. While HA is the only GAG not normally substituted with sulfate groups, sulfated hyaluronan (sHA) has previously been used in studies with promising antitumor results. The aim of the present study was to evaluate the effects sHA fragments have on breast cancer cells with different estrogen receptor (ER) status. To this end, ERα-positive MCF-7, and ERβ-positive MDA-MB-231 cells were treated with non-sulfated HA or sHA fragments of 50 kDa. The functional properties of the breast cancer cells and the expression of key matrix effectors were investigated. According to the results, sHA attenuates cell proliferation, migration, and invasion, while increasing adhesion on collagen type I. Furthermore, sHA modulates the expression of epithelial-to-mesenchymal transition (EMT) markers, such as E-cadherin and snail2/slug. Additionally, sHA downregulates matrix remodeling enzymes such as the matrix metalloproteinases MT1-MMP, MMP2, and MMP9. Notably, sHA exhibits a stronger effect on the breast cancer cell properties compared to the non-sulfated counterpart, dependent also on the type of cancer cell type. Consequently, a deeper understanding of the mechanism by which sHA facilitate these processes could contribute to the development of novel therapeutic strategies.

Keywords: sulfated hyaluronan; extracellular matrix; breast cancer; estrogen receptors; epithelial-to-mesenchymal transition; matrix metalloproteinases

Citation: Koutsakis, C.; Tavianatou, A.-G.; Kokoretsis, D.; Baroutas, G.; Karamanos, N.K. Sulfated Hyaluronan Modulates the Functional Properties and Matrix Effectors Expression of Breast Cancer Cells with Different Estrogen Receptor Status. *Biomolecules* **2021**, *11*, 1916. https://doi.org/10.3390/biom11121916

Academic Editors: George Tzanakakis and Dragana Nikitovic

Received: 15 November 2021
Accepted: 16 December 2021
Published: 20 December 2021

Publisher's Note: MDPI stays neutral with regard to jurisdictional claims in published maps and institutional affiliations.

1. Introduction

Breast cancer is characterized by high heterogeneity and constitutes one of the most commonly reported types of cancer globally [1]. Throughout disease development, the expression patterns of the estrogen receptors (ERs) are crucial for regulating breast cancer cell properties, morphology, as well the expression of several effectors associated with aggressiveness of the malignancy [2–4]. Depending on the ER status, breast cancer cells are categorized into ERα-positive, with epithelial characteristics and low metastatic potential, and ERα-negative, which exhibit higher metastatic potential and are associated with more aggressive phenotypes [5]. During breast cancer progression, cellular phenotypes can be altered due to epithelial-to-mesenchymal transition (EMT), which leads to loss of cell polarity and cell-to-cell junctions, thus facilitating cellular migration and metastasis [6,7]. As a result, the expression of epithelial markers, such as E-cadherin, is significantly downregulated, whereas the expression levels of mesenchymal markers, such as snail2/slug, are elevated [8–10]. Additionally, overexpression of matrix enzymes, such as matrix metalloproteinases (MMPs) accounts for the extensive reorganization of the extracellular matrix (ECM), further contributing to cell migration and invasion [11].

ECM is a complex, well-organized, three-dimensional network of macromolecules, which provides structural support to the cells and at the same time regulates their morphology and function [12,13]. One major component of ECM is the polysaccharide hyaluronan (HA), a linear non-sulfated GAG composed of disaccharide repeating units of D-glucuronic acid (GlcA) and *N*-acetyl-D-glucosamine (GlcNAc) [13]. HA is renowned for its role and participation in various physiological processes, like embryogenesis, wound healing and inflammation, as well as being linked to tumor initiation and development [14–16]. After binding to its receptors, HA can induce intracellular signaling pathways, thus regulating cell behavior. The most well-characterized HA receptor is CD44, which is implicated in various cellular processes [17–19]. Three enzymes are responsible for HA synthesis, the HA synthases HAS1–3, each producing HA of different sizes [20–22]. Moreover, HA degradation into low molecular fragments is carried out by the action of the specific hyaluronidases [23].

Depending on its molecular size, HA exerts different biological functions and regulates specific cellular behavior. For instance, HA fragments of <10, 30, and 200 kDa affect breast cancer cells' functional properties and morphology, as well as the expression patterns of several molecules in a size-dependent manner [24,25]. Moreover, high molecular weight (HMW) HA is known for its anti-inflammatory, anti-angiogenic, and anti-proliferative properties [26]. Although elevated levels of endogenous HA have been linked with an aggressive phenotype in breast cancer cells, several studies have reported that low molecular weight (LMW) HA (100–300 kDa) can attenuate cancer cell growth and metastasis [27–29]. Recently, the effects of modified HA molecules have come under focus. These molecules have undergone modifications on their structure, such as the addition of sulfate groups. The inhibitory role of sulfated HA (sHA) in cancer has been shown in prostate cancer cells, where it attenuates the activity of HYAL1, decreasing cancer cell proliferation and invasion [30]. Moreover, the antitumor capabilities of sHA have also been shown in pre-clinical models of bladder cancer, where sHA fragments inhibited the proliferation, migration and invasion of cancer cells, as well as angiogenesis [31]. Together, these observations suggest the potential use of HA fragments as novel therapeutic approaches in malignancies. However, studies regarding the direct comparison between the effects exerted by non- and sulfated HA of the same molecular size are not available, according to our knowledge.

The aim of the present study was therefore to investigate the role of sHA on breast cancer cells with different ER status. To this end, non-sulfated 50 kDA HA and sHA fragments of the same molecular size were used to treat two breast cancer cell lines of different ER status: the less invasive ERα-positive MCF-7 and the highly aggressive ERα-negative/ERβ-positive MDA-MB-231 cells. Here, we demonstrate that sHA fragments regulate the functional properties of breast cancer cells and modify the expression of EMT markers, as well as key matrix effectors, such as MMPs and hyaluronan synthases.

2. Materials and Methods

2.1. HA and sHA Fragments

The HA fragments of 50 kDa were isolated at Fidia Farmaceutici S.p.A as described in a previous study of our research group [25]. The sulfated HA (MW = 50 kDa, degree of sulfation: 2.6) was kindly provided by Fidia Farmaceutici S.p.A (Abano Terme, Italy).

2.2. Cell Cultures and Reagents

The MCF-7 (ERα-positive, low metastatic potential) and MDA-MB-231 (ERα-negative/ERβ-positive, high metastatic potential) breast cancer cell lines were obtained from the American Type Culture Collection (ATCC, Manassas, VA, USA). The cells were cultured at 37 °C in a humidified atmosphere of 5% CO_2 and 95% air. The cell culture medium used was Dulbecco's Modified Eagle Medium (DMEM, LM-D1110/500, Biosera, Nuaillé, France), supplemented with 10% fetal bovine serum (FBS, FB-1000/500, Biosera, Nuaillé, France), antimicrobial agents (100 IU/mL penicillin, 100 µg/mL streptomycin, 10 µg/mL gentamycin sulfate and 2.5 µg/mL amphotericin B), 1 mM sodium pyruvate, and 2 mM

L-glutamine (Biosera, Nuaillé, France). The cells were harvested using trypsin-EDTA 1× in PBS (LM-T1706/500, Biosera, Nuaillé, France) at approximately 80–85% cell confluency. All experiments were conducted in serum-free conditions using three separate biological replicates. The cytostatic agent cytarabine was purchased from Sigma-Aldrich (Saint Louis, MO, USA). All other chemicals used were of the best commercially available grade.

2.3. Proliferation Assay

MCF-7 and MDA-MB-231 cells were seeded on 96-well plates at a density of 7500 and 5000 cells per well respectively. After a 24 h incubation in complete cell culture medium, the cells were serum starved overnight. The HA and sHA fragments were added at a concentration of 200 µg/mL and the cells were incubated for an additional 24 h. In order to evaluate the effects on breast cancer cell proliferation a crystal violet assay was performed, as described in a previous study [32]. The cells were washed twice with PBS and stained with 0.5% (*w*/*v*) crystal violet in 20% methanol/distilled water solution. After a 20 min incubation on a bench rocker, the staining solution was aspirated, and the cells were washed three times with distilled water. The plates were left at room temperature to air dry overnight. The next day, methanol was added to solubilize the dye, followed by a 20 min incubation on a bench rocker. Finally, the optical density of each well was measured at 570 nm using a TECAN photometer.

2.4. Wound Healing Assay

MCF-7 and MDA-MB-231 cells were seeded on 12-well plates at a density of 3×10^5 and 2.5×10^5 cells per well respectively. The cells were cultured in complete cell culture medium for 24 h and were serum starved overnight. The next day, the cell monolayer was wounded by scratching with a sterile 100 µL pipette tip. The wells were then washed twice with PBS to remove the detached cells. To minimize possible contribution of cell proliferation to the migration results, serum-free medium containing the cytostatic cytarabine (10 µM) was added and remained until the end of the assay. After a 40 min incubation, the HA and sHA fragments were added at a final concentration of 200 µg/mL. The wound closure was captured at two time points of 0 and 24 h using a digital camera connected to a phase-contrast microscope. The cell migration was determined by quantification of the wound surface area difference between the two time points using image analysis (Fiji v1.52p) [33].

2.5. Collagen Type I Cell Adhesion Assay

MCF-7 and MDA-MB-231 cells were seeded on 6-well plates at a density of 3×10^5 and 2.5×10^5 cells per well respectively. The cells were cultured in complete cell culture medium for 24 h and were serum starved overnight. The next day, the HA and sHA fragments were added at a final concentration of 200 µg/mL in serum-free medium. Meanwhile, 96-well plates were prepared with 40 µg/mL collagen type I in PBS and were stored at 4 °C overnight. The next day the solution was aspirated, the plates were washed twice with PBS, and blocked with 1% BSA in PBS for 30 min. After a 24 h incubation, the cells in the 6-well plates were harvested using 4 mM EDTA in PBS, before being centrifuged and resuspended in serum-free medium containing 0.1% BSA. The cells were then seeded in the pre-coated collagen type I 96-well plates at a density of 2×10^4 per well for the MCF-7 and 1×10^4 cells per well for the MDA-MB-231. After seeding, the cells were incubated at 37 °C for 40 min to adhere to the collagen type I. Subsequently, the cells were washed twice with PBS to remove the non-adherent ones. To determine the adhesion rate, the adherent cells were stained following the crystal violet assay, as described above (Section 2.3).

2.6. Collagen Type I Invasion Assay

To evaluate the HA and sHA fragments effects on breast cancer cells invasive capacity, a collagen type I invasion assay was used. The cells were cultured in complete cell culture medium for 24 h and were serum starved overnight. A collagen type I solution was prepared as follows: 5 volumes of CMF-HBSS were mixed with 2.65 volumes of complete medium, 1 volume of MEM 10×, 1 volume of 0.25 M NaHCO3, 0.3 volumes of 1 M NaOH and 4 volumes of collagen type I (stock concentration 5 mg/mL). The final concentration of collagen type I was 1 mg/mL. The solution was spread homogeneously in 12-well plates and the plates were left to gellify at 37 °C, 5% CO_2 for 1 h. The cells were then seeded in the plates at a density of 6×10^4 cells per well. The HA and sHA fragments were added at a final concentration of 200 µg/mL and the cells were incubated for 24 h, after which, images were taken using a digital camera connected to a phase-contrast microscope. The quantification was performed using an image analysis software (Fiji v1.52p), as described in a previous study [34].

2.7. Wound Healing and Adhesion Assays Using Hyaluronidase Pre-Treatment

The processes followed were the same as the wound healing and adhesion assays (Sections 2.4 and 2.5, respectively), with the distinctive step of adding hyaluronidase from Streptomyces hyalurolyticus (Sigma-Aldrich, Saint Louis, MO, USA, H1136-1AMP) in serum free-medium at a final concentration of 1 U/mL. After a 1 h incubation at 37 °C, 5% CO_2, the assays were conducted with the addition of the cytostatic cytarabine (wound healing assay) or the seeding of the cells on collagen type I substrate (adhesion assay), as described above.

2.8. RNA Isolation, cDNA Synthesis and Real-Time PCR

MCF-7 and MDA-MB-231 cells were cultured in petri dishes at a density of 60×10^4 cells for 24 h and were serum starved overnight. In order to evaluate the effects hyaluronidase in the expression of MT1-MMP, cells were first pre-treated with hyaluronidase for a 1 h incubation at 37 °C, 5% CO_2. Then, the HA and sHA fractions were added to the cultures at a final concentration of 200 µg/mL. Following a 24 h incubation, the cells were collected, and RNA isolation was carried out using the NucleoSpin® RNA II Kit (Macherey-Nagel, Allentown, PA, USA). To quantify the isolated RNA, the absorbance of each sample was measured at 260 nm and RNA purity was determined by evaluating the 260/280 nm and 260/230 nm ratios. For the cDNA synthesis, the PrimeScript™ 1st strand cDNA synthesis kit perfect real time (Takara Bio Inc., Kusatsu, Japan) was used. Real-time PCR was conducted using KAPA Taq ReadyMix DNA Polymerase (KAPA BIOSYSTEMS, Wilmington, MA, USA) according to the manufacturer's instructions in 20 µL reaction mixture. The amplification was performed utilizing Rotor Gene Q (Qiagen, Hilden, Germany). All the reactions were performed in triplicate and a standard curve was included for each pair of primers for assay validation. In addition, a melting curve analysis was performed for detecting the SYBR Green-based objective amplicon. To provide quantification, the point of product accumulation in the early logarithmic phase of the amplification plot was defined by assigning a fluorescence threshold above the background, defined as the threshold cycle (Ct) number. Relative expression of different gene transcripts was calculated using the ΔΔCt method. The Ct of any gene of interest was normalized to the Ct of the normalizer (GAPDH). Fold changes (arbitrary units) were determined as $2^{-\Delta\Delta Ct}$. The genes of interest and the primers used are presented in Table 1.

2.9. Immunofluorescence

MCF-7 and MDA-MB-231 cells were seeded on glass coverslips in 24-well plates at a density of 6×10^4 and 5×10^4 cells per well respectively. The cells were cultured in complete cell culture medium for 24 h and then serum starved overnight. The next day, the HA and sHA fragments were added at a final concentration of 200 µg/mL and the cells were incubated for 24 h. Following a PBS wash, the cells were fixed in cold methanol and acetone for 5 min each. Afterwards, the cells were washed three times with PBS-Tween 0.01% and were permeabilized with 0.05% Triton X-100/PBS-Tween 0.01%. The coverslips were blocked with 5% BSA in PBS-Tween 0.01% for 1 h, and were subsequently stained with the primary antibody against E-cadherin (Takara, Kusatsu, Japan, ECCD-2, 1:200) and

CD44 (Hermes-3) (Abcam, Cambridge, UK, 1:1000) in 1% BSA/PBS, overnight at 4 °C. The secondary antibody (anti-mouse Alexa Fluor-594, 1:1000) (Biotium, Fremont, CA, USA) was used in 1% BSA/PBS-Tween 0.01% for a 1-h incubation in the dark, and the coverslips were mounted with DAPI on microscope slides. The stained slides were observed through a 60× objective using a fluorescence microscope (OLYMPUS CKX41, Waltham, MA, USA) and images were captured using the QImaging MicroPublisher 3.3RTV digital camera (Adept Turnkey, Perth, Australia). Additionally, the intensity of the immunofluorescence was calculated and quantified as corrected total cell fluorescence (CTCF) with the formula: CTCF = Integrated Density— (Area of selected cell × Mean fluorescence of background readings) using image analysis software (Fiji v1.52p).

Table 1. Primer sequences used for the genes of interest in real-time PCR.

Gene		Primer Sequence (5′–3′)	Annealing T (°C)
HAS2	F	TCGCAACACGTAACGCAAT	60 °C
	R	ACTTCTCTTTTTCCACCCCATTT	
HAS3	F	AACAAGTACGACTCATGGATTTCCT	60 °C
	R	GCCCGCTCCACGTTGA	
E-cadherin	F	TACGCCTGGGACTCCACCTA	60 °C
	R	CCAGAAACGGAGGCCTGAT	
snail2/slug	F	AGACCCTGGTTGCTTCAAGGA	60 °C
	R	CTCAGATTTGACCTGTCTGCAAA	
MT1-MMP	F	CATGGGCAGCGATGAAGTCT	60 °C
	R	CCAGTATTTGTTCCCCTTGTAGAAGTA	
MMP2	F	CGTCTGTCCCAGGATGACATC	62 °C
	R	ATGTCAGGAGAGGCCCCATA	
GAPDH	F	AGGCTGTTGTCATACTTCTCAT	60 °C
	R	GGAGTCCACTGGCGTCTT	
MMP9	F	TTCCAGTACCGAGAGAAAGCCTAT	60 °C
	R	GGTCACGTAGCCCACTTGGT	
CD44	F	ATAATAAAGGAGCAGCACTTCAGGA	60 °C
	R	ATAATTTGTGTCTTGGTCTCTGGTAGC	

2.10. Western Blot Analysis

MCF-7 and MDA-MB-231 cells were seeded in petri dishes at a confluency of 90%. The cells were cultured in complete cell culture medium for 24 h and then serum starved overnight. The next day, the HA and sHA fragments were added at a final concentration of 200 µg/mL and the cells were incubated for 24 h. Following a triple wash with cold PBS, the cells were lysed using lysis buffer (25 mM HEPES, 150 mM NaCl, 5 mM EDTA, 10% glycerol, 1% Triton X-100, 0.5 mM sodium orthovanadate (Sigma-Aldrich) and protease inhibitor cocktail 1× (Chemicon, Millipore, Burlington, MA, USA) at 4 °C for 30 min. Total protein quantity was calculated using Bradford assay (Thermo Scientific, Waltham, MO, USA) and equal protein samples were reduced with β-mercaptoethanol and boiled at 100 °C for 3 min. The samples were then separated by 10% SDS-PAGE and transferred to PVDF membranes (Macherey-Nagel). The membranes were blocked with 5% BSA in TBS 0.1% Tween for 2 h at room temperature and incubated overnight with the Hermes-3 antibody (Abcam, 1:1000). The following day the membranes were incubated with HRP-conjugated secondary goat anti-mouse IgG (Sigma-Aldrich) for 2 h at room temperature and the subsequent protein detection was performed by Pierce ECL Western Blotting Substrate (Thermo Scientific), according to the manufacturer's instructions.

2.11. Statistical Analysis

Reported values are expressed as mean $\pm$ standard deviation (SD) of experiments in triplicate. Statistically significant differences were evaluated using the analysis of variance (ANOVA) test followed by Tukey's test to determine statistical differences between each data set of the three groups (control, 50 kDa HA, and sHA treatments). Differences were considered statistically significant at the level of $p \leq 0.05$, indicated by an asterisk (*) for the treatment and control group comparison and by a hash sign (#) for the comparison between the two treatments. Statistical analysis and graphs were made using GraphPad Prism 8 (GraphPad Software, San Diego, CA, USA).

3. Results

3.1. Sulfated HA Affects Breast Cancer Cells' Functional Properties

In order to evaluate the effects of sHA on the ERα-positive MCF-7 and the ERα-negative/ERβ-positive MDA-MB-231 cell lines, the cells were treated with non-sulfated HA and sHA fragments of 50 kDa at a final concentration of 200 μg/mL, based on a previous study of our research group [25]. The first step was to determine whether sHA affects cell proliferation. The obtained results indicate a slight decrease in the proliferation capability of the MCF-7 cells for both HA fragments tested (Figure 1A). On the other hand, the MDA-MB-231 cell proliferation exhibits a significant decrease, both following treatment with the sHA fragments (ca 31%) and the non-sulfated 50 kDa HA (ca 22%) (Figure 1B).

A hallmark of cancer progression is the ability of cancer cells to migrate and metastasize [35]. For this reason, the migration, adhesion, and invasive properties of the breast cancer cells were further investigated. Notably, treatment with sHA results in lower migratory capacity for both MCF-7 (ca 20%) and MDA-MB-231 cells (ca 30%), while non-sulfated HA of the same size do not exhibit a similar effect (Figure 1C,D,I,K). Next, the adhesiveness of the cells on collagen type I was determined. According to the obtained data, treatment with sHA leads to an increase of ca 35% for the MCF-7 (Figure 1E) and ca 76% for the MDA-MB-231 cells (Figure 1F). Lastly, the effects of cellular invasion on collagen type I were investigated. Both the non-sulfated and sHA fragments show a decreased the invasiveness (ca 38%) of the MCF-7 cells (Figure 1G), whereas in the MDA-MB-231 cells, the invasiveness is affected significantly only after the sHA treatment, with a decrease of ca 30% (Figure 1H), indicating that the various effects of sHA depend on the breast cancer cell types bearing different ERs among other biomolecules.

3.2. Sulfated HA Competition with Endogenous HA Affects Cell Migration and Adhesion

As shown above, the sHA fragments significantly attenuate the migratory potential of MDA-MB-231 cells and slightly the MCF-7 ones. To investigate whether this effect could be attributed to competition between the exogenously added sHA fragments and the endogenous HA, the breast cancer cells were pre-treated with hyaluronidase as to remove the existing pericellular HA coating, and a wound healing assay following treatments with HA and sHA treatments was subsequently performed. Notably, the obtained data show an alteration of the previously observed effects, with no significant changes appearing in the migratory capabilities in both cell lines following hyaluronidase treatment (Figure 2A,B). To further evaluate whether hyaluronidase could affect adhesion, another major functional property, the same setup with hyaluronidase was applied. It is noted that no significant differences were observed upon such a treatment (Figure 2C,D). It is therefore plausible to suggest that the effects of sHA fragments on the migratory as well as the adhesive potential of both breast cancer cell lines could at least in part be attributed to competition with endogenous HA.

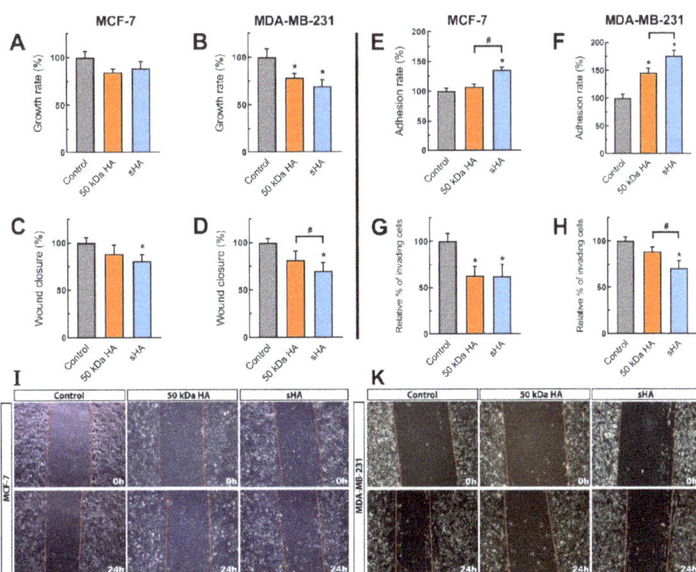

Figure 1. The effects of HA and sHA 50 kDa fragments on the functional properties of MCF-7 and MDA-MB-231 breast cancer cells. (**A,B**) Cell proliferation after 24 h of treatment at a concentration of 200 µg/mL. (**C,D**) Cell migration after 24 h of treatment at a concentration of 200 µg/mL. (**E,F**) Cell adhesion on collagen type I after 24 h of treatment at a concentration of 200 µg/mL. (**G,H**) Cell invasion after 24 h of treatment at a concentration of 200 µg/mL. Each bar represents mean ± SD values from triplicate samples. (**I,K**) Photos of the wound healing assays at time point 0 and 24 h. An asterisk (*) indicates statistically significant differences ($p < 0.05$) compared to the control groups and (#) indicates statistically significant differences ($p < 0.05$) between the treatments.

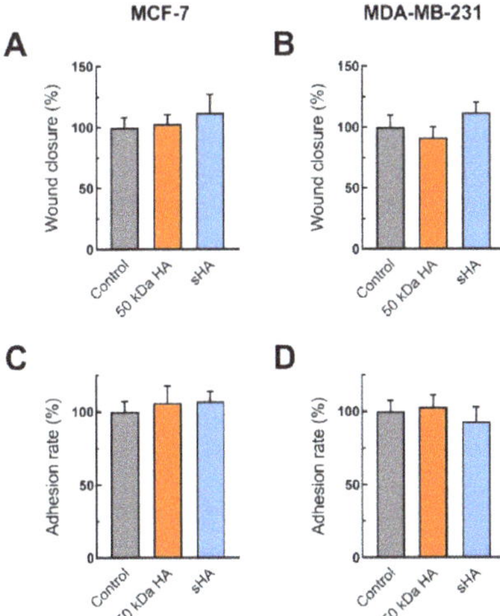

Figure 2. The effects of HA and sHA 50 kDa fragments on the migratory and adhesive potential of MCF-7 [(**A,C**), respectively] and MDA-MB-231 [(**B,D**), respectively] following treatment with hyaluronidase at a final concentration of 1 U/mL for 1 h. Each bar represents mean ± SD values from triplicate samples.

3.3. Sulfated HA Modulates the Expression of EMT Markers

A vital process in cancer progression is the EMT, during which epithelial markers such as E-cadherin are suppressed, while mesenchymal markers such as snail2/slug are elevated [8]. For this reason, the potential implication of sHA in the expression of EMT markers was investigated. Treatment with either HA or sHA fragments resulted in upregulation of E-cadherin of the mesenchymal-like MDA-MB-231 cells, with non-sulfated HA yielding a 1.3-fold change in the mRNA levels of the marker and sHA showing a 1.6-fold change (Figure 3B). On the other hand, no differences are observed in the mRNA levels of E-cadherin for the epithelial-like MCF-7 cells following the same treatment (Figure 3A). The above results are further corroborated by immunofluorescence for the E-cadherin protein (Figure 3E) and its subsequent intensity quantification analysis, calculated as corrected total cell fluorescence (CTCF) (Figure 3G,F). Additionally, the expression levels of the mesenchymal marker snail2/slug were studied. Following treatment with non-sulfated and sHA a significant downregulation of snail2/slug in the MCF-7 cells was observed, both by non-sulfated HA (ca 66%) and sHA (ca 60%) (Figure 3C). Similarly, the mRNA levels of snail2/slug are also diminished after treatments, with non-sulfated HA and sHA causing a significant downregulation (ca 31% and 37%, respectively) (Figure 3D).

3.4. Sulfated HA Downregulates the Expression of Matrix Metalloproteinases

To further evaluate the changes caused by sHA in the functional properties of the breast cancer cells, the expression of matrix metalloproteinases was studied. MMPs are enzymes involved in ECM remodeling and are found significantly modulated in breast cancer. Firstly, the expression of MT1-MMP, which has been linked with migration and cell invasion in breast cancer cells, was investigated [23]. Treatment with non-sulfated HA and sHA significantly decreased the mRNA levels of MT1-MMP in both breast cancer cell lines. For the MCF-7 cells a decrease of ca 20% and ca 40% is caused by the non-sulfated HA and sHA fragments, respectively (Figure 4A). In MDA-MB-231 cells these suppressive effects were more profound (31 and 47%, respectively) (Figure 4B). These data were further confirmed at the protein level using Western blotting (Figure 4E). It is worth noting that the observed effects were abolished using pre-treatment with hyaluronidase, demonstrating the importance of competition between the added HA and sHA fragments with the endogenous HA (Figure 4C,D), even in the more profound inhibitory effect exerted by sHA.

Moreover, the expression of MMP2 and MMP9, two key enzymes in ECM degradation, was also investigated [11]. The mRNA levels of MMP2 are significantly altered upon treatment with fragments, with non-sulfated HA causing downregulation in both MCF-7 (ca 33%) and MDA-MB-231 cells (ca 40%). sHA decreased the levels of MMP2 in the respective cell lines ca 48% and ca 23% (Figure 4F,G). Following a similar pattern, the expression levels of MMP9 were also downregulated after treatment, with non-sulfated HA (ca 35% decrease in both MCF-7 and MDA-MB-231 cells), while the sHA fragments caused a more profound decrease (ca 42% and ca 60% in MCF-7 and MDA-MB-231, respectively) (Figure 4H,I).

3.5. Sulfated HA Modulates the Expression of HAS but Not CD44

Considering the effects sHA of on the breast cancer cells' functional properties, the correlation between these changes and modulations to the expression of HA partners, such as HA synthases (HAS2 and -3) and CD44, was investigated. HAS2 is the main enzyme responsible for HA synthesis in MDA-MB-231 cells and its elevated levels have been linked to EMT, invasive potential and metastasis, whereas HAS3 is the main HA active synthase in MCF-7 cells [36–39]. Our data indicate that treatments with non-sulfated or sHA fragments lead to significant downregulation of HAS3 and HAS2 in MCF-7 and MDA-MB-231 cells, respectively. Specifically, non-sulfated HA causes a decrease in HAS3 mRNA levels of ca 20%, and sHA decreases HAS-3 expression by ca 30% (Figure 5A). HAS2 mRNA levels in MDA-MB-231 cells were significantly downregulated by ca 38% upon treatment with non-sulfated HA and ca 65% upon treatment with sHA fragments (Figure 5B).

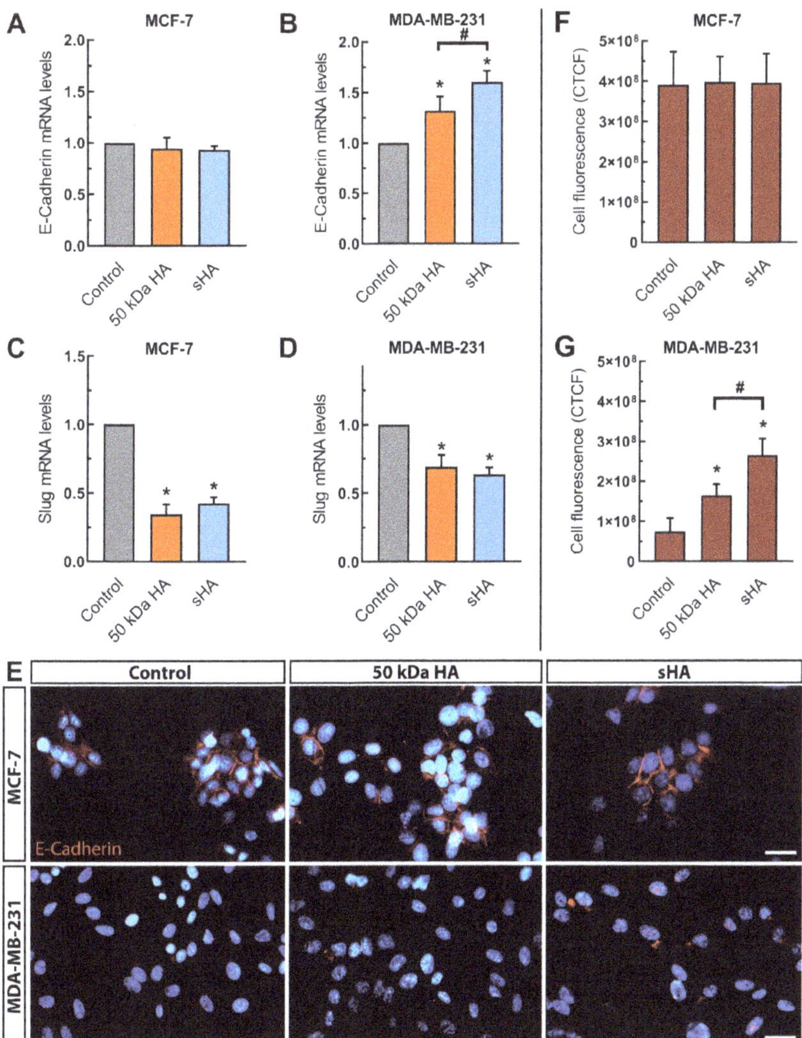

Figure 3. The effects of HA and sHA 50 kDa fragments on the expression of EMT markers. (**A,B**) Quantitative RT-PCR analysis of E-cadherin mRNA levels in MCF-7 (**A**) and MDA-MB-231 cells (**B**) following a 24 h treatment at a final concentration of 200 µg/mL. (**C,D**) Quantitative RT-PCR analysis of snail2/slug mRNA levels following a 24 h treatment at a final concentration of 200 µg/mL in MCF-7 (**C**) and MDA-MB-231 cells (**D**). (**E**) Immunofluorescence imaging for E-cadherin (red) after treatment with HA and sHA fragments for 24 h at a final concentration of 200 µg/mL. Nuclei are shown in blue (DAPI). Scale bars: 20 µm. (**F,G**) Quantification of E-cadherin fluorescence, calculated as corrected total cell fluorescence (CTCF) in both breast cancer cell lines following treatment. Each bar represents mean ± SD values from triplicate samples. An asterisk (*) indicates statistically significant differences ($p < 0.05$) compared to the control groups and (#) indicates statistically significant differences ($p < 0.05$) between the treatments.

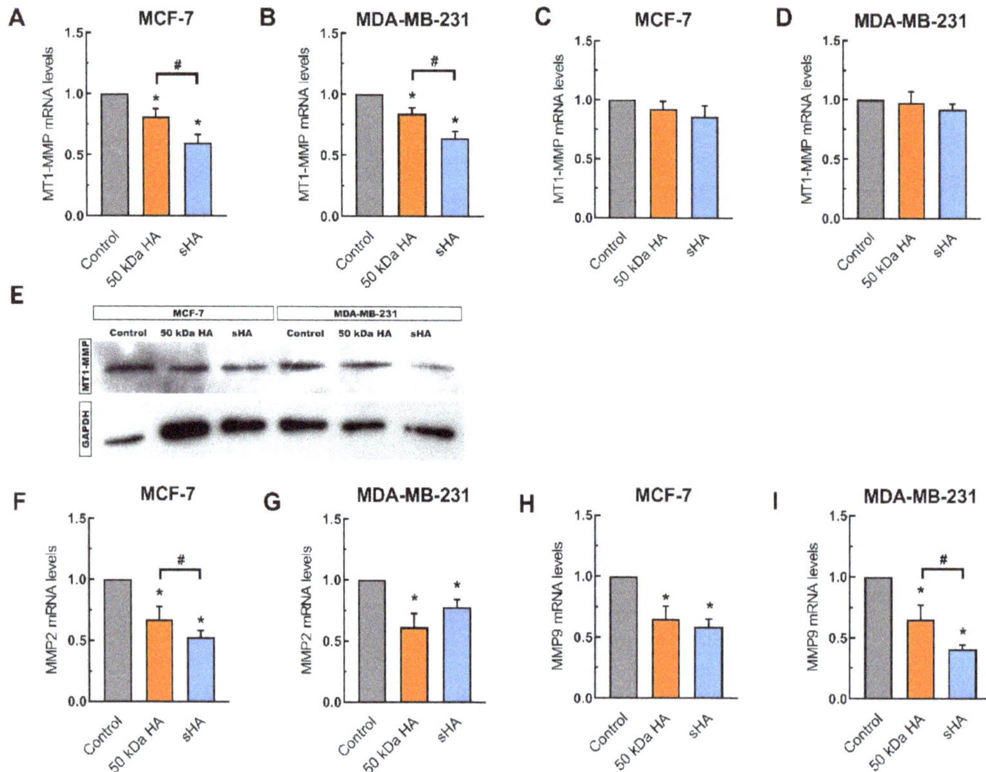

Figure 4. The effects of HA and sHA 50 kDa fragments on the expression of matrix remodeling enzymes. (**A,B**) Quantitative RT-PCR analysis of MT1-MMP mRNA levels in MCF-7 (**A**) and MDA-MB-231 cells (**B**) following a 24 h treatment at a final concentration of 200 µg/mL. (**C,D**) Quantitative RT-PCR analysis of MT1-MMP mRNA levels following hyaluronidase pre-treatment in MCF-7 (**C**) and MDA-MB-231 cells (**D**). (**E**) Western blot analysis of MT1-MMP protein levels in MCF-7 and MDA-MB-231 cells. (**F,G**) Quantitative RT-PCR analysis of MMP2 mRNA levels following a 24 h treatment at a final concentration of 200 µg/mL in MCF-7 (**F**) and MDA-MB-231 cells (**G**). (**H,I**) Quantitative RT-PCR analysis of MMP9 mRNA levels following a 24 h treatment at a final concentration of 200 µg/mL in MCF-7 (**H**) and MDA-MB-231 cells (**I**). Each bar represents mean $\pm$ SD values from triplicate samples. An asterisk (*) indicates statistically significant differences ($p < 0.05$) compared to the control groups and (#) indicates statistically significant differences ($p < 0.05$) between the treatments.

CD44 is one of the main HA receptors, and its overexpression correlates with cell migration, EMT and metastasis in breast cancer [18,40]. Therefore, we evaluated whether sHA plays a role in altering the expression of CD44. Following a 24 h treatment with 200 µg/mL HA and sHA in MCF-7 and MDA-MB-231 cells, no significant differences were observed in the expression (mRNA and protein) levels of CD44, except a slight decrease seen upon treatments with sHA, in both breast cancer cell line tested (Figure 5C–F). These data indicate that CD44 is sufficiently available for the extra signaling or other receptors may be implicated.

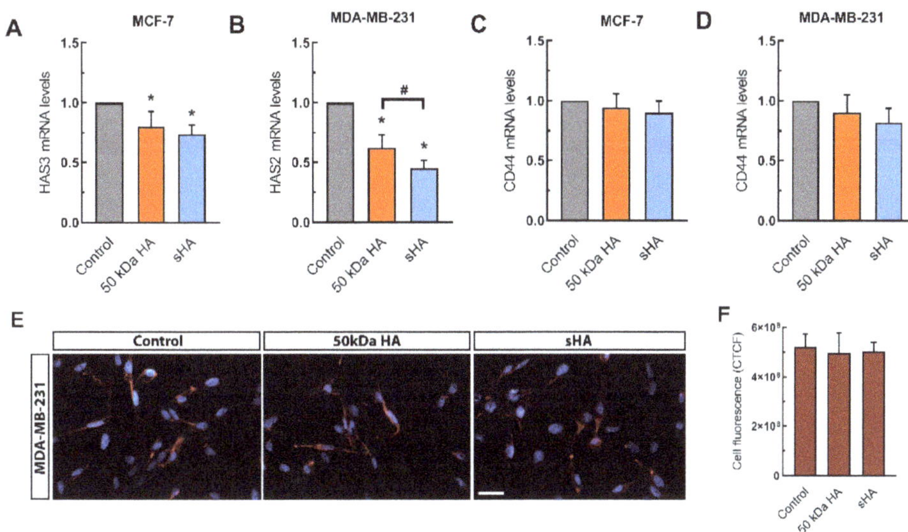

Figure 5. The effects of HA and sHA 50 kDa fragments on the expression of HA partners. (**A,B**) Quantitative RT-PCR analysis of HAS3 mRNA levels in MCF-7 (**A**) and HAS2 mRNA MDA-MB-231 cells (**B**) following a 24 h treatment at a final concentration of 200 μg/mL. (**C,D**) Quantitative RT–PCR analysis of mRNA levels of CD44 mRNA levels following a 24 h treatment at a final concentration of 200 μg/mL in MCF-7 (**C**) and MDA-MB-231 cells (**D**). (**E**) Immunofluorescence imaging for CD44 (red) after treatment with HA and sHA fragments for 24 h at a final concentration of 200 μg/mL. Nuclei are shown in blue (DAPI). Scale bars: 20 μm. (**F**) Quantification of CD44 fluorescence, calculated as corrected total cell fluorescence (CTCF) in MDA-MB-231 cells. Each bar represents mean $\pm$ SD values from triplicate samples. An asterisk (*) indicates statistically significant differences ($p < 0.05$) compared to the control groups and (#) indicates statistically significant differences ($p < 0.05$) between the treatments.

4. Discussion

HA is a major macromolecule of the three-dimensional ECM network, playing a crucial role in a variety of biological processes, including cancer development and progression. Through its binding to the receptor CD44, HA regulates intracellular signaling pathways and cell functional properties in a size-dependent manner [26]. Recent studies have shown the role of HA fragments of different molecular size in modulating breast cancer cells proliferation, migration, and invasion [24,25]. Furthermore, treatment with such HA fragments results in significant alterations in the morphology and aggressive phenotype of breast cancer cells. HA is the only GAG not physically modified by the addition of sulfate groups [13]. Research studies, however, have previously demonstrated the potential of sHA fragments to exhibit antitumor effects on prostate and bladder cancer cells [30,31]. In the present study, we investigated the effects of sHA on breast cancer cells with different ER status; the low metastatic ERα-positive MCF-7 and the more aggressive ERα-negative/ERβ-positive MDA-MB-231 cells. To this end, non-sulfated HA and the sHA fragment of the same molecular weight (50 kDa) were evaluated regarding their effects on the breast cancer cell functional properties. Moreover, the expression of key ECM effectors was also studied.

Our data demonstrate that treatment with sHA leads to an attenuation of breast cancer cell proliferation, migration, and invasion, while also increasing cell adhesion on collagen type I. Notably, sHA exhibited a more profound effect compared to the non-sulfated HA counterpart of the same molecular size, thus suggesting an anticancer regulatory capability for sHA. Additionally, differences in the effects of sHA between the two breast cancer cell lines suggest the dependency of its action on the type of cells, among other parameters, the ER status is considered a key difference. These observations are further supported by modulations to the expression of matrix metalloproteinases, which are typically overexpressed during breast cancer progression. Treatment with sHA, however, lead to the downregulation of MT1-MMP, known to be involved in breast cancer cell migration and invasion, as well as the downregulation of MMP2 and MMP9, two MMPs with pivotal role in ECM degradation [23]. Additionally, sHA caused a significant downregulation in the expression of HA synthases, which have been linked to breast cancer invasion and metastasis [36–39].

In the study by Li et al. [41], the authors demonstrated that the CD44-mediated TGF-β1 and EGF signaling, and the co-localization of CD44/EGFR influenced the activation of EGF signaling induced by TGF-β1 in lung and breast cancer cells. The effects were abolished following inhibition of HAS2 by 4-MU. The 50 kDa sHA effects, as clearly demonstrated here, may well be at least in part attributed to the competition with the endogenous HA.

An important process aiding in breast cancer metastasis is the epithelial-to-mesenchymal transition [7]. Our findings show that sHA alters the expression of EMT markers, causing an upregulation of the epithelial E-cadherin for the mesenchymal-like MDA-MB-231 cells and downregulating the mesenchymal marker snail2/slug in both breast cancer cell lines. Taking these changes in EMT markers expression into account, it would be of high interest to examine whether sHA also affects the cell morphology, using scanning electron microscopy. Based on the results obtained for the functional properties (migration and adhesion) as well as in the expression of MT1-MMP using hyaluronidase, it is plausible to suggest that the effects of sHA fragment could at least in part be attributed to competition with endogenous HA. Adding to this, no significant changes in the expression of CD44 was found following sHA treatment, possibly since the CD44 is sufficiently available and/or other changes in intracellular signaling pathways play a role. This, however, requires further elucidation.

In conclusion, our findings demonstrate that sHA regulates breast cancer cells' functional properties and the gene expression profile of ECM effectors. The exact mechanism by which the sHA fragments drive these actions has to be further elucidated in the future. Consequently, our novel data open new avenues for further research in respect to breast cancer targeting and designate sHA as a useful molecule in the development of therapeutic strategies.

Author Contributions: Conceptualization, A.-G.T. and N.K.K.; Investigation C.K., A.-G.T., D.K. and G.B.; Visualization, C.K.; Writing-original draft preparation, C.K., A.-G.T. and D.K.; Writing-review and editing, C.K., A.-G.T. and N.K.K.; Formal analysis, C.K.; Resources, N.K.K.; Funding acquisition, A.-G.T. and N.K.K.; Project administration, N.K.K.; Supervision, N.K.K. All authors have read and agreed to the published version of the manuscript.

Funding: This work was supported by the EU Horizon 2020 project RISE-2014, action No. 645756 "GLYCANC-Matrix glycans as multifunctional pathogenesis factors and therapeutic targets in cancer". A.-G.T. was supported by General Secretariat for Research and Technology (GSRT) & Hellenic Foundation of Research & Innovation (H.F.R.I.).

Institutional Review Board Statement: No applicable.

Informed Consent Statement: No applicable.

Data Availability Statement: The data presented in this study are available on request.

Acknowledgments: We wish to thank the Research & Development department of Fidia Farmaceutici S.p.A. for kindly providing us with the sHA fragments, as well as the materials and resources necessary for the 50 kDa HA synthesis. Additionally, we thank Spyros Skandalis (University of Patras) for providing us with the CD44 and HAS primers.

Conflicts of Interest: The authors declare no conflict of interest.

References

1. Sung, H.; Ferlay, J.; Siegel, R.L.; Laversanne, M.; Soerjomataram, I.; Jemal, A.; Bray, F. Global Cancer Statistics 2020: GLOBOCAN Estimates of Incidence and Mortality Worldwide for 36 Cancers in 185 Countries. *CA Cancer J. Clin.* **2021**, *71*, 209–249. [CrossRef] [PubMed]
2. Piperigkou, Z.; Karamanos, N.K. Estrogen Receptor-Mediated Targeting of the Extracellular Matrix Network in Cancer. *Semin. Cancer Biol.* **2020**, *62*, 116–124. [CrossRef] [PubMed]
3. Bouris, P.; Skandalis, S.S.; Piperigkou, Z.; Afratis, N.; Karamanou, K.; Aletras, A.J.; Moustakas, A.; Theocharis, A.D.; Karamanos, N.K. Estrogen Receptor Alpha Mediates Epithelial to Mesenchymal Transition, Expression of Specific Matrix Effectors and Functional Properties of Breast Cancer Cells. *Matrix Biol.* **2015**, *43*, 42–60. [CrossRef] [PubMed]
4. Chekhun, S.; Bezdenezhnykh, N.; Shvets, J.; Lukianova, N. Expression of Biomarkers Related to Cell Adhesion, Metastasis and Invasion of Breast Cancer Cell Lines of Different Molecular Subtype. *Exp. Oncol.* **2013**, *35*, 174–179.
5. Nicolini, A.; Ferrari, P.; Duffy, M.J. Prognostic and Predictive Biomarkers in Breast Cancer: Past, Present and Future. *Semin. Cancer Biol.* **2018**, *52*, 56–73. [CrossRef]
6. Gritsenko, P.G.; Ilina, O.; Friedl, P. Interstitial Guidance of Cancer Invasion. *J. Pathol.* **2012**, *226*, 185–199. [CrossRef]
7. Jo, S.J.; Park, P.-G.; Cha, H.-R.; Ahn, S.G.; Kim, M.J.; Kim, H.; Koo, J.S.; Jeong, J.; Park, J.H.; Dong, S.M.; et al. Cellular Inhibitor of Apoptosis Protein 2 Promotes the Epithelial-Mesenchymal Transition in Triple-Negative Breast Cancer Cells through Activation of the AKT Signaling Pathway. *Oncotarget* **2017**, *8*, 78781–78795. [CrossRef]
8. Medici, D.; Hay, E.D.; Olsen, B.R. Snail and Slug Promote Epithelial-Mesenchymal Transition through β-Catenin–T-Cell Factor-4-Dependent Expression of Transforming Growth Factor-B3. *MBoC* **2008**, *19*, 4875–4887. [CrossRef]
9. Thiery, J.P. Epithelial–Mesenchymal Transitions in Tumour Progression. *Nat. Rev. Cancer* **2002**, *2*, 442–454. [CrossRef]
10. Sciacovelli, M.; Frezza, C. Metabolic Reprogramming and Epithelial-to-Mesenchymal Transition in Cancer. *FEBS J.* **2017**, *284*, 3132–3144. [CrossRef]
11. Gialeli, C.; Theocharis, A.D.; Karamanos, N.K. Roles of Matrix Metalloproteinases in Cancer Progression and Their Pharmacological Targeting: MMPs as Potential Targets in Malignancy. *FEBS J.* **2011**, *278*, 16–27. [CrossRef]
12. Karamanos, N.K.; Theocharis, A.D.; Neill, T.; Iozzo, R.V. Matrix Modeling and Remodeling: A Biological Interplay Regulating Tissue Homeostasis and Diseases. *Matrix Biol.* **2019**, *75–76*, 1–11. [CrossRef]
13. Theocharis, A.D.; Skandalis, S.S.; Gialeli, C.; Karamanos, N.K. Extracellular Matrix Structure. *Adv. Drug Deliv. Rev.* **2016**, *97*, 4–27. [CrossRef]
14. Oikari, S.; Kettunen, T.; Tiainen, S.; Häyrinen, J.; Masarwah, A.; Sudah, M.; Sutela, A.; Vanninen, R.; Tammi, M.; Auvinen, P. UDP-Sugar Accumulation Drives Hyaluronan Synthesis in Breast Cancer. *Matrix Biol.* **2018**, *67*, 63–74. [CrossRef]
15. Heldin, P.; Lin, C.-Y.; Kolliopoulos, C.; Chen, Y.-H.; Skandalis, S.S. Regulation of Hyaluronan Biosynthesis and Clinical Impact of Excessive Hyaluronan Production. *Matrix Biol.* **2019**, *78–79*, 100–117. [CrossRef]
16. Passi, A.; Vigetti, D.; Buraschi, S.; Iozzo, R.V. Dissecting the Role of Hyaluronan Synthases in the Tumor Microenvironment. *FEBS J.* **2019**, *286*, 2937–2949. [CrossRef]
17. Ouhtit, A.; Rizeq, B.; Saleh, H.A.; Rahman, M.M.; Zayed, H. Novel CD44-Downstream Signaling Pathways Mediating Breast Tumor Invasion. *Int. J. Biol. Sci.* **2018**, *14*, 1782–1790. [CrossRef]
18. Misra, S.; Heldin, P.; Hascall, V.C.; Karamanos, N.K.; Skandalis, S.S.; Markwald, R.R.; Ghatak, S. Hyaluronan-CD44 Interactions as Potential Targets for Cancer Therapy. *FEBS J.* **2011**, *278*, 1429–1443. [CrossRef]
19. Skandalis, S.S.; Karalis, T.T.; Chatzopoulos, A.; Karamanos, N.K. Hyaluronan-CD44 Axis Orchestrates Cancer Stem Cell Functions. *Cell. Signal.* **2019**, *63*, 109377. [CrossRef]
20. Garantziotis, S.; Savani, R.C. Hyaluronan Biology: A Complex Balancing Act of Structure, Function, Location and Context. *Matrix Biol.* **2019**, *78–79*, 1–10. [CrossRef]
21. Velesiotis, C.; Vasileiou, S.; Vynios, D.H. A Guide to Hyaluronan and Related Enzymes in Breast Cancer: Biological Significance and Diagnostic Value. *FEBS J.* **2019**, *286*, 3057–3074. [CrossRef]
22. Tammi, M.I.; Oikari, S.; Pasonen-Seppänen, S.; Rilla, K.; Auvinen, P.; Tammi, R.H. Activated Hyaluronan Metabolism in the Tumor Matrix—Causes and Consequences. *Matrix Biol.* **2019**, *78–79*, 147–164. [CrossRef]
23. Piperigkou, Z.; Kyriakopoulou, K.; Koutsakis, C.; Mastronikolis, S.; Karamanos, N.K. Key Matrix Remodeling Enzymes: Functions and Targeting in Cancer. *Cancers* **2021**, *13*, 1441. [CrossRef]
24. Tavianatou, A.; Piperigkou, Z.; Koutsakis, C.; Barbera, C.; Beninatto, R.; Franchi, M.; Karamanos, N.K. The Action of Hyaluronan in Functional Properties, Morphology and Expression of Matrix Effectors in Mammary Cancer Cells Depends on Its Molecular Size. *FEBS J.* **2021**. [CrossRef]
25. Tavianatou, A.-G.; Piperigkou, Z.; Barbera, C.; Beninatto, R.; Masola, V.; Caon, I.; Onisto, M.; Franchi, M.; Galesso, D.; Karamanos, N.K. Molecular Size-Dependent Specificity of Hyaluronan on Functional Properties, Morphology and Matrix Composition of Mammary Cancer Cells. *Matrix Biol. Plus* **2019**, *3*, 100008. [CrossRef]
26. Tavianatou, A.G.; Caon, I.; Franchi, M.; Piperigkou, Z.; Galesso, D.; Karamanos, N.K. Hyaluronan: Molecular Size-dependent Signaling and Biological Functions in Inflammation and Cancer. *FEBS J.* **2019**, *286*, 2883–2908. [CrossRef]
27. Toole, B.; Ghatak, S.; Misra, S. Hyaluronan Oligosaccharides as a Potential Anticancer Therapeutic. *Curr. Pharm. Biotechnol.* **2008**, *9*, 249–252. [CrossRef]

28. Urakawa, H.; Nishida, Y.; Knudson, W.; Knudson, C.B.; Arai, E.; Kozawa, E.; Futamura, N.; Wasa, J.; Ishiguro, N. Therapeutic Potential of Hyaluronan Oligosaccharides for Bone Metastasis of Breast Cancer. *J. Orthop. Res.* **2012**, *30*, 662–672. [CrossRef]

29. Zhao, Y.; Qiao, S.; Shi, S.; Yao, L.; Hou, X.; Li, C.; Lin, F.-H.; Guo, K.; Acharya, A.; Chen, X.; et al. Modulating Three-Dimensional Microenvironment with Hyaluronan of Different Molecular Weights Alters Breast Cancer Cell Invasion Behavior. *ACS Appl. Mater. Interfaces* **2017**, *9*, 9327–9338. [CrossRef]

30. Benitez, A.; Yates, T.J.; Lopez, L.E.; Cerwinka, W.H.; Bakkar, A.; Lokeshwar, V.B. Targeting Hyaluronidase for Cancer Therapy: Antitumor Activity of Sulfated Hyaluronic Acid in Prostate Cancer Cells. *Cancer Res.* **2011**, *71*, 4085–4095. [CrossRef]

31. Jordan, A.R.; Lokeshwar, S.D.; Lopez, L.E.; Hennig, M.; Chipollini, J.; Yates, T.; Hupe, M.C.; Merseburger, A.S.; Shiedlin, A.; Cerwinka, W.H.; et al. Antitumor Activity of Sulfated Hyaluronic Acid Fragments in Pre-Clinical Models of Bladder Cancer. *Oncotarget* **2017**, *8*, 24262–24274. [CrossRef] [PubMed]

32. Feoktistova, M.; Geserick, P.; Leverkus, M. Crystal Violet Assay for Determining Viability of Cultured Cells. *Cold Spring Harb. Protoc.* **2016**, *2016*. [CrossRef] [PubMed]

33. Schindelin, J.; Arganda-Carreras, I.; Frise, E.; Kaynig, V.; Longair, M.; Pietzsch, T.; Preibisch, S.; Rueden, C.; Saalfeld, S.; Schmid, B.; et al. Fiji: An Open-Source Platform for Biological-Image Analysis. *Nat. Methods* **2012**, *9*, 676–682. [CrossRef] [PubMed]

34. De Wever, O.; Hendrix, A.; De Boeck, A.; Westbroek, W.; Braems, G.; Emami, S.; Sabbah, M.; Gespach, C.; Bracke, M. Modeling and Quantification of Cancer Cell Invasion through Collagen Type I Matrices. *Int. J. Dev. Biol.* **2010**, *54*, 887–896. [CrossRef]

35. Hanahan, D.; Weinberg, R.A. Hallmarks of Cancer: The Next Generation. *Cell* **2011**, *144*, 646–674. [CrossRef]

36. Kolliopoulos, C.; Lin, C.-Y.; Heldin, C.-H.; Moustakas, A.; Heldin, P. Has2 Natural Antisense RNA and Hmga2 Promote Has2 Expression during TGFβ-Induced EMT in Breast Cancer. *Matrix Biol.* **2019**, *80*, 29–45. [CrossRef]

37. Heldin, P.; Basu, K.; Kozlova, I.; Porsch, H. HAS2 and CD44 in Breast Tumorigenesis. In *Advances in Cancer Research*; Elsevier: Amsterdam, The Netherlands, 2014; Volume 123, pp. 211–229. ISBN 978-0-12-800092-2.

38. Bernert, B.; Porsch, H.; Heldin, P. Hyaluronan Synthase 2 (HAS2) Promotes Breast Cancer Cell Invasion by Suppression of Tissue Metalloproteinase Inhibitor 1 (TIMP-1). *J. Biol. Chem.* **2011**, *286*, 42349–42359. [CrossRef]

39. Bohaumilitzky, L.; Huber, A.-K.; Stork, E.M.; Wengert, S.; Woelfl, F.; Boehm, H. A Trickster in Disguise: Hyaluronan's Ambivalent Roles in the Matrix. *Front. Oncol.* **2017**, *7*, 242. [CrossRef]

40. Hamilton, S.R.; Fard, S.F.; Paiwand, F.F.; Tolg, C.; Veiseh, M.; Wang, C.; McCarthy, J.B.; Bissell, M.J.; Koropatnick, J.; Turley, E.A. The Hyaluronan Receptors CD44 and Rhamm (CD168) Form Complexes with ERK1,2 That Sustain High Basal Motility in Breast Cancer Cells. *J. Biol. Chem.* **2007**, *282*, 16667–16680. [CrossRef]

41. Li, L.; Qi, L.; Liang, Z.; Song, W.; Liu, Y.; Wang, Y.; Sun, B.; Zhang, B.; Cao, W. Transforming Growth Factor-B1 Induces EMT by the Transactivation of Epidermal Growth Factor Signaling through HA/CD44 in Lung and Breast Cancer Cells. *Int. J. Mol. Med.* **2015**, *36*, 113–122. [CrossRef]

biomolecules

MDPI

Review

Current Perspectives on the Role of Matrix Metalloproteinases in the Pathogenesis of Basal Cell Carcinoma

Mircea Tampa [1,2], Simona Roxana Georgescu [1,2,*], Madalina Irina Mitran [3], Cristina Iulia Mitran [3], Clara Matei [1], Ana Caruntu [4,5,*], Cristian Scheau [6], Ilinca Nicolae [2], Andreea Matei [6], Constantin Caruntu [6,7], Carolina Constantin [8,9] and Monica Neagu [8,9,10]

[1] Department of Dermatology, Carol Davila University of Medicine and Pharmacy, 020021 Bucharest, Romania; tampa_mircea@yahoo.com (M.T.); matei_clara@yahoo.com (C.M.)

[2] Department of Dermatology, Victor Babes Clinical Hospital for Infectious Diseases, 030303 Bucharest, Romania; drnicolaei@yahoo.ro

[3] Department of Microbiology, Carol Davila University of Medicine and Pharmacy, 020021 Bucharest, Romania; madalina.irina.mitran@gmail.com (M.I.M.); cristina.iulia.mitran@gmail.com (C.I.M.)

[4] Department of Oral and Maxillofacial Surgery, "Carol Davila" Central Military Emergency Hospital, 010825 Bucharest, Romania

[5] Faculty of Dental Medicine, Titu Maiorescu University, 031593 Bucharest, Romania

[6] Department of Physiology, Carol Davila University of Medicine and Pharmacy, 050474 Bucharest, Romania; cristian.scheau@umfcd.ro (C.S.); andreea.matei@drd.umfcd.ro (A.M.); costin.caruntu@gmail.com (C.C.)

[7] Department of Dermatology, Prof. N.C. Paulescu National Institute of Diabetes, Nutrition and Metabolic Diseases, 011233 Bucharest, Romania

[8] Immunology Department, Victor Babes National Institute of Pathology, 050096 Bucharest, Romania; caroconstantin@gmail.com (C.C.); neagu.monica@gmail.com (M.N.)

[9] Department of Pathology, Colentina University Hospital, Bucharest 020125, Romania

[10] Faculty of Biology, University of Bucharest, Bucharest 76201, Romania

* Correspondence: srg.dermatology@gmail.com (S.R.G.); ana.caruntu@gmail.com (A.C.)

Citation: Tampa, M.; Georgescu, S.R.; Mitran, M.I.; Mitran, C.I.; Matei, C.; Caruntu, A.; Scheau, C.; Nicolae, I.; Matei, A.; Caruntu, C.; et al. Current Perspectives on the Role of Matrix Metalloproteinases in the Pathogenesis of Basal Cell Carcinoma. *Biomolecules* 2021, 11, 903. https://doi.org/10.3390/biom11060903

Academic Editors: Dragana Nikitovic and George Tzanakakis

Received: 15 May 2021
Accepted: 15 June 2021
Published: 17 June 2021

Abstract: Basal cell carcinoma (BCC) is the most common skin malignancy, which rarely metastasizes but has a great ability to infiltrate and invade the surrounding tissues. One of the molecular players involved in the metastatic process are matrix metalloproteinases (MMPs). MMPs are enzymes that can degrade various components of the extracellular matrix. In the skin, the expression of MMPs is increased in response to various stimuli, including ultraviolet (UV) radiation, one of the main factors involved in the development of BCC. By modulating various processes that are linked to tumor growth, such as invasion and angiogenesis, MMPs have been associated with UV-related carcinogenesis. The sources of MMPs are multiple, as they can be released by both neoplastic and tumor microenvironment cells. Inhibiting the action of MMPs could be a useful therapeutic option in BCC management. In this review that reunites the latest advances in this domain, we discuss the role of MMPs in the pathogenesis and evolution of BCC, as molecules involved in tumor aggressiveness and risk of recurrence, in order to offer a fresh and updated perspective on this field.

Keywords: BCC; MMP; TIMP; invasion; tumor progression

1. Introduction

Basal cell carcinoma (BCC) is the most common form of skin cancer in humans, displaying a worldwide increase in incidence. BCC may be considered the result of a complex interaction between genetic and environmental factors, with exposure to ultraviolet (UV) light being a key player in its pathogenesis. The incidence of BCC starts to increase within the fourth decade of life, while young people are rarely affected. An exception to this tendency is constituted by the patients with either genodermatoses (such as xeroderma pigmentosum, Gorlin-Goltz syndrome, Bazex or Rombo syndrome) or different degrees of immunosuppression [1,2]. BCC most commonly appears on the face in individuals with fair skin; other possible locations are the trunk and extremities. According to some recent theories, BCC originates in hair follicles; therefore, it is rarely, if ever, diagnosed on non-hair-bearing sites such as the mucous membranes (e.g. oral or genital mucosa) [3].

In spite of the fact that it is a slow-growing tumor rarely displaying local invasiveness and metastasis, BCC causes, due to its ability to invade and infiltrate the surrounding tissue, considerable morbidity; altogether, due to its high incidence, it represents an important public health issue [1,2]. Hence, numerous efforts are made to discover new noninvasive diagnostic techniques [4,5] and new candidate molecules that can be used as both biomarkers of progression and future therapy targets in BCC.

The pathogenesis of BCC is complex and incompletely deciphered. Matrix metalloproteinases (MMPs) create a suitable microenvironment for tumor development, representing key molecules in tumor progression [6]. In this review, we aim to assemble novel data on the involvement of MMPs in the pathogenesis and progression of BCC and underline the role of these proteolytic enzymes in tumor aggressiveness, the risk of recurrence and as a valuable source for scouting new BCC therapeutic approaches.

2. MMPs as Molecular Promoters in Carcinogenesis

MMPs are members of the metzincin protease superfamily of zinc-endopeptidases and have traditionally been described as molecules that primarily degrade extracellular matrix (ECM) proteins [7]. Other members of the superfamily are A disintegrin and metalloproteinases (ADAMs) and ADAMs with thrombospondin motifs (ADAMTSs), which present, in their structure, a conserved methionine residue adjacent to the active site [7]. Nowadays, it is well-known that MMPs act on a wide range of substrates, including membrane receptors, cytokines, growth factors, signaling molecules and ligands [8]. Recent experimental in vivo studies have suggested that the main substrates of MMPs are nonmatrix molecules [8–11]. Therefore, MMPs can be considered as cell signal regulators rather than as destructive enzymes [8,9]. In the human body, MMPs are synthesized by numerous cellular types, such as fibroblasts, macrophages, endothelial cells, vascular smooth muscle, osteoblasts, etc. [12,13]. MMPs are grouped into several classes depending on the organization mode of their structural domains (Table 1). The structure of MMPs includes a propeptide, a metalloproteinase domain with catalytic action, a linker peptide of variable length and a hemopexin domain [12].

MMPs can be grouped into soluble MMPs and membrane-bound MT-MMPs. Soluble MMPs are released as inactive zymogens and, subsequently, are activated in the extracellular space by other MMPs and/or other proteases. MT-MMPs display a basic amino acid motif of RX(K/R)R at the C-terminal site of their prodomain, which is cleaved by proprotein convertases (PCs) (e.g., furin), resulting in the activation of the enzyme [14–16]. Their activity is modulated by general protease inhibitors, including α2-macroglobulin, and by specific inhibitors, such as tissue inhibitors of metalloproteinases (TIMPs) [17].

Recent research has revealed new in vivo substrates for these proteases. MMPs are involved in angiogenesis by modulating the bioavailability of angiogenic factors that are sequestered by the ECM or the basement membrane. For example, VEGF is mobilized by MMP-1, MMP-3, MMP-7, MMP-9, MMP-16 and MMP-19 through the cleavage of VEGF-binding ECM proteins [8]. Another interesting observation is that MMP-14 overexpression is associated with increased VEGF-A transcriptional activation [18]. MMPs increase cell migration both by degrading the ECM that impedes cell motility and by breaking down the proteins of adherens junctions, maintaining cellular cohesion [19]. In this case, E-cadherin is an important target. Another MMP substrate related to cell migration is the γ2 chain of laminin-5 [20]. MMPs also participate in the inflammatory process. The action of MMPs on the ECM components leads to the release of chemotactic molecules. For example, the fragments of elastin that results through MMP-12 cleavage may represent a chemoattractant factor for the influx of macrophages. In addition, MMP-12 can act on the serine protease inhibitor α1-PI, a non-ECM protein, resulting in components that act as chemoattractants for neutrophils [8]. The discovery of new substrates for MMPs offers novel perspectives on the role of MMPs in disease pathogenesis. MMPs are deeply involved in cell differentiation and proliferation, as well as in apoptosis, angiogenesis and the immune response [12,13].

The increased expression of MMPs in tumor cells and adjacent tumor tissue has been associated with a more aggressive tumor behavior [21–23]. The sources of MMPs are multiple; therefore, MMPs can be released by stromal cells, tumor cells or circulating cells [24]. The degradation of the ECM and basement membrane are important events in tumor invasion and metastasis [25]. Certain proteins in the ECM structure, such as fibronectin and laminin, promote cell migration and angiogenesis [26]. Basement membrane (BM) disruption is a key event in tumor invasion. Cells employ various mechanisms to cross the basement membrane; some of these events still remain incompletely elucidated. Cells may use protease-dependent or -independent invasion programs. Protease-dependent transmigration relies on the activity of membrane-type matrix metalloproteinases [27]. Using COS cells (i.e., an epithelial cell type that displays no BM degradative or invasive activity), it has been shown that MMP-2, MMP-3, MMP-7, MMP-9, MMP-11 and MMP-13 do not enable COS cells to degrade the BM. Conversely MMP-14, MMP-15 and MMP-16 endow COS cells with the ability to remodel the BM [27]. Collagens are most abundant in the ECM. The degradation of type I and III collagens is catalyzed by MMP-1, MMP-8, MMP-13, MMP-14 and MMP-16. MMP-2 degrades the solubilized monomers of collagens I, II and III [28]. Regarding MMP-2 and MMP-9, Rowe and Weiss point out that MMP-2 or MMP-9 can degrade in vitro type IV collagen; however, this ability is limited in vivo [27]. It should be considered that some substrates identified in vitro do not necessarily predict the activity of these enzymes in vivo; therefore, the studies performed in vivo should represent the basis for understanding the physio/pathological roles of MMPs.

Moreover, MMPs modulate several signaling pathways that contribute to tumor progression, activating cell signaling molecules, such as focal adhesion kinase (FAK), mitogen-activated protein kinase (MAPK), Rous sarcoma oncogene (SRC), rat sarcoma viral oncogene homologous (RAS) and phosphatidylinositol 3-kinase (PI3K) [24]. It is known that MMPs are involved in all stages of tumor progression and that they modulate crucial signaling pathways [29,30]. For example TIMP-2 binds to MMP-14 on the tumor cell surface, and this binding induces cell proliferation mediated through the subsequent activation of extracellular-regulated kinases (ERK) 1/2 [31]. Similarly, MMP-7 and ADAM10 induce antiapoptotic signaling events in cancer cells through the cleavage of the Fas ligand from the cell's surface [32]. In other types of cancers, MMPs in the absence of their inhibitors activate canonical Wnt-signaling pathways, favoring the invasive capacity of cells [33].

In addition, MMPs can induce the shedding of the transmembrane precursors of the growth factors and growth factor-binding proteins (GF-BP). When MMPs degrade the insulin-like growth factor-binding protein (IGF-BP), IGFs are released [34]. Fibroblast growth factors are released when the basement membrane-specific heparan sulfate proteoglycan core protein is cleaved by MMPs. Through these processes, MMPs contribute to a pro-tumoral milieu [35].

MMPs also indirectly participate in the regulation of proliferative signals by integrins [26]. MMPs can be considered angiomodulators, as they are involved in the activation of proangiogenic factors and participate in the formation of new vessels, important structures sustaining tumor growth and progression [7]. Thus, many recent studies have highlighted that MMPs contribute to the main carcinogenesis steps, such as tumor progression, angiogenesis, invasion and metastasis; therefore, these biomolecules may also be regarded as potential target molecules in the prevention and treatment of neoplasms [36,37].

Table 1. MMP classifications and their substrates.

Groups	Substrates and Targets	MMPs as Activators for Other MMPs
Collagenases [8,9,12,38,39]		
MMP-1 (collagenase-1)	type I, II, III, VII, VIII, X and XI collagens [40,41], gelatin, nidogen [42], casein, aggrecan [43], perlecan, serpins, tenascin-C [44], versican, vitronectin, fibronectin, L-selectin, ovostatin, myelin basic protein, SDF-1 [45], pentraxin-3 [46], IGFBP [47], TNF precursor [48], VEGF-binding ECM proteins [49]	MMP-1 activates pro-MMP-2 [50] and pro-MMP-9 [51]
MMP-8 (collagenase-2)	type I, II, III, V, VII, VIII, and X collagens [52], gelatin, aggrecan [53], elastin, laminin [54], nidogen, fibronectin [55], ovostatin	MMP-8 activates pro-MMP-8
MMP-13 (collagenase-3)	type I-IV, IX, X, and XIV collagens [56–58], gelatin, plasminogen, fibronectin [58], osteonectin, aggrecan [59], perlecan [35], laminin, tenascin [58], casein	MMP-13 activates pro-MMP-2 and pro-MMP-9 [60]
MMP-18 (collagenase-4)	type I-III collagens, gelatin	
Gelatinases [8,9,11,12,46,61]		
MMP-2 (gelatinase A)	gelatin, type I-V, VII, X and XI collagens [40,62,63], elastin [63], aggrecan, laminin [64], fibronectin, nidogen, versican [51], tenascin, vitronectin, myelin basic protein, IGFBP-5 [65], follistatin-like 1 protein [66], follistatin-like 3 protein [46], mHB-EGF, CCL7/MCP-3 [67], CX3CL1/fractalkine, galectin-1, galectin-3 [68], transglutaminase [69], osteopontin, big endothelin-1 [70], TNF precursor, TGF beta, thrombospondin-2 and pyruvate kinase M1/M2 [71]	
MMP-9 (gelatinase B)	gelatin, type IV, V, VII, X and XIV collagens [40,63,72,73], aggrecan [43], elastin [63], fibronectin, laminin, nidogen, versican [51], decorin, myelin basic protein, casein, vitronectin, cytokines, chemokines, mHB-EGF [74], interleukin-8 [75], galectin-3 [68], interleukin-2 receptor-α [76], GCP-2/LIX [77], IGFBP [71], TNF precursor, TGF beta, VEGF-binding ECM proteins [49], thrombospondin-2 and pyruvate kinase M1/M2 [71]	
Stromelysins [8,9,12,26,78,79]		
MMP-3 (stromelysin-1)	type I-V [63] and IX-XI collagens, gelatin, aggrecan [80], ovostatin, nidogen [81], laminin, elastin, casein [82], osteonectin [60], decorin [44], fibronectin, perlecan [35], proteoglycans, versican [51], tenascin, myelin basic protein, osteopontin, plasminogen [83], IGFBP-3 [84], TNF precursor, VEGF-binding ECM proteins [49]	MMP-3 activates pro-MMP-1 [85], pro-MMP-8, pro-MMP-13 [86] and gelatinases [87,88]
MMP-10 (stromelysin-2)	type III-V [63,89], IX and X collagens, gelatin [86], aggrecan [80], fibronectin, casein [82], elastin [63], laminin, nidogen, proteoglycans, fibrilin-10	MMP-10 activates pro-MMP-1 [82], pro-MMP-8 [90] and pro-MMP-10
MMP-11 (stromelysin-3)	gelatin, fibronectin [91,92], aggrecan, laminin receptor [78]	
Matrilysins [8,9,12,61]		
MMP-7 (matrilysin-1)	type IV and X collagens [41,93], gelatin [93], aggrecan [43], decorin [94], elastin [63], entactin [73], casein [93], transferrin [95], fibronectin [93], laminin [93], plasminogen [96], vitronectin, tenascin [97], myelin, proteoglycans, β4-integrin, mHB-EGF [98], E-cadherin [19], osteopontin, syndecan [99], FasL [100]	MMP-7 activates pro-MMP-2 [101], pro-MMP-7 and pro-MMP-9 [102]
MMP-26 (matrilysin-2)	type IV collagen [103], gelatin [103], fibronectin [103], fibrinogen, vitronectin [104], casein, α2-macroglobulin	MMP-26 activates pro-MMP-2 and pro-MMP-9
Trans-membrane [8,9,12,46,105]		
MMP-14 (MT1-MMP)	type I-III collagens [106,107], gelatin, casein, fibronectin, laminin, nidogen [107], aggrecan, elastin, fibrin, perlecan, tenascin, vitronectin and proteoglycans [106–108], fibrilin-1, α2-macroglobulin [106], dickkopf-1 and cysteine-rich motor neuron-1 [109], galectin-1 [110], galectin-3 [111], syndecan-1, follistatin-like 3 protein [46], cyclophilin A [46], transglutaminase [112], pentraxin-3 [46]	MMP-14 activates pro-MMP-2 [113], pro-MMP-8 and pro-MMP-13 [114]
MMP-15 (MT2-MMP)	type I collagen, gelatin, fibronectin [107], laminin [107], aggrecan, perlecan [107], nidogen [107], tenascin, vitronectin, transglutaminase [112]	MMP-15 activates pro-MMP-2 [115] and pro-MMP-13
MMP-16 (MT3-MMP)	type I and III collagens [116], gelatin, casein [117], fibronectin, aggrecan, laminin, perlecan, vitronectin, syndecan [118], transglutaminase [112], VEGF-binding ECM proteins [49]	MMP-16 activates pro-MMP-2 [115], pro-MMP-9 and pro-MMP-13

Table 1. *Cont.*

Groups	Substrates and Targets	MMPs as Activators for Other MMPs
MMP-24 (MT5-MMP)	gelatin, fibronectin [119], proteoglycans [119], N-cadherin [120]	MMP-24 activates pro-MMP-2 [121] and pro-MMP-13
GP1-anchored [8,9,11,12,26,122,123]		
MMP-17 (MT4-MMP)	fibrinogen [124], fibrin [124], gelatin [125], TNF precursor	
MMP-25 (MT6-MMP)	type IV collagen, proteoglycans, gelatin [126], fibronectin [127], vimentin [122], cystatin C [122], galectin-1 [122]	MMP-25 activates pro-MMP-2 [128]
Other enzymes [8,9,12,38]		
MMP-12 (macrophage metalloelastase)	type I, IV [129] and V collagens, gelatin [129], elastin [130], fibronectin [129], laminin [129], vitronectin [129], proteoglycans [131], elastin, entactin [132], osteonectin, aggrecan, myelin, fibrinogen [131], α1-antitripsin [133], serine protease inhibitor α1-PI, TNF precursor	
MMP-19 (RASI-1)	type I and IV collagens, gelatin [134], aggrecan [135], fibronectin, casein, laminin, nidogen [136], tenascin, cartilage oligomeric matrix protein [135]	MMP-19 activates pro-MMP-9
MMP-20 (enamelysin)	type V collagen, aggrecan [135], amelogenin [121]	
MMP-21 (Xenopus-MMP)	-	
MMP-23 (CA-MMP)	gelatin	
MMP-27 (human MMP-22 homolog)	gelatin	
MMP-28 (epylisin)	casein	

IGFB: insulin-like growth factor-binding proteins, SDF-1: stromal cell-derived factor 1, TNF: tumor necrosis factor, VEGF: vascular endothelial growth factor, mHB-EGF: heparin-binding EGF-like growth factor, CCL7/MCP: 3-chemokine (C-C motif) ligand 7/monocyte chemotactic protein 3, CX3CL1: C-X3-C motif chemokine ligand 1 and TGF beta: transforming growth factor beta.

3. UV Radiation as a Trigger for MMP Production

It is well-known that UV radiation causes genetic alterations in keratinocytes that are responsible for skin carcinogenesis [1]. However, there are multiple mechanisms by which UV light generates skin carcinogenesis [137,138]. There is a link between MMPs, UV radiation, skin aging and carcinogenesis [139]. Sun exposure leads to the alteration of numerous signaling pathways, such as the mitogen-activated protein kinase (MAPK), the nuclear factor-kappa beta (NF-kB), the JAK/STAT (signal transduction and activation of transcription) and the nuclear factor erythroid 2-related factor 2 (Nrf2), as critical networks for modulating inflammation and cancer [140]. Epidermal keratinocytes exposed to UVB secrete a plethora of mediators that, in turn, activate the release of MMPs by dermal fibroblasts [141–143]. An important role in photooxidative damage is played by proinflammatory cytokines such as interleukin (IL)-1, IL-6, IL-8 and tumor necrosis (TNF)-alpha released by irradiated keratinocytes, which will activate the MAPK pathway in dermal fibroblasts [144,145]. TNF and MMPs are important aggressiveness drivers in skin cancers. In melanoma cell lines, an aggressiveness score based on TNF and MMP-2 has been described [146]. Moreover, we have shown that the regression process in melanoma is clearly associated with a reduced tissue expression of MMPs and that TIMPs modulate their activity [147,148]. Moreover, TNF is a proinflammatory cytokine secreted by macrophages, T-lymphocytes and mastocytes, inducing an overexpression of MMP-2, MMP-3, MMP-7 and MMP-9 in the tumor microenvironment, contributing to the invasive capacity of malignant cells [149].

The protein-directed Ser/Thr kinases, extracellular signal-regulated kinase (ERK), cJun N-terminal kinase (JNK) and p38 kinase are activated by proinflammatory cytokines and further promote the expression of activator protein 1 (AP 1), which is a positive regulator of MMP production [144]. The AP-1 transcription factor results from the dimerization of Jun (c-Jun, JunB and JunD) and Fos (c-Fos, FosB, Fra-1 and Fra-2) proteins and is activated through phosphorylation under the action of various stimuli, such as growth factors, cytokines, UV radiation, etc. [150,151]. UV-induced apoptosis in dermal fibroblasts requires JNK for the release of cytochrome C from the mitochondria [152]. The P38 kinase promotes COX-2 expression and, additionally, regulates the stabilization of genes that encode for IL-8 and IL-6, inducing a proinflammatory milieu [153].

Increased AP-1 and NF-κB expression is associated with large amounts of proinflammatory cytokines, such as IL-8, that would induce tumor cell proliferation and survival in squamous cell carcinoma of the head and neck [154,155]. AP-1 promotes the transcription of certain MMPs involved in ECM degradation (MMP-1, MMP-2, MMP-3 and MMP-9) and inhibits the production of type I collagen. MMP expression and activity can be regulated at many levels (gene transcription, proenzyme activation, post-translational modifications, extracellular inhibition and degradation) [156]. Primarily, the expression of MMPs is regulated at the transcriptional level, resulting in low levels of these enzymes in physiological conditions. MMPs share cis-regulatory elements in their promoter sequences, which allow the stimulation/inhibition of their expression. There are many molecules that could modulate the MMP expression, such as hormones, cytokines and growth factors. Post-transcriptional regulatory processes have been also described to be involved in the modulation of MMP gene expression (methylation, phosphorylation, acetylation, etc.) [156,157].

UVA promotes the activation of the epidermal growth factor receptor (EGFR)-signaling pathway and the tyrosine phosphorylation of β-catenin. Subsequently, β-catenin is translocated into the nucleus and binds to T-cell transcription factor 4 (TCF-4), resulting in the stimulation of MMP gene transcription [158]. The Wnt/β-catenin pathway regulates numerous processes, including cell proliferation, migration and invasion [159]. Moreover, B-catenin mediates several genetic transcripts, such as cyclin D, matrilysin metalloproteinase, survivin, etc., involved in tumor development [160]. UVA-induced EGFR expression promotes p70 (S6K)/p90 (RSK) activation by PI-3 and ERK kinases. The activation of EGFR, in conjunction with PI-3 kinase or MAPK signaling, leads to the expression of downstream effector molecules that affect the function of certain components of the mRNA translation apparatus. The activation of the two molecules, p70S6K and p90RSK, induces AP-1 expression and promotes tumor development [161]. EGFR overexpression has been identified in various tumors, being involved in the modulation of various signaling pathways responsible for cell proliferation, invasion and metastasis. In BCC, the crosstalk between the EGFR and Hedgehog pathways induces the activation of RAS/MEK/ERK and JUN/AP-1 signaling [162].

4. Tumor Microenvironment—An Important Source of MMPs in BCC

MMPs are important components of the tumor microenvironment, having essential roles in cancer progression by modulating cell growth, differentiation and migration. MMPs mediate the release and activity of numerous molecules, such as cytokines, growth factors and adhesion molecules, which regulate the function of the cells encountered in the tumor microenvironment [163]. MMPs are involved in ECM degradation when the primary tumor is initiated, but correspondingly, MMPs are involved in the metastatic process when a new tumor niche in the secondary organ/tissue is established [123]. Cancer-associated fibroblasts (CAFs) actively participate in the remodeling of the ECM, and besides secreting collagen and fibronectin, CAFs actively produce MMPs (e.g., MMP-1, MMP-3, MMP-7, MMP-9 and MMP-13); these MMPs concomitantly release growth factors (e.g., VEGF) from the ECM, facilitating neoplastic cell migration together with the pro-tumoral action of CAFs [164]. At the molecular level, MMPs process adhesion and cytoskeletal proteins, favoring cancer progression [165]. The intracellular roles of MMPs should not be neglected. MMP-2 is representative for intracellular MMPs, and its roles are ancient and conserved [166]. It is important to point out that both extra- and intracellular MMPs are involved in carcinogenesis. MMP-1 was shown to promote tumor growth and chemoresistance. This can be explained by the fact that MMP-1 was identified in the mitochondrial membrane and in the nucleus and may inhibit caspase activation, conferring resistance to apoptosis [167]. In osteosarcoma cells, nucleolar MMP-2 induces ribosomal RNA transcription and proliferation of malignant cells through the cleavage of the N-terminal tail of histone H3 [16].

Nuclear MMP-3 is correlated with cancer progression in glioma. MMP-3 promotes the transcription of the connective tissue growth factor (CTGF) gene, involved in cell migration and tumor growth, by interacting with heterochromatin protein gamma [16,168]. MMP-9 accelerates the extravasation of VEGFR-1+ cells in the tumor niche. VEGFR-1+ cells will activate integrin and chymosin to promote adhesion, survival and growth in tumor cells. In the lungs of patients diagnosed with esophageal cancer, melanoma and ovarian cancer, a high expression of MMP-9 was found, which suggests that primary tumors can stimulate the production of MMP-9 in premetastatic areas [169].

Nevertheless, fibroblasts and inflammatory cells are important sources of MMPs rather than malignant cells, thus demonstrating the clear role of inflammatory infiltrates in tumor progression [170]. Plasminogen binds to its receptors and is converted to plasmin by plasminogen activators, such as the urokinase plasminogen activator (uPA). Plasmin may activate various proMMPs, including MMP-1, MMP-3 and MMP-9, and certain growth factors, such as VEGF and transforming growth factor beta (TGF-β). Therefore, plasmin has a proteolytic activity, being involved in the release of ECM components, but may also activate intracellular signaling pathways [171] (Figure 1—parts of the figure were drawn by using and modifying pictures from Servier Medical Art under https://creativecommons.org/licenses/by/3.0/, accessed on 16 June 2021). The tumor microenvironment represents the obvious results of several signaling pathway activations that further induced a plethora of cytokine and growth factor releases, promoting tumor progression, invasion and metastasis [172]. These mediators are released by stromal cells (fibroblasts, macrophages, endothelial cells, lymphocytes, etc.), as well as by neoplastic cells [173]. In the tumor microenvironment, two main types of macrophages have been identified: anti-inflammatory M1 and proangiogenic M2 polarized macrophages, with a predominance of M2 macrophages in aggressive tumors. Kaiser et al. pointed out that, in BCC, the ratio between M1 and M2 macrophages is not involved in tumor aggression, but other factors such as the expression of cyclooxygenase (COX)-2 would influence the tumor aggressiveness [174]. However, Tiju et al. observed that, in aggressive forms of BCC compared to those with milder evolution, there is a higher number of macrophages, which can induce the secretion of MMP-9 in BCC cells and enhance tumor invasion through the activation of a p38 MAPK/NF-kB/COX-2 cascade [175]. On the contrary, Padoveze et al. did not reveal a relationship between tumor-associated macrophages and recurrent BCC [176]. At least, in other cancers, in lung adenocarcinoma (LUAD), a comprehensive proteogenomic characterization of the tumors in comparison to matched normal adjacent tissues showed that, besides other molecules, MMP-8, MMP-9 and MMP-12 are correlated with rich macrophage infiltration [177]. In pancreatic cancer, a recent work has shown an association between the MMP-28 gene and immune cells infiltrating a tumor [178].

Data on tumor-infiltrating lymphocytes in BCC are scarce; however, histopathological examinations have revealed a dense infiltrate that seems to correlate with a better course of the disease [179]. CD4 + T-helper cells and FoxP3 + T-regulatory (Tregs) cells are predominant in the intra- and peritumoral inflammatory infiltrates in BCC samples; the number of CD8 + T cells, NK and immature dendritic cells is variable, depending on several factors [180]. A high amount of CD8 + T cells represents a host antitumor response, and a reduction in the number of CD8 + T cells is associated with an enhanced risk of tumor recurrence [181,182]. Additionally, an increased number of Tregs is associated with an unfavorable outcome [172]. MMPs have the ability to cleave IL-2Rα on the T-cell surface and impede their proliferation. Additionally, MMP-11 can indirectly increase the survival of malignant cells in the presence of cytotoxic molecules released by NK cells [17,183].

In the BCC stroma, as well as in the peritumoral area, numerous CAFs that exhibit morphological alterations compared to normal fibroblasts and a loss of contact inhibition are present [172]. As previously stated, CAFs can secrete various MMPs, including MMP-1, MMP-2, MMP-3, MMP-9, MMP-11, MMP-13, MMP-14 and MMP-19, which promote ECM remodeling [123]. CAFs may also be involved in tumor escapes from the immune response by releasing MMP-2, MMP-9 and MMP-14 [163].

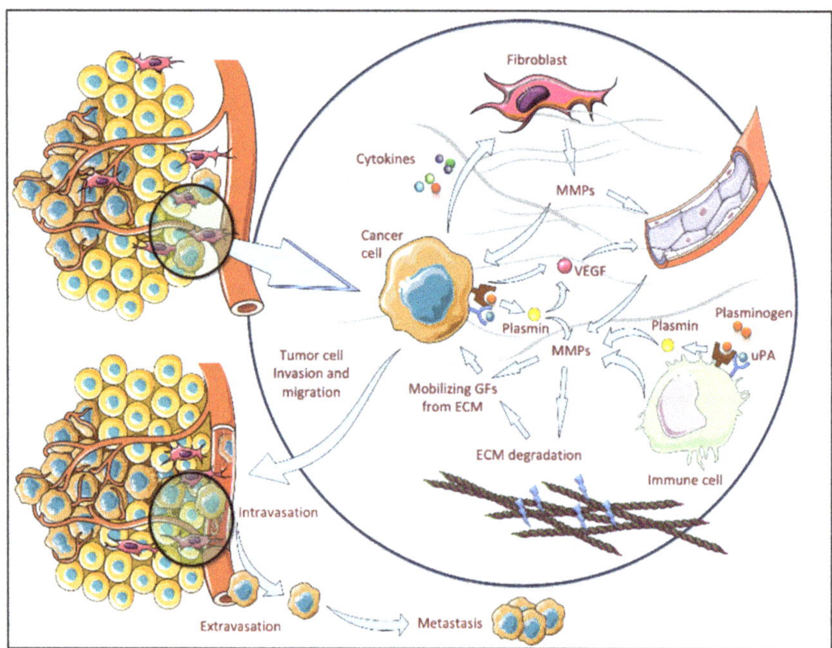

Figure 1. The interplay between MMPs and the tumor microenvironment. In the tumor microenvironment, MMPs are released by fibroblasts and immune cells, as well as by tumor cells. Plasminogen binds to its receptors and is converted to plasmin by urokinase plasminogen activator (uPA). In turn, plasmin may activate various MMPs and certain growth factors. All these molecules create a suitable microenvironment for ECM degradation, tumor cell invasion and migration.

Tumor cells mitigate the immune response by releasing MMP-2, MMP-9, MMP-13 and MMP-14, resulting in the inhibition of T-cell proliferation and antigen presentation [163]. Additionally, in BCC, there is an immunosuppressive micro medium characterized by immature dendritic cells and Th2-type cytokines [184]. In contrast, in regressing BCC, Th1-type cytokines are predominant [180].

5. The Role of MMPs in BCC Pathogenesis

Notwithstanding with the fact that numerous studies assessing MMP expression are available, many of them provide limited data regarding the biologically relevant activity of MMPs. Assays specifically designed to detect the activation, exposed active sites and enzymatic activity of MMPs should be employed. Crawford et al. employed fluorescent MMP substrates (both in vitro and in vivo) to characterize the patterns of MMP activity in zebrafish embryos. By using conventional gelatin zymography, MMPs were identified in the embryos as early as 3 somites, whereas in vivo techniques revealed type IV collagen degradation at the somite boundaries in as early as 4 somites [185]. The association between immunostaining and in vivo activity-based protein profiling ("a method that uses a chemical probe that targets active MMPs and becomes covalently bound to the protease") can characterize MMP localization and MMP activity [186]. Most studies evaluating MMP activation are based on tissue culture and biochemical assays. However, the native tissue context influences the activity of MMPs [187].

Significant progress has been made in deciphering the pathogenesis of BCC, but there are still numerous gaps in understanding the processes involved in the occurrence and evolution of BCC. In recent years, numerous studies have deepened our understanding of the role of MMPs in BCC.

5.1. The Expression of MMPs in BCC

A study conducted by Yucel et al. suggested that MMP-1 plays the most important role in the degradation of collagen in BCC [188]. The degradation of intact type I collagen results in high molecular weight collagen fragments; these fragments are not further degraded, and as MMP-2 and MMP-9 continue to degrade the intact collagen, high molecular fragments accumulate in a high rate. The accumulation of high molecular weight collagen fragments impairs the function of stromal fibroblasts. These events may underlie the increase in MMP production associated with BCC development [188]. When the intact collagen is fragmented, high molecular weight breakdown products are biologically active. Fibroblasts in contact with large collagen fragments change from a pro-synthesis phenotype that implies increased procollagen synthesis in association with a low MMP expression to a degradative phenotype characterized by an increased MMP expression and a reduced collagen synthesis [188,189]. The high expression of MMP-1 in the tumor stroma induces structural changes at the tumor periphery, with a loss of the palisading arrangement, which suggests a poor differentiation and a histological aspect that is correlated with an unfavorable prognosis [190].

Chen et al., using RT-PCR, evaluated the expression of MMP-2 in independent cultures and observed that MMP-2 is overexpressed in fibroblasts and melanoma cells and, to a much lesser extent, in keratinocytes and BCC cells. In contrast, when the noncontact cocultivation of fibroblasts with keratinocytes or BCC cells was performed, a decrease in MMP-2 expression by fibroblasts was observed. When the cocultivation with melanoma cells was performed, MMP-2 expression by fibroblasts was increased. Performing the same steps for MMP-1, noncontact cocultivation with keratinocytes, BCC cells and melanoma cells led to an increased expression of MMP-1 by fibroblasts. These results suggest the role of epidermal–mesenchymal interactions and the host immune response in the progression of BCC [191].

Monhian et al. showed a higher expression of MMP-1 and MMP-9 in the peritumoral tissue compared to the areas of the skin located distal to the tumor and a significant correlation between the presence of active gelatinolytic enzymes and broad fragmentation of the collagen substrate. Several factors could be involved in the increased MMP expression in tumor-associated tissue. EGFR agonists such as heparin-binding epidermal growth factor and amphiregulin could promote cell proliferation and stimulate MMP expression. A role was also attributed to proinflammatory cytokines such as IL-1, a significant amount of IL-1 identified in the skin, which acts as an inducer of gene transcription for MMPs [192]. Manola et al. evaluated the role of MMP-2 and MMP-9 in the peritumoral cleft present in the BCC samples, but they did not find a statistically significant correlation between the MMP expression and the presence of the peritumoral cleft [193]. Varani et al. showed, by analyzing 54 histological samples of BCC, different patterns of MMP expression. MMP-1 and MMP-2 were expressed both in tumor cells and normal epithelial cells, but their activity was higher in BCC compared to normal skin. MMP-8 was expressed only in the stroma. In contrast, MMP-9 and MMP-13 were expressed mainly in the healthy skin adjacent to the tumor [189]. Zlatarova et al. conducted a study on eyelid BCCs that revealed an increased expression of MMP-1, MMP-9 and MMP-13 in malignant epithelial cells but, also, in the surrounding stroma (inflammatory cells, fibroblasts and endothelial cells). Moreover, a significantly increased expression of MMPs was detected at the periphery of the tumor, an area with local invasion potential [194]. Petterson et al. showed that the depletion of cell surface CD44 was correlated with a high MMP-7 expression in BCC and SCC samples [195].

The study reported by Ciążyńska et al. was the first one to highlight the MMP-8 expression in BCC, using RT-PCR and a Western blot analysis. They identified the overexpression of MMP-8 mRNA in BCC; however, the Western blot analysis revealed only a slight increase in the MMP-8 protein expression. The study also showed important differences in the levels of mRNA and protein overexpression of MMP-1, MMP-3 and MMP-9 in the tumor tissue compared to normal skin [196].

Hattori et al. revealed that MMP-13 was expressed by the endothelial cells in 17 of the 20 BCC analyzed samples, at both mRNA and protein level, whereas MMP-1 was identified in only two cases. MMP-13 was also detected in the microvessels of normal skin from the edge of the surgical wound. These data suggest the role of MMP-13 in angiogenesis and indicate that the endothelial cells in the skin represent a significant source. The study pinpointed that MMP-13 expression is upregulated by IL-1alpha and, to a lesser extent, by phorbol myristate acetate, a tumor promoter, and by tumor necrosis factor alpha [197]. El-Havary et al. immunohistochemically evaluated the expression of MMP-13 and the cellular marker of Ki 67 proliferation in BCC and SCC specimens from patients with and without xeroderma pigmentosum. They did not observe differences in the MMP-13 expression between the two groups, but when analyzing the Ki67 expression, they detected higher levels in those with xeroderma pigmentosum, which may explain the more aggressive behavior of these tumors in patients with xeroderma pigmentosum [198].

Boyd et al. compared the expression of several MMPs (MMPs-1, -7, -8, -9, -10, -13 and -26) and their tissue inhibitors (TIMPs-1 and -3) in BCC samples from kidney transplant recipients and a control group of immunocompetent individuals. The immunohistochemical analysis did not reveal significant differences between the two groups regarding the expression of MMPs by BCC cells. However, MMP-1, MMP-9 and TIMP-1 were more frequently expressed by stromal macrophages in BCC samples from immunocompetent individuals, emphasizing the role of the tumor microenvironment in BCC behavior in association with the patient's immune status [199].

5.2. The Link between MMPs and BCC Invasiveness and Recurrence

The release of MMPs is one of the first events in the complex process of tumor invasion, resulting in important cytoskeleton changes that will allow cell migration. This process is governed by molecular interactions, cell-to-cell adhesion molecules such as E cadherin and β-catenin and chemokine receptor ligands such as CXCR4 [6]. MMPs could play a role not only in the aggressiveness of the tumor but, also, in its recurrence.

Histopathologically, BCC can be divided into six types: nodular, superficial, infiltrative, morpheaform, micronodular and mixed. The most common types encountered in medical practice are the nodular and superficial BCCs [179,200]. Based on its invasive behavior and recurrence risk, BCC can be considered as high-risk BCC (the morpheaform, infiltrative or basosquamous types) and low-risk BCC (the nodular and superficial types) [201,202]. Thus, depending on the histopathological type, BCC can be considered a more aggressive or less aggressive tumor. However, in many cases, areas with aggressive growth patterns and areas with a less aggressive pattern have been present in BCC samples [203].

Poswar et al. revealed that the protein expression of MMP-2 in the tumor stroma was more extensive in high-risk BCCs compared to low-risk BCCs [201]. In line with this, the same study showed that the protein expression of MMP-2 was higher in SCC (parenchyma and stroma) compared to BCC, suggesting that MMP-2 could play a defining role in the invasive nature of tumors [201]. However, a recent study did not find differences regarding the expression of MMP-2 mRNA in infiltrative BCC compared to nodular BCC. Interestingly, a higher expression of mRNA for MMP-2 and a lower expression of mRNA for type IV collagen were observed in the tissue adjacent to the nodular type compared to the infiltrative one. In the infiltrative type, the level of type IV collagen mRNA was increased in the surrounding tissue, probably as a defense mechanism against tumor infiltration [204]. Lower amounts of type IV collagen, a major component of the basement membrane, were identified in SCC samples compared to BCC, which may explain the higher invasiveness of SCC and can be considered as a marker of aggressiveness [105,205]. A study conducted by Orimoto et al. did not show significant differences in MMP-2 mRNA expression when compared between the nodular, superficial and sclerosing BCC types; however, they suggested that MMP-2 could be regarded as a marker for the differentiation between BCC and the surrounding normal tissue [206].

Zhu et al., using an immunohistochemical analysis, showed an increased expression of MMP-9 in both primary and metastatic SCC compared to BCC and normal skin tissue, highlighting the role of MMP-9 in tumor invasion and metastasis: MMP-9 degrades collagen and elastin, allowing tumor cell migration [207]. Gozdzialska et al. detected a higher expression of mRNA for MMP-9 in infiltrative BCC compared to nodular BCC [204]. Kadeh et al. analyzed the MMP-10 expression in BCC compared to SCC samples and observed a higher MMP-10 expression in both the tumor epithelium and stroma in SCC. In addition, in the case of SCC, there was a positive correlation between the MMP-10 expression and tumor grade. The authors concluded that MMP-10 may have a role in the different invasive patterns observed in BCC and SCC, contributing to the tumor aggressive behavior [208]. Cribier et al. indicated that MMP-11 expression is elevated in high-risk BCC [209]. Consistent with this, Greco et al. detected that HMGA1 and MMP-11 mRNA expressions were higher, as well as the HMGA1 and MMP-11 protein expression levels in the BCC and SCC samples compared to healthy tissue, and the levels of the two markers were higher in SCC compared to BCC, suggesting their involvement in tumor aggressiveness [210]. Greco et al. proposed that MMP-11 as a differentiation marker between self-limiting skin tumors and more aggressive ones in nonmelanoma skin cancer [210].

Kerkela et al. detected MMP-14 mRNA in stromal fibroblasts in fibrosing and keratotic BCC samples but not in adenoid BCC samples, being the first study to evaluate the expression of this MMP in BCC. In addition, MMP-10 was expressed only in epithelial laminin 5-positive cancer cells [39]. Laminin 5, after being cleaved by MMP-2 and MMP-14, promotes cell migration and invasion [105]. In the study by Kerkela et al., the release of laminin 5 by tumor cells was observed only in sclerodermiform BCC [105]. Oh et al. suggested that MMP-14 should be considered as a marker for high-risk BCC. The mechanism by which MMP-14 could be involved in the pathogenesis of high-risk BCC is the degradation of the E-cadherin/β-catenin complex and the activation of other MMPs, such as MMP-2 [211]. Oh et al. attributed an important role to β-catenin in increasing MMP-14 expression [211]. MMP-14 degrades type I collagen, the most abundant ECM component, and modulates cell–ECM interactions, promoting cell migration and tumor invasion [27]. ECM proteolysis mechanisms are involved in regulating epithelial cells, as well as carcinoma cell trafficking in vivo. MMP-14 was identified as an important effector of the matrix-remodeling processes in breast cancer. Feinberg et al. showed that MMP-14 regulates physiologic processes such as normal mammary gland branching morphogenesis, but also, MMP-14 induces local invasion and metastasis, and they concluded that there is a differential regulation of the normal and malignant mammary epithelial cell invasion programs [212].

MMPs participate in the invasion process by degrading E cadherin, a molecule with a pivotal role in cell–cell adhesion in epithelial tissues [213]. The role of the E-cadherin/β-catenin protein complex in mesenchymal–epithelial transition is well-known and is strongly implicated in cancer progression [214].

Rogosic et al. performed a study on 64 BCC samples and evaluated the expression of MMP-1, MMP-2, MMP-9, MMP-13 and E-cadherin using immunohistochemical staining. They revealed that the expression of MMP-1 in tumor cells was five times higher in morpheaform and recurrent BCC than in superficial, cystic or micronodular BCC. In addition, the expression of MMP-9 and MMP-13 in stromal cells was associated with morpheaform and recurrent BCC. Moreover, in morpheaform and recurrent BCC, E-cadherin was absent [215].

Chu et al. analyzed 19 recurrent BCCs and observed an increased expression of CXCR4. The study showed that stromal cell-derived factor 1a (SDF-1a), the CXCR4 ligand, is involved in the invasive behavior of BCC. This mechanism is based on the upregulation of mRNA expression and activity of MMP-13 by SDF-1a through the phosphorylation of ERK1/2 and induction of the expression of the AP-1 component c-Jun [216]. CXCR4-transfected BCC cells were administered to nude mice that developed aggressive BCC with an increased expression of CXCR4 and MMP-13 [216]. Ciurea et al. investigated the expression of CXCR4, MMP-13 and β-catenin in samples of facial BCC as following: metatypical (six cases), infiltrative–morpheaform (eight cases), micronodular (six cases) and superficial (five cases). CXCR4 expression was the highest in metatypical BCC compared to the other subtypes, especially in the areas with squamous transformation. The same observations were made for MMP-13 and β-catenin. In all cases, MMP-13 expression was higher in the stroma (fibroblasts and inflammatory cells) compared to tumor cells [160]. As previously acknowledged, MMP-13 can be regarded as a marker of malignant transformation in keratinocytes and has a role in tumor invasion by degrading ECM [199].

El-Khalawany et al. performed a study on 22 samples of recurrent BCCs and analyzed the expression of COX-2, ezrin—a cytoplasmic peripheral membrane protein—and MMP-9 compared to nonrecurrent BCCs. COX-2 expression was identified in 90.9% of recurrent BCCs and 59.1% of nonrecurrent BCCs, the found difference being statistically significant. In contrast, regarding the expression of ezrin and MMP-9, there were no statistically significant differences between the two groups [217]. Karahan et al. evaluated 30 BCC samples and did not observe any significant differences between the histological types of primary BCCs with respect to MMP-2 and MMP-9 expression. Moreover, the differences were not significant between the primary BCC and recurrent BCC. However, a positive correlation was identified between the COX-2 and MMP-9 expression. The COX-2 expression was higher in recurrent BCC compared to primary BCC [218].

6. TIMPs—Tissue Inhibitors of Metalloproteinases—Potential New Agents in the Management of BCC

TIMPs, endogenous inhibitors of MMPs that bind to their catalytic site, represent a family of polypeptides that includes four members: TIMP-1 TIMP-2, TIMP-3 and TIMP-4 [219].

The balance between TIMPs and MMPs plays an important role in cell homeostasis and in the ECM remodeling process. The destruction of this balance leads to the appearance of numerous pathological processes involved in various diseases, including malignant tumors [220]. Structurally, TIMPs are composed of a large N-terminal domain that inhibits MMP function and a smaller C-terminal domain. The N-terminal domain of TIMPs is able to bind the majority of MMPs, while the C-terminal site of TIMP-1 and TIMP-2 binds to the hemopexin domain of pro-MMP-2 and pro-MMP-9 [21,219]. The tissue expression of TIMPs is both constitutive and inducible and is modulated by various molecules and growth factors [221]. In addition to their role in maintaining ECM homeostasis, TIMPs have many functions independent of metalloproteinases, being multifunctional proteins. They are involved in cell migration, exhibit pro and antiapoptotic activity, are antiangiogenic factors, etc. Recent studies have shown that TIMPs, in addition to their inhibitory effect on MMPs, exert an inhibitory effect on integrin-metalloproteinases, ADAMs and ADAMTSs [222].

Of note, only a few studies in the literature have evaluated the expression of TIMPs in BCC. The expression of TIMP-1 and TIMP-2 was detected in stromal cells, and TIMP-3 expression in tumor epithelial cells, in the infiltrative area [105]. Regarding TIMP-1, an increased expression was observed primarily in morpheaform BCC, which was associated with an unfavorable prognosis. Furthermore, an association between the increased expression of TIMP-1 in tumor cells and/or stroma and the recurrence rate of eyelid BCC was identified [194]. TIMPs prevent tumor development by forming a fibrotic capsule and by inhibiting angiogenesis; however, it should be emphasized that TIMPs can also act as activators of MMPs; therefore, in some cancers, there may be an increase in the expression of MMPs in association with an increase in the expression of TIMPs [194,223]. Fu et al. revealed a higher expression of MMP-2 and MMP-9 and a lower expression of TIMP-1 and TIMP-2 in SCC compared to BCC [224]. Regarding TIMP-3 expression, no differences were observed when the samples of BCC, SCC and actinic keratoses were analyzed [201].

Recent research has suggested that TIMPs could be used as therapeutic agents in carcinogenesis and influence the outcome and prognosis of neoplasms [223].. Mutagenesis studies have revealed mutations that could improve the selectivity of TIMPs for MMPs.

There are some studies with encouraging results; for instance, the affinity of TIMP-1 toward MMP-14 was increased by replacing a single amino acid in the binding interface [225]. In addition, in another study of selective MMP-14 inhibition, the authors created a mutant TIMP that blocked the collagenase activity of MMP-14 in cell culture models of breast cancer and fibrosarcoma [226].

7. Conclusions

Most of the studies in this field show that MMPs are overexpressed in BCC, where they promote the release of numerous molecules, such as cytokines, growth factors and adhesion molecules, which provide a favorable environment for tumor development and progression. The majority of the studies investigating MMPs in BCC are correlative analyses of expression data that is often based on mRNA levels or protein levels, and very few studies have examined the biologically relevant activity of these proteases in the context of their possible roles in BCC. Knowledge of the MMP profile in BCC samples could be the basis for predicting the characteristics of tumor behaviors, particularly with respect to aggressiveness and amenability to specific clinical interventions. However, data on the substrates of these enzymes in vivo and their roles in processes other than ECM remodeling are scarce; therefore, correlations between MMP expression and tumor aggressiveness are difficult to interpret. This review gathered up and organized the newest studies on the matter, an endeavor that has not been undertaken in recent years, although data on the MMP involvement in BCC has been building up. Based on the aforementioned studies, we believe that MMP-targeted therapies could find their place in BCC management; however, further studies are needed.

Author Contributions: All authors equally contributed to this paper. Conceptualization, M.T., S.R.G. and M.N.; methodology, C.I.M., M.I.M., I.N. and A.C.; investigation, C.M., I.N., A.M. and C.S.; data curation C.C. (Constantin Caruntu), C.S., A.C. and C.M.; writing—original draft preparation M.T., C.I.M., M.I.M. and A.M.; writing—review and editing, C.C. (Constantin Caruntu), C.C. (Carolina Constantin), S.R.G. and M.N. and funding acquisition, C.C. (Carolina Constantin), C.C. (Constantin Caruntu) and M.N. All authors have read and agreed to the published version of the manuscript.

Funding: This study was partially supported by the Core Program, implemented with the support of NASR, project PN 19.29.01.01, and a grant from the Romanian Ministry of Research and Innovation, CCCDI-UEFISCDI, project number 61PCCDI/2018 PN-III-P1-1.2-PCCDI-2017-0341 within PNCDI-III, and the article processing charge was funded by Carol Davila University of Medicine and Pharmacy.

Conflicts of Interest: The authors declare no conflict of interest.

References

1. Wong, C.S.; Strange, R.C.; Lear, J.T. Basal cell carcinoma. *Bio Med. J.* **2003**, *327*, 794–798. [CrossRef]
2. Kim, D.P.; Kus, K.J.; Ruiz, E. Basal Cell Carcinoma Review. *Hematol. Oncol. Clin. N. Am.* **2019**, *33*, 13–24. [CrossRef]
3. Marzuka, A.G.; Book, S.E. Basal Cell Carcinoma: Pathogenesis, Epidemiology, Clinical Features, Diagnosis, Histopathology, and Management. *Yale J. Biol. Med.* **2015**, *88*, 167–179. [PubMed]
4. Lupu, M.; Popa, I.M.; Voiculescu, V.M.; Caruntu, A.; Caruntu, C. A Systematic Review and Meta-Analysis of the Accuracy of in VivoReflectance Confocal Microscopy for the Diagnosis of Primary Basal Cell Carcinoma. *J. Clin. Med.* **2019**, *8*, 1462. [CrossRef] [PubMed]
5. Ilie, M.A.; Caruntu, C.; Lupu, M.; Lixandru, D.; Tampa, M.; Georgescu, S.-R.; Bastian, A.; Constantin, C.; Neagu, M.; Zurac, S.A.; et al. Current and Future Applications of Confocal Laser Scanning Microscopy Imaging in Skin Oncology. *Oncol. Lett.* **2019**, *17*, 4102–4111. [CrossRef]
6. Pittayapruek, P.; Meephansan, J.; Prapapan, O.; Komine, M.; Ohtsuki, M. Role of Matrix Metalloproteinases in Photoaging and Photocarcinogenesis. *Int. J. Mol. Sci.* **2016**, *17*, 868. [CrossRef]
7. Quintero-Fabián, S.; Arreola, R.; Becerril-Villanueva, E.; Torres-Romero, J.C.; Arana-Argáez, V.; Lara-Riegos, J.; Ramírez-Camacho, M.A.; Alvarez-Sánchez, M.E. Role of Matrix Metalloproteinases in Angiogenesis and Cancer. *Front. Oncol.* **2019**, *9*, 1370. [CrossRef] [PubMed]
8. Rodríguez, D.; Morrison, C.J.; Overall, C.M. Matrix Metalloproteinases: What Do They Not Do? New Substrates and Biological Roles Identified by Murine Models and Proteomics. *Biochim. Biophys. Acta* **2010**, *1803*, 39–54. [CrossRef]
9. Laronha, H.; Caldeira, J. Structure and Function of Human Matrix Metalloproteinases. *Cells* **2020**, *9*, 1076. [CrossRef]

10. Jobin, P.G.; Butler, G.S.; Overall, C.M. New Intracellular Activities of Matrix Metalloproteinases Shine in the Moonlight. *Biochim. Biophys. Acta Mol. Cell Res.* **2017**, *1864*, 2043–2055. [CrossRef]

11. Das, N.; Benko, C.; Gill, S.E.; Dufour, A. The Pharmacological TAILS of Matrix Metalloproteinases and Their Inhibitors. *Pharmaceuticals* **2020**, *14*, 31. [CrossRef]

12. Cui, N.; Hu, M.; Khalil, R.A. Biochemical and Biological Attributes of Matrix Metalloproteinases. *Prog. Mol. Biol. Transl. Sci.* **2017**, *147*, 1–73. [CrossRef]

13. Neagu, M.; Constantin, C.; Bostan, M.; Caruntu, C.; Ignat, S.R.; Dinescu, S.; Costache, M. Proteomic Technology "Lens" for Epithelial-Mesenchymal Transition Process Identification in Oncology. *Anal. Cell. Pathol.* **2019**, *2019*, 3565970. [CrossRef]

14. Stawowy, P.; Meyborg, H.; Stibenz, D.; Stawowy, N.B.P.; Roser, M.; Thanabalasingam, U.; Veinot, J.P.; Chrétien, M.; Seidah, N.G.; Fleck, E.; et al. Furin-Like Proprotein Convertases Are Central Regulators of the Membrane Type Matrix Metalloproteinase–Pro-Matrix Metalloproteinase-2 Proteolytic Cascade in Atherosclerosis. *Circulation* **2005**, *111*, 2820–2827. [CrossRef]

15. Itoh, Y. Membrane-Type Matrix Metalloproteinases: Their Functions and Regulations. *Matrix Biol.* **2015**, *44–46*, 207–223. [CrossRef]

16. Bassiouni, W.; Ali, M.A.M.; Schulz, R. Multifunctional Intracellular Matrix Metalloproteinases: Implications in Disease. *FEBS J.* **2021**. [CrossRef]

17. Nissinen, L.; Kähäri, V.-M. Matrix Metalloproteinases in Inflammation. *Biochim. Biophys. Acta BBA Gen. Subj.* **2014**, *1840*, 2571–2580. [CrossRef]

18. Sounni, N.E.; Roghi, C.; Chabottaux, V.; Janssen, M.; Munaut, C.; Maquoi, E.; Galvez, B.G.; Gilles, C.; Frankenne, F.; Murphy, G.; et al. Up-Regulation of Vascular Endothelial Growth Factor-A by Active Membrane-Type 1 Matrix Metalloproteinase through Activation of Src-Tyrosine Kinases. *J. Biol. Chem.* **2004**, *279*, 13564–13574. [CrossRef]

19. McGuire, J.K.; Li, Q.; Parks, W.C. Matrilysin (Matrix Metalloproteinase-7) Mediates E-Cadherin Ectodomain Shedding in Injured Lung Epithelium. *Am. J. Pathol.* **2003**, *162*, 1831–1843. [CrossRef]

20. Gilles, C.; Polette, M.; Coraux, C.; Tournier, J.M.; Meneguzzi, G.; Munaut, C.; Volders, L.; Rousselle, P.; Birembaut, P.; Foidart, J.M. Contribution of MT1-MMP and of Human Laminin-5 Gamma2 Chain Degradation to Mammary Epithelial Cell Migration. *J. Cell Sci.* **2001**, *114*, 2967–2976. [CrossRef]

21. Vargová, V.; Pytliak, M.; Mechírová, V. Matrix Metalloproteinases. In *Matrix Metalloproteinase Inhibitors*; Experientia Supplementum; Gupta, S.P., Ed.; Springer: Basel, Switzerland, 2012; Volume 103, pp. 1–33. [CrossRef]

22. Scheau, C.; Badarau, I.A.; Costache, R.; Caruntu, C.; Mihai, G.L.; Didilescu, A.C.; Constantin, C.; Neagu, M. The Role of Matrix Metalloproteinases in the Epithelial-Mesenchymal Transition of Hepatocellular Carcinoma. *Anal. Cell. Pathol.* **2019**, *2019*, 9423907. [CrossRef] [PubMed]

23. Badea, M.; Baroş, A.; Bohîlţea, R.E.; Julea, I.E.; Furtunescu, F.; Istrate-Ofiţeru, A.M.; Iovan, L.; Cîrstoiu, M.M.; Burcin, M.R.; Turcan, N.; et al. Modern Interdisciplinary Monitoring of Cervical Cancer Risk. *Rom. J. Morphol. Embryol.* **2019**, *60*, 469–478. [PubMed]

24. Alaseem, A.; Alhazzani, K.; Dondapati, P.; Alobid, S.; Bishayee, A.; Rathinavelu, A. Matrix Metalloproteinases: A Challenging Paradigm of Cancer Management. *Semin. Cancer Biol.* **2019**, *56*, 100–115. [CrossRef]

25. Berdiaki, A.; Neagu, M.; Giatagana, E.-M.; Kuskov, A.; Tsatsakis, A.M.; Tzanakakis, G.N.; Nikitovic, D. Glycosaminoglycans: Carriers and Targets for Tailored Anti-Cancer Therapy. *Biomolecules* **2021**, *11*, 395. [CrossRef] [PubMed]

26. Jabłońska-Trypuć, A.; Matejczyk, M.; Rosochacki, S. Matrix Metalloproteinases (MMPs), the Main Extracellular Matrix (ECM) Enzymes in Collagen Degradation, as a Target for Anticancer Drugs. *J. Enzym. Inhib. Med. Chem.* **2016**, *31*, 177–183. [CrossRef] [PubMed]

27. Rowe, R.G.; Weiss, S.J. Breaching the Basement Membrane: Who, When and How? *Trends Cell Biol.* **2008**, *18*, 560–574. [CrossRef]

28. Van Doren, S.R. Matrix Metalloproteinase Interactions with Collagen and Elastin. *Matrix Biol.* **2015**, *44–46*, 224–231. [CrossRef]

29. Dufour, A.; Overall, C.M. Missing the Target: Matrix Metalloproteinase Antitargets in Inflammation and Cancer. *Trends Pharmacol. Sci.* **2013**, *34*, 233–242. [CrossRef]

30. Cabral-Pacheco, G.A.; Garza-Veloz, I.; Castruita-De la Rosa, C.; Ramirez-Acuña, J.M.; Perez-Romero, B.A.; Guerrero-Rodriguez, J.F.; Martinez-Avila, N.; Martinez-Fierro, M.L. The Roles of Matrix Metalloproteinases and Their Inhibitors in Human Diseases. *Int. J. Mol. Sci.* **2020**, *21*, 9739. [CrossRef]

31. D'Alessio, S.; Ferrari, G.; Cinnante, K.; Scheerer, W.; Galloway, A.C.; Roses, D.F.; Rozanov, D.V.; Remacle, A.G.; Oh, E.-S.; Shiryaev, S.A.; et al. Tissue Inhibitor of Metalloproteinases-2 Binding to Membrane-Type 1 Matrix Metalloproteinase Induces MAPK Activation and Cell Growth by a Non-Proteolytic Mechanism. *J. Biol. Chem.* **2008**, *283*, 87–99. [CrossRef]

32. Gialeli, C.; Theocharis, A.D.; Karamanos, N.K. Roles of Matrix Metalloproteinases in Cancer Progression and Their Pharmacological Targeting. *FEBS J.* **2011**, *278*, 16–27. [CrossRef] [PubMed]

33. Hojilla, C.V.; Kim, I.; Kassiri, Z.; Fata, J.E.; Fang, H.; Khokha, R. Metalloproteinase Axes Increase Beta-Catenin Signaling in Primary Mouse Mammary Epithelial Cells Lacking TIMP3. *J. Cell Sci.* **2007**, *120*, 1050–1060. [CrossRef] [PubMed]

34. Sounni, N.E.; Noel, A. Membrane Type-Matrix Metalloproteinases and Tumor Progression. *Biochimie* **2005**, *87*, 329–342. [CrossRef]

35. Whitelock, J.M.; Murdoch, A.D.; Iozzo, R.V.; Underwood, P.A. The Degradation of Human Endothelial Cell-Derived Perlecan and Release of Bound Basic Fibroblast Growth Factor by Stromelysin, Collagenase, Plasmin, and Heparanases. *J. Biol. Chem.* **1996**, *271*, 10079–10086. [CrossRef]

36. Sun, J. Matrix Metalloproteinases and Tissue Inhibitor of Metalloproteinases Are Essential for the Inflammatory Response in Cancer Cells. *J. Signal Transduct.* **2010**, *2010*, 985132. [CrossRef] [PubMed]

37. Tampa, M.; Caruntu, C.; Mitran, M.; Mitran, C.; Sarbu, I.; Rusu, L.-C.; Matei, C.; Constantin, C.; Neagu, M.; Georgescu, S.-R. Markers of Oral Lichen Planus Malignant Transformation. *Dis. Markers* **2018**, *2018*, 1959506. [CrossRef]

38. Herouy, Y.; Simkin, R.; Ulloa, J.; Pearson, I.C.; Mortimer, P.S.; Ciucci, J.L. Metalloproteinases (MMPs) and Their Inhibitors in Venous Leg Ulcer Healing. *Phlebo Lymphol.* **2004**, *44*, 231–267.

39. Kerkelä, E.; Ala-aho, R.; Jeskanen, L.; Lohi, J.; Grénman, R.; Kähäri, M.-V.; Saarialho-Kere, U. Differential Patterns of Stromelysin-2 (MMP-10) and MT1-MMP (MMP-14) Expression in Epithelial Skin Cancers. *Br. J. Cancer* **2001**, *84*, 659–669. [CrossRef]

40. Welgus, H.G.; Fliszar, C.J.; Seltzer, J.L.; Schmid, T.M.; Jeffrey, J.J. Differential Susceptibility of Type X Collagen to Cleavage by Two Mammalian Interstitial Collagenases and 72-KDa Type IV Collagenase. *J. Biol. Chem.* **1990**, *265*, 13521–13527. [CrossRef]

41. Sires, U.I.; Schmid, T.M.; Fliszar, C.J.; Wang, Z.Q.; Gluck, S.L.; Welgus, H.G. Complete Degradation of Type X Collagen Requires the Combined Action of Interstitial Collagenase and Osteoclast-Derived Cathepsin-B. *J. Clin. Invest.* **1995**, *95*, 2089–2095. [CrossRef]

42. Sires, U.I.; Griffin, G.L.; Broekelmann, T.J.; Mecham, R.P.; Murphy, G.; Chung, A.E.; Welgus, H.G.; Senior, R.M. Degradation of Entactin by Matrix Metalloproteinases. Susceptibility to Matrilysin and Identification of Cleavage Sites. *J. Biol. Chem.* **1993**, *268*, 2069–2074. [CrossRef]

43. Fosang, A.J.; Last, K.; Knäuper, V.; Neame, P.; Murphy, G.; Hardingham, T.; Tschesche, H.; Hamilton, J. Fibroblast and Neutrophil Collagenases Cleave at Two Sites in the Cartilage Aggrecan Interglobular Domain. *Biochem. J.* **1993**, *295*, 273–276. [CrossRef]

44. Imai, K.; Kusakabe, M.; Sakakura, T.; Nakanishi, I.; Okada, Y. Susceptibility of tenascin to degradation by matrix metalloproteinases and serine proteinases. *FEBS Lett.* **1994**, *352*, 216–218. [CrossRef]

45. Egeblad, M.; Werb, Z. New Functions for the Matrix Metalloproteinases in Cancer Progression. *Nat. Rev. Cancer* **2002**, *2*, 161–174. [CrossRef] [PubMed]

46. Butler, G.S.; Dean, R.A.; Tam, E.M.; Overall, C.M. Pharmacoproteomics of a Metalloproteinase Hydroxamate Inhibitor in Breast Cancer Cells: Dynamics of Membrane Type 1 Matrix Metalloproteinase-Mediated Membrane Protein Shedding. *Mol. Cell Biol.* **2008**, *28*, 4896–4914. [CrossRef] [PubMed]

47. Fowlkes, J.L.; Serra, D.M.; Nagase, H.; Thrailkill, K.M. MMPs are IGFBP-degrading proteinases: Implications for cell proliferation and tissue growth. *Ann. N. Y. Acad. Sci.* **1999**, *878*, 696–699. [CrossRef]

48. Gearing, A.J.H.; Beckett, P.; Christodoulou, M.; Churchill, M.; Clements, J.; Davidson, A.H.; Drummond, A.H.; Galloway, W.A.; Gilbert, R.; Gordon, J.L.; et al. Processing of Tumour Necrosis Factor-α Precursor by Metalloproteinases. *Nature* **1994**, *370*, 555–557. [CrossRef] [PubMed]

49. Lee, S.; Jilani, S.M.; Nikolova, G.V.; Carpizo, D.; Iruela-Arispe, M.L. Processing of VEGF-A by Matrix Metalloproteinases Regulates Bioavailability and Vascular Patterning in Tumors. *J. Cell Biol.* **2005**, *169*, 681–691. [CrossRef]

50. Crabbe, T.; O'Connell, J.P.; Smith, B.J.; Docherty, A.J. Reciprocated matrix metalloproteinase activation: A process performed by interstitial collagenase and progelatinase A. *Biochemistry* **1994**, *33*, 14419–14425. [CrossRef]

51. Perides, G.; Asher, R.A.; Lark, M.W.; Lane, W.S.; Robinson, R.A.; Bignami, A. Glial Hyaluronate-Binding Protein: A Product of Metalloproteinase Digestion of Versican? *Biochem. J.* **1995**, *312*, 377–384. [CrossRef]

52. Visse, R.; Nagase, H. Matrix Metalloproteinases and Tissue Inhibitors of Metalloproteinases: Structure, Function, and Biochemistry. *Circ. Res.* **2003**, *92*, 827–839. [CrossRef] [PubMed]

53. Fosang, A.J.; Last, K.; Neame, P.J.; Murphy, G.; Knäuper, V.; Tschesche, H.; Hughes, C.E.; Caterson, B.; Hardingham, T.E. Neutrophil Collagenase (MMP-8) Cleaves at the Aggrecanase Site E373-A374 in the Interglobular Domain of Cartilage Aggrecan. *Biochem. J.* **1994**, *304*, 347–351. [CrossRef]

54. Gutiérrez-Fernández, A.; Fueyo, A.; Folgueras, A.R.; Garabaya, C.; Pennington, C.J.; Pilgrim, S.; Edwards, D.R.; Holliday, D.L.; Jones, J.L.; Span, P.N.; et al. Matrix Metalloproteinase-8 Functions as a Metastasis Suppressor through Modulation of Tumor Cell Adhesion and Invasion. *Cancer Res.* **2008**, *68*, 2755–2763. [CrossRef] [PubMed]

55. Tschesche, H.; Knäuper, V.; Krämer, S.; Michaelis, J.; Oberhoff, R.; Reinke, H. Latent Collagenase and Gelatinase from Human Neutrophils and Their Activation. *Matrix Suppl.* **1992**, *1*, 245–255. [PubMed]

56. Freije, J.M.; Díez-Itza, I.; Balbín, M.; Sánchez, L.M.; Blasco, R.; Tolivia, J.; López-Otín, C. Molecular Cloning and Expression of Collagenase-3, a Novel Human Matrix Metalloproteinase Produced by Breast Carcinomas. *J. Biol. Chem.* **1994**, *269*, 16766–16773. [CrossRef]

57. Knäuper, V.; López-Otin, C.; Smith, B.; Knight, G.; Murphy, G. Biochemical Characterization of Human Collagenase-3 (*). *J. Biol. Chem.* **1996**, *271*, 1544–1550. [CrossRef]

58. Knäuper, V.; Cowell, S.; Smith, B.; López-Otin, C.; O'Shea, M.; Morris, H.; Zardi, L.; Murphy, G. The Role of the C-Terminal Domain of Human Collagenase-3 (MMP-13) in the Activation of Procollagenase-3, Substrate Specificity, and Tissue Inhibitor of Metalloproteinase Interaction. *J. Biol. Chem.* **1997**, *272*, 7608–7616. [CrossRef]

59. Fosang, A.J.; Last, K.; Knäuper, V.; Murphy, G.; Neame, P.J. Degradation of Cartilage Aggrecan by Collagenase-3 (MMP-13). *FEBS Lett.* **1996**, *380*, 17–20. [CrossRef]

60. Sasaki, T.; Göhring, W.; Mann, K.; Maurer, P.; Hohenester, E.; Knäuper, V.; Murphy, G.; Timpl, R. Limited Cleavage of Extracellular Matrix Protein BM-40 by Matrix Metalloproteinases Increases Its Affinity for Collagens. *J. Biol. Chem.* **1997**, *272*, 9237–9243. [CrossRef]

61. Groblewska, M.; Siewko, M.; Mroczko, B.; Szmitkowski, M. The Role of Matrix Metalloproteinases (MMPs) and Their Inhibitors (TIMPs) in the Development of Esophageal Cancer. *Folia Histochem. Cytobiol.* **2012**, *50*, 12–19. [CrossRef] [PubMed]

62. Aimes, R.T.; Quigley, J.P. Matrix Metalloproteinase-2 Is an Interstitial Collagenase: Inhibitor-Free Enzyme Catalyzes the Cleavage of Collagen Fibrils and Soluble Native Type I Collagen Generating the Specific $\frac{3}{4}$-and $\frac{1}{4}$-Length Fragments. *J. Biol. Chem.* **1995**, *270*, 5872–5876. [CrossRef] [PubMed]

63. Murphy, G.; Cockett, M.I.; Ward, R.; Docherty, A.J. Matrix Metalloproteinase Degradation of Elastin, Type IV Collagen and Proteoglycan. A Quantitative Comparison of the Activities of 95 KDa and 72 KDa Gelatinases, Stromelysins-1 and-2 and Punctuated Metalloproteinase (PUMP). *Biochem. J.* **1991**, *277*, 277–279. [CrossRef] [PubMed]

64. Giannelli, G.; Falk-Marzillier, J.; Schiraldi, O.; Stetler-Stevenson, W.G.; Quaranta, V. Induction of Cell Migration by Matrix Metalloprotease-2 Cleavage of Laminin-5. *Science* **1997**, *277*, 225–228. [CrossRef] [PubMed]

65. Thrailkill, K.M.; Quarles, L.D.; Nagase, H.; Suzuki, K.; Serra, D.M.; Fowlkes, J.L. Characterization of Insulin-like Growth Factor-Binding Protein 5-Degrading Proteases Produced throughout Murine Osteoblast Differentiation. *Endocrinology* **1995**, *136*, 3527–3533. [CrossRef] [PubMed]

66. Dean, R.A.; Butler, G.S.; Hamma-Kourbali, Y.; Delbé, J.; Brigstock, D.R.; Courty, J.; Overall, C.M. Identification of Candidate Angiogenic Inhibitors Processed by Matrix Metalloproteinase 2 (MMP-2) in Cell-Based Proteomic Screens: Disruption of Vascular Endothelial Growth Factor (VEGF)/Heparin Affin Regulatory Peptide (Pleiotrophin) and VEGF/Connective Tissue Growth Factor Angiogenic Inhibitory Complexes by MMP-2 Proteolysis. *Mol Cell. Biol.* **2007**, *27*, 8454. [CrossRef]

67. McQuibban, G.A.; Gong, J.H.; Tam, E.M.; McCulloch, C.A.; Clark-Lewis, I.; Overall, C.M. Inflammation Dampened by Gelatinase A Cleavage of Monocyte Chemoattractant Protein-3. *Science* **2000**, *289*, 1202–1206. [CrossRef] [PubMed]

68. Ochieng, J.; Fridman, R.; Nangia-Makker, P.; Kleiner, D.E.; Liotta, L.A.; Stetler-Stevenson, W.G.; Raz, A. Galectin-3 Is a Novel Substrate for Human Matrix Metalloproteinases-2 and-9. *Biochemistry* **1994**, *33*, 14109–14114. [CrossRef] [PubMed]

69. Belkin, A.M.; Zemskov, E.A.; Hang, J.; Akimov, S.S.; Sikora, S.; Strongin, A.Y. Cell-Surface-Associated Tissue Transglutaminase Is a Target of MMP-2 Proteolysis. *Biochemistry* **2004**, *43*, 11760–11769. [CrossRef]

70. Fernandez-Patron, C.; Zouki, C.; Whittal, R.; Chan, J.S.; Davidge, S.T.; Filep, J.G. Matrix Metalloproteinases Regulate Neutrophil-Endothelial Cell Adhesion through Generation of Endothelin-1[1-32]. *FASEB J.* **2001**, *15*, 2230–2240. [CrossRef]

71. Prudova, A.; auf dem Keller, U.; Butler, G.S.; Overall, C.M. Multiplex N-Terminome Analysis of MMP-2 and MMP-9 Substrate Degradomes by ITRAQ-TAILS Quantitative Proteomics. *Mol. Cell. Proteom.* **2010**, *9*, 894–911. [CrossRef]

72. Murphy, G.; Reynolds, J.; Bretz, U.; Baggiolini, M. Partial Purification of Collagenase and Gelatinase from Human Polymorphonuclear Leucocytes. Analysis of Their Actions on Soluble and Insoluble Collagens. *Biochem. J.* **1982**, *203*, 209–221. [CrossRef]

73. Sires, U.I.; Dublet, B.; Aubert-Foucher, E.; van der Rest, M.; Welgus, H.G. Degradation of the COL1 Domain of Type XIV Collagen by 92-KDa Gelatinase. *J. Biol. Chem.* **1995**, *270*, 1062–1067. [CrossRef]

74. Higashiyama, S. Metalloproteinase-Mediated Shedding of Heparin-Binding EGF-like Growth Factor and Its Pathophysiological Roles. *Protein Pept. Lett.* **2004**, *11*, 443–450. [CrossRef] [PubMed]

75. Van den Steen, P.E.; Proost, P.; Wuyts, A.; Van Damme, J.; Opdenakker, G. Neutrophil Gelatinase B Potentiates Interleukin-8 Tenfold by Aminoterminal Processing, Whereas It Degrades CTAP-III, PF-4, and GRO-α and Leaves RANTES and MCP-2 Intact. *Blood J. Am. Soc. Hematol.* **2000**, *96*, 2673–2681. [CrossRef]

76. Sheu, B.-C.; Hsu, S.-M.; Ho, H.-N.; Lien, H.-C.; Huang, S.-C.; Lin, R.-H. A Novel Role of Metalloproteinase in Cancer-Mediated Immunosuppression. *Cancer Res.* **2001**, *61*, 237–242. [PubMed]

77. Van den Steen, P.E.; Wuyts, A.; Husson, S.J.; Proost, P.; Van Damme, J.; Opdenakker, G. Gelatinase B/MMP-9 and Neutrophil Collagenase/MMP-8 Process the Chemokines Human GCP-2/CXCL6, ENA-78/CXCL5 and Mouse GCP-2/LIX and Modulate Their Physiological Activities. *Eur. J. Biochem.* **2003**, *270*, 3739–3749. [CrossRef]

78. Amano, T.; Kwak, O.; Fu, L.; Marshak, A.; Shi, Y.-B. The Matrix Metalloproteinase Stromelysin-3 Cleaves Laminin Receptor at Two Distinct Sites between the Transmembrane Domain and Laminin Binding Sequence within the Extracellular Domain. *Cell Res.* **2005**, *15*, 150–159. [CrossRef] [PubMed]

79. Mathew, S.; Fu, L.; Fiorentino, M.; Matsuda, H.; Das, B.; Shi, Y.-B. Differential Regulation of Cell Type-Specific Apoptosis by Stromelysin-3: A Potential Mechanism via the Cleavage of the Laminin Receptor during Tail Resorption in *Xenopus laevis*. *J. Biol. Chem.* **2009**, *284*, 18545–18556. [CrossRef]

80. Fosang, A.J.; Neame, P.J.; Hardingham, T.E.; Murphy, G.; Hamilton, J.A. Cleavage of Cartilage Proteoglycan between G1 and G2 Domains by Stromelysins. *J. Biol. Chem.* **1991**, *266*, 15579–15582. [CrossRef]

81. Alexander, C.M.; Howard, E.W.; Bissell, M.J.; Werb, Z. Rescue of Mammary Epithelial Cell Apoptosis and Entactin Degradation by a Tissue Inhibitor of Metalloproteinases-1 Transgene. *J. Cell Biol.* **1996**, *135*, 1669–1677. [CrossRef]

82. Windsor, L.J.; Grenett, H.; Birkedal-Hansen, B.; Bodden, M.; Engler, J.; Birkedal-Hansen, H. Cell Type-Specific Regulation of SL-1 and SL-2 Genes. Induction of the SL-2 Gene but Not the SL-1 Gene by Human Keratinocytes in Response to Cytokines and Phorbolesters. *J. Biol. Chem.* **1993**, *268*, 17341–17347. [CrossRef]

83. Lijnen, H.R.; Ugwu, F.; Bini, A.; Collen, D. Generation of an Angiostatin-like Fragment from Plasminogen by Stromelysin-1 (MMP-3). *Biochemistry* **1998**, *37*, 4699–4702. [CrossRef] [PubMed]

84. Fowlkes, J.L.; Enghild, J.J.; Suzuki, K.; Nagase, H. Matrix Metalloproteinases Degrade Insulin-like Growth Factor-Binding Protein-3 in Dermal Fibroblast Cultures. *J. Biol. Chem.* **1994**, *269*, 25742–25746. [CrossRef]

85. Murphy, G.; Cockett, M.I.; Stephens, P.E.; Smith, B.J.; Docherty, A.J. Stromelysin Is an Activator of Procollagenase. A Study with Natural and Recombinant Enzymes. *Biochem. J.* **1987**, *248*, 265–268. [CrossRef] [PubMed]

86. Knäuper, V.; Wilhelm, S.M.; Seperack, P.K.; DeClerck, Y.A.; Langley, K.E.; Osthues, A.; Tschesche, H. Direct Activation of Human Neutrophil Procollagenase by Recombinant Stromelysin. *Biochem. J.* **1993**, *295*, 581–586. [CrossRef] [PubMed]

87. Shapiro, S.D.; Fliszar, C.J.; Broekelmann, T.J.; Mecham, R.P.; Senior, R.M.; Welgus, H.G. Activation of the 92-KDa Gelatinase by Stromelysin and 4-Aminophenylmercuric Acetate: Differential processing and stabilization of the carboxyl-terminal domain by tissue inhibitor of metalloproteinases (TIMP). *J. Biol. Chem.* **1995**, *270*, 6351–6356. [CrossRef]

88. Miyazaki, K.; Umenishi, F.; Funahashi, K.; Koshikawa, N.; Yasumitsu, H.; Umeda, M. Activation of TIMP-2/Progelatinase a Complex by Stromelysin. *Biochem. Biophys. Res. Commun.* **1992**, *185*, 852–859. [CrossRef]

89. Nicholson, R.; Murphy, G.; Breathnach, R. Human and rat malignant-tumor-associated mRNAs encode stromelysin-like metallo-proteinases. *Biochemistry* **1989**, *28*, 5195–5203. [CrossRef]

90. Knäuper, V.; Murphy, G.; Tschesche, H. Activation of Human Neutrophil Procollagenase by Stromelysin 2. *Eur. J. Biochem.* **1996**, *235*, 187–191. [CrossRef]

91. Matchett, E.F.; Wang, S.; Crawford, B.D. Paralogues of Mmp11 and Timp4 Interact during the Development of the Myotendinous Junction in the Zebrafish Embryo. *J. Dev. Biol.* **2019**, *7*, 22. [CrossRef]

92. Jenkins, M.H.; Alrowaished, S.S.; Goody, M.F.; Crawford, B.D.; Henry, C.A. Laminin and Matrix Metalloproteinase 11 Regulate Fibronectin Levels in the Zebrafish Myotendinous Junction. *Skelet Muscle* **2016**, *6*, 18. [CrossRef] [PubMed]

93. Miyazaki, K.; Hattori, Y.; Umenishi, F.; Yasumitsu, H.; Umeda, M. Purification and Characterization of Extracellular Matrix-Degrading Metalloproteinase, Matrin (Pump-1), Secreted from Human Rectal Carcinoma Cell Line. *Cancer Res.* **1990**, *50*, 7758–7764. [PubMed]

94. Imai, K.; Hiramatsu, A.; Fukushima, D.; Pierschbacher, M.D.; Okada, Y. Degradation of Decorin by Matrix Metalloproteinases: Identification of the Cleavage Sites, Kinetic Analyses and Transforming Growth Factor-B1 Release. *Biochem. J.* **1997**, *322*, 809–814. [CrossRef] [PubMed]

95. Abramson, S.R.; Conner, G.E.; Nagase, H.; Neuhaus, I.; Woessner, J.F. Characterization of Rat Uterine Matrilysin and Its cDNA: Relationship To Human Pump-1 And Activation Of Procollagenases. *J. Biol. Chem.* **1995**, *270*, 16016–16022. [CrossRef]

96. Patterson, B.C.; Sang, Q.A. Angiostatin-Converting Enzyme Activities of Human Matrilysin (MMP-7) and Gelatinase B/Type IV Collagenase (MMP-9). *J. Biol. Chem.* **1997**, *272*, 28823–28825. [CrossRef]

97. Siri, A.; Knäuper, V.; Veirana, N.; Caocci, F.; Murphy, G.; Zardi, L. Different Susceptibility of Small and Large Human Tenascin-C Isoforms to Degradation by Matrix Metalloproteinases. *J. Biol. Chem.* **1995**, *270*, 8650–8654. [CrossRef]

98. Hao, L.; Du, M.; Lopez-Campistrous, A.; Fernandez-Patron, C. Agonist-Induced Activation of Matrix Metalloproteinase-7 Promotes Vasoconstriction through the Epidermal Growth Factor–Receptor Pathway. *Circ. Res.* **2004**, *94*, 68–76. [CrossRef]

99. Li, Q.; Park, P.W.; Wilson, C.L.; Parks, W.C. Matrilysin Shedding of Syndecan-1 Regulates Chemokine Mobilization and Transepithelial Efflux of Neutrophils in Acute Lung Injury. *Cell* **2002**, *111*, 635–646. [CrossRef]

100. Strand, S.; Vollmer, P.; van den Abeelen, L.; Gottfried, D.; Alla, V.; Heid, H.; Kuball, J.; Theobald, M.; Galle, P.R.; Strand, D. Cleavage of CD95 by Matrix Metalloproteinase-7 Induces Apoptosis Resistance in Tumour Cells. *Oncogene* **2004**, *23*, 3732–3736. [CrossRef]

101. Crabbe, T.; Smith, B.; O'Connell, J.; Docherty, A. Human Progelatinase A Can Be Activated by Matrilysin. *FEBS Lett.* **1994**, *345*, 14–16. [CrossRef]

102. Imai, K.; Yokohama, Y.; Nakanishi, I.; Ohuchi, E.; Fujii, Y.; Nakai, N.; Okada, Y. Matrix Metalloproteinase 7 (Matrilysin) from Human Rectal Carcinoma Cells: Activation of the Precursor, Interaction with Other Matrix Metalloproteinases and Enzymic Properties. *J. Biol. Chem.* **1995**, *270*, 6691–6697. [CrossRef]

103. Uría, J.A.; López-Otín, C. Matrilysin-2, a New Matrix Metalloproteinase Expressed in Human Tumors and Showing the Minimal Domain Organization Required for Secretion, Latency, and Activity. *Cancer Res.* **2000**, *60*, 4745–4751. [PubMed]

104. Marchenko, G.N.; Ratnikov, B.I.; Rozanov, D.V.; Godzik, A.; Deryugina, E.I.; Strongin, A.Y. Characterization of Matrix Metalloproteinase-26, a Novel Metalloproteinase Widely Expressed in Cancer Cells of Epithelial Origin. *Biochem. J.* **2001**, *356*, 705–718. [CrossRef] [PubMed]

105. Kerkelä, E.; Saarialho-Kere, U. Matrix Metalloproteinases in Tumor Progression: Focus on Basal and Squamous Cell Skin Cancer. *Exp. Dermatol.* **2003**, *12*, 109–125. [CrossRef]

106. Ohuchi, E.; Imai, K.; Fujii, Y.; Sato, H.; Seiki, M.; Okada, Y. Membrane Type 1 Matrix Metalloproteinase Digests Interstitial Collagens and Other Extracellular Matrix Macromolecules. *J. Biol. Chem.* **1997**, *272*, 2446–2451. [CrossRef] [PubMed]

107. D'ortho, M.; Will, H.; Atkinson, S.; Butler, G.; Messent, A.; Gavrilovic, J.; Smith, B.; Timpl, R.; Zardi, L.; Murphy, G. Membrane-type Matrix Metalloproteinases 1 and 2 Exhibit Broad-spectrum Proteolytic Capacities Comparable to Many Matrix Metalloproteinases. *Eur. J. Biochem.* **1997**, *250*, 751–757. [CrossRef] [PubMed]

108. Pei, D.; Weiss, S.J. Transmembrane-Deletion Mutants of the Membrane-Type Matrix Metalloproteinase-1 Process Progelatinase A and Express Intrinsic Matrix-Degrading Activity. *J. Biol. Chem.* **1996**, *271*, 9135–9140. [CrossRef] [PubMed]

109. Pinzone, J.J.; Hall, B.M.; Thudi, N.K.; Vonau, M.; Qiang, Y.-W.; Rosol, T.J.; Shaughnessy, J.D. The Role of Dickkopf-1 in Bone Development, Homeostasis, and Disease. *Blood* **2009**, *113*, 517–525. [CrossRef]

110. Dean, R.A.; Overall, C.M. Proteomics Discovery of Metalloproteinase Substrates in the Cellular Context by ITRAQ™ Labeling Reveals a Diverse MMP-2 Substrate Degradome. *Mol. Cell. Proteom.* **2007**, *6*, 611–623. [CrossRef] [PubMed]

111. Ochieng, J.; Green, B.; Evans, S.; James, O.; Warfield, P. Modulation of the Biological Functions of Galectin-3 by Matrix Metalloproteinases. *Biochim. Biophys. Acta* **1998**, *1379*, 97–106. [CrossRef]

112. Belkin, A.M.; Akimov, S.S.; Zaritskaya, L.S.; Ratnikov, B.I.; Deryugina, E.I.; Strongin, A.Y. Matrix-Dependent Proteolysis of Surface Transglutaminase by Membrane-Type Metalloproteinase Regulates Cancer Cell Adhesion and Locomotion. *J. Biol. Chem.* **2001**, *276*, 18415–18422. [CrossRef] [PubMed]

113. Sato, H.; Takino, T.; Okada, Y.; Cao, J.; Shinagawa, A.; Yamamoto, E.; Seiki, M. A Matrix Metalloproteinase Expressed on the Surface of Invasive Tumour Cells. *Nature* **1994**, *370*, 61–65. [CrossRef]

114. Knäuper, V.; Will, H.; López-Otin, C.; Smith, B.; Atkinson, S.J.; Stanton, H.; Hembry, R.M.; Murphy, G. Cellular Mechanisms for Human Procollagenase-3 (MMP-13) Activation: Evidence That MT1-MMP (MMP-14) and Gelatinase A (MMP-2) Are Able to Generate Active Enzyme. *J. Biol. Chem.* **1996**, *271*, 17124–17131. [CrossRef] [PubMed]

115. Tanaka, M.; Sato, H.; Takino, T.; Iwata, K.; Inoue, M.; Seiki, M. Isolation of a Mouse MT2-MMP Gene from a Lung CDNA Library and Identification of Its Product. *FEBS Lett.* **1997**, *402*, 219–222. [CrossRef]

116. Matsumoto, S.; Katoh, M.; Saito, S.; Watanabe, T.; Masuho, Y. Identification of Soluble Type of Membrane-Type Matrix Metalloproteinase-3 Formed by Alternatively Spliced MRNA. *Biochim. Biophys. Acta BBA Gene Struct. Expr.* **1997**, *1354*, 159–170. [CrossRef]

117. Shofuda, K.; Yasumitsu, H.; Nishihashi, A.; Miki, K.; Miyazaki, K. Expression of Three Membrane-Type Matrix Metalloproteinases (MT-MMPs) in Rat Vascular Smooth Muscle Cells and Characterization of MT3-MMPs with and without Transmembrane Domain. *J. Biol. Chem.* **1997**, *272*, 9749–9754. [CrossRef] [PubMed]

118. Asundi, V.K.; Erdman, R.; Stahl, R.C.; Carey, D.J. Matrix Metalloproteinase-dependent Shedding of Syndecan-3, a Transmembrane Heparan Sulfate Proteoglycan, in Schwann Cells. *J. Neurosci. Res.* **2003**, *73*, 593–602. [CrossRef]

119. Wang, X.; Yi, J.; Lei, J.; Pei, D. Expression, Purification and Charaterization of Recombinant Mouse MT5-MMP Protein Products. *FEBS Lett.* **1999**, *462*, 261–266. [CrossRef]

120. Porlan, E.; Martí-Prado, B.; Morante-Redolat, J.M.; Consiglio, A.; Delgado, A.C.; Kypta, R.; López-Otín, C.; Kirstein, M.; Fariñas, I. MT5-MMP Regulates Adult Neural Stem Cell Functional Quiescence through the Cleavage of N-Cadherin. *Nat. Cell Biol.* **2014**, *16*, 629–638. [CrossRef]

121. Llano, E.; Pendás, A.M.; Knäuper, V.; Sorsa, T.; Salo, T.; Salido, E.; Murphy, G.; Simmer, J.P.; Bartlett, J.D.; López-Otín, C. Identification and Structural and Functional Characterization of Human Enamelysin (MMP-20). *Biochemistry* **1997**, *36*, 15101–15108. [CrossRef]

122. Starr, A.E.; Bellac, C.L.; Dufour, A.; Goebeler, V.; Overall, C.M. Biochemical Characterization and N-Terminomics Analysis of Leukolysin, the Membrane-Type 6 Matrix Metalloprotease (MMP25): Chemokine and Vimentin Cleavages Enhance Cell Migration and Macrophage Phagocytic Activities. *J. Biol. Chem.* **2012**, *287*, 13382–13395. [CrossRef] [PubMed]

123. Gonzalez-Avila, G.; Sommer, B.; Mendoza-Posada, D.A.; Ramos, C.; Garcia-Hernandez, A.A.; Falfan-Valencia, R. Matrix Metalloproteinases Participation in the Metastatic Process and Their Diagnostic and Therapeutic Applications in Cancer. *Crit. Rev. Oncol. Hematol.* **2019**, *137*, 57–83. [CrossRef] [PubMed]

124. English, W.R.; Puente, X.S.; Freije, J.M.; Knäuper, V.; Amour, A.; Merryweather, A.; López-Otın, C.; Murphy, G. Membrane Type 4 Matrix Metalloproteinase (MMP17) Has Tumor Necrosis Factor-α Convertase Activity but Does Not Activate pro-MMP2. *J. Biol. Chem.* **2000**, *275*, 14046–14055. [CrossRef] [PubMed]

125. Wang, Y.; Johnson, A.R.; Ye, Q.-Z.; Dyer, R.D. Catalytic Activities and Substrate Specificity of the Human Membrane Type 4 Matrix Metalloproteinase Catalytic Domain. *J. Biol. Chem.* **1999**, *274*, 33043–33049. [CrossRef]

126. English, W.R.; Velasco, G.; Stracke, J.O.; Knäuper, V.; Murphy, G. Catalytic Activities of Membrane-Type 6 Matrix Metalloproteinase (MMP25). *FEBS Lett.* **2001**, *491*, 137–142. [CrossRef]

127. Kang, T.; Yi, J.; Guo, A.; Wang, X.; Overall, C.M.; Jiang, W.; Elde, R.; Borregaard, N.; Pei, D. Subcellular Distribution and Cytokine- and Chemokine-Regulated Secretion of Leukolysin/MT6-MMP/MMP-25 in Neutrophils. *J. Biol. Chem.* **2001**, *276*, 21960–21968. [CrossRef]

128. Velasco, G.; Cal, S.; Merlos-Suárez, A.; Ferrando, A.A.; Alvarez, S.; Nakano, A.; Arribas, J.; López-Otín, C. Human MT6-Matrix Metalloproteinase: Identification, Progelatinase A Activation, and Expression in Brain Tumors. *Cancer Res.* **2000**, *60*, 877–882.

129. Chandler, S.; Cossins, J.; Lury, J.; Wells, G. Macrophage Metalloelastase Degrades Matrix and Myelin Proteins and Processes a Tumour Necrosis Factor-α Fusion Protein. *Biochem. Biophys. Res. Commun.* **1996**, *228*, 421–429. [CrossRef]

130. Panchenko, M.V.; Stetler-Stevenson, W.G.; Trubetskoy, O.V.; Gacheru, S.N.; Kagan, H.M. Metalloproteinase Activity Secreted by Fibrogenic Cells in the Processing of Prolysyl Oxidase: Potential Role of Procollagen C-Proteinase. *J. Biol. Chem.* **1996**, *271*, 7113–7119. [CrossRef]

131. Banda, M.; Werb, Z. The Role of Macrophage Elastase in the Proteolysis of Fibrinogen, Plasminogen, and Fibronectin. *Fed. Am. Soc. Exp. Biol.* **1980**, *39*, 1756.

132. Gronski, T.J., Jr.; Martin, R.L.; Kobayashi, D.K.; Walsh, B.C.; Holman, M.C.; Huber, M.; Van Wart, H.E.; Shapiro, S.D. Hydrolysis of a Broad Spectrum of Extracellular Matrix Proteins by Human Macrophage Elastase. *J. Biol. Chem.* **1997**, *272*, 12189–12194. [CrossRef]

133. Banda, M.; Clark, E.; Sinha, S.; Travis, J. Interaction of Mouse Macrophage Elastase with Native and Oxidized Human Alpha 1-Proteinase Inhibitor. *J. Clin. Investig.* **1987**, *79*, 1314–1317. [CrossRef] [PubMed]

134. Kolb, C.; Mauch, S.; Peter, H.-H.; Krawinkel, U.; Sedlacek, R. The Matrix Metalloproteinase RASI-1 Is Expressed in Synovial Blood Vessels of a Rheumatoid Arthritis Patient. *Immunol. Lett.* **1997**, *57*, 83–88. [CrossRef]

135. Stracke, J.O.; Fosang, A.J.; Last, K.; Mercuri, F.A.; Pendás, A.M.; Llano, E.; Perris, R.; Di Cesare, P.E.; Murphy, G.; Knäuper, V. Matrix Metalloproteinases 19 and 20 Cleave Aggrecan and Cartilage Oligomeric Matrix Protein (COMP). *FEBS Lett.* **2000**, *478*, 52–56. [CrossRef]
136. Titz, B.; Dietrich, S.; Sadowski, T.; Beck, C.; Petersen, A.; Sedlacek, R. Activity of MMP-19 Inhibits Capillary-like Formation Due to Processing of Nidogen-1. *Cell. Mol. Life Sci.* **2004**, *61*, 1826–1833. [CrossRef]
137. Liu-Smith, F.; Jia, J.; Zheng, Y. UV-Induced Molecular Signaling Differences in Melanoma and Non-Melanoma Skin Cancer. *Adv. Exp. Med. Biol.* **2017**, *996*, 27–40. [CrossRef]
138. Caruntu, C.; Mirica, A.; Roşca, A.E.; Mirica, R.; Caruntu, A.; Tampa, M.; Matei, C.; Constantin, C.; Neagu, M.; Badarau, A.I.; et al. The role of estrogens and estrogen receptors in melanoma development and progression. *Acta Endocrinol.* **2016**, *12*, 234–241. [CrossRef] [PubMed]
139. Gupta, A.; Kaur, C.D.; Jangdey, M.; Saraf, S. Matrix Metalloproteinase Enzymes and Their Naturally Derived Inhibitors: Novel Targets in Photocarcinoma Therapy. *Ageing Res. Rev.* **2014**, *13*, 65–74. [CrossRef] [PubMed]
140. Bosch, R.; Philips, N.; Suárez-Pérez, J.; Juarranz, A.; Devmurari, A.; Chalensouk-Khaosaat, J.; González, S. Mechanisms of Photoaging and Cutaneous Photocarcinogenesis, and Photoprotective Strategies with Phytochemicals. *Antioxidants* **2015**, *4*, 248–268. [CrossRef] [PubMed]
141. Wang, X. IL-1 Receptor Antagonist Attenuates MAP Kinase/AP-1 Activation and MMPl Expression in UVA-Irradiated Human Fibroblasts Induced by Culture Medium from UVB-Irradiated Human Skin Keratinocytes. *Int. J. Mol. Med.* **2005**, *16*, 1117–1124. [CrossRef]
142. Lan, C.-C.E.; Hung, Y.-T.; Fang, A.-H.; Ching-Shuang, W. Effects of Irradiance on UVA-Induced Skin Aging. *J. Dermatol. Sci.* **2019**, *94*, 220–228. [CrossRef]
143. Turcan, N.; Bohiltea, R.E.; Ionita-Radu, F.; Furtunescu, F.; Navolan, D.; Berceanu, C.; Nemescu, D.; Cirstoiu, M.M. Unfavorable Influence of Prematurity on the Neonatal Prognostic of Small for Gestational Age Fetuses. *Exp. Ther. Med.* **2020**, *20*. [CrossRef] [PubMed]
144. Kim, H.; Woo, S.; Choi, W.; Kim, H.; Yi, C.; Kim, K.; Cheng, J.; Yang, S.; Suh, J. Scopoletin Downregulates MMP-1 Expression in Human Fibroblasts via Inhibition of P38 Phosphorylation. *Int. J. Mol. Med.* **2018**. [CrossRef] [PubMed]
145. Nicolae, I.; Ene, C.D.; Georgescu, S.R.; Tampa, M.; Matei, C.; Ceausu, E. Effects of UV Radiation and Oxidative DNA Adduct 8-Hydroxy-2′-Deoxiguanosine on the Skin Diseases. *Rev. Chim.* **2014**, *9*, 1036–1041.
146. Rossi, S.; Cordella, M.; Tabolacci, C.; Nassa, G.; D'Arcangelo, D.; Senatore, C.; Pagnotto, P.; Magliozzi, R.; Salvati, A.; Weisz, A.; et al. TNF-Alpha and Metalloproteases as Key Players in Melanoma Cells Aggressiveness. *J. Exp. Clin. Cancer Res.* **2018**, *37*, 326. [CrossRef]
147. Zurac, S.; Negroiu, G.; Petrescu, S.; Tudose, I.; Andrei, R.; Caius, S.; Neagu, M.; Constantin, C.; Staniceanu, F. Matrix Metalloproteinases Underexpression in Melanoma with Regression. In Proceedings of the 24th European Congress of Pathology, Prague, Czech Republic, 8–12 September 2012.
148. Zurac, S.; Neagu, M.; Constantin, C.; Cioplea, M.; Nedelcu, R.; Bastian, A.; Popp, C.; Nichita, L.; Andrei, R.; Tebeica, T.; et al. Variations in the Expression of TIMP1, TIMP2 and TIMP3 in Cutaneous Melanoma with Regression and Their Possible Function as Prognostic Predictors. *Oncol. Lett.* **2016**, *11*, 3354–3360. [CrossRef] [PubMed]
149. Hagemann, T.; Robinson, S.C.; Schulz, M.; Trümper, L.; Balkwill, F.R.; Binder, C. Enhanced Invasiveness of Breast Cancer Cell Lines upon Co-Cultivation with Macrophages Is Due to TNF-Alpha Dependent up-Regulation of Matrix Metalloproteases. *Carcinogenesis* **2004**, *25*, 1543–1549. [CrossRef]
150. Cooper, S.; Ranger-Moore, J.; Bowden, T.G. Differential Inhibition of UVB-Induced AP-1 and NF-KappaB Transactivation by Components of the Jun BZIP Domain. *Mol. Carcinog.* **2005**, *43*, 108–116. [CrossRef] [PubMed]
151. Neagu, M.; Constantin, C.; Caruntu, C.; Dumitru, C.; Surcel, M.; Zurac, S. Inflammation: A Key Process in Skin Tumorigenesis. *Oncol. Lett.* **2019**, *17*, 4068–4084. [CrossRef]
152. Zhang, W.; Liu, H.T. MAPK Signal Pathways in the Regulation of Cell Proliferation in Mammalian Cells. *Cell Res.* **2002**, *12*, 9–18. [CrossRef] [PubMed]
153. Bachelor, M.A.; Bowden, G.T. UVA-Mediated Activation of Signaling Pathways Involved in Skin Tumor Promotion and Progression. *Semin. Cancer Biol.* **2004**, *14*, 131–138. [CrossRef]
154. Wolf, J.S.; Chen, Z.; Dong, G.; Sunwoo, J.B.; Bancroft, C.C.; Capo, D.E.; Yeh, N.T.; Mukaida, N.; Van Waes, C. (Interleukin)-1alpha Promotes Nuclear Factor-KappaB and AP-1-Induced IL-8 Expression, Cell Survival, and Proliferation in Head and Neck Squamous Cell Carcinomas. *Clin. Cancer Res.* **2001**, *7*, 1812–1820.
155. Tampa, M.; Mitran, M.I.; Mitran, C.I.; Sarbu, M.I.; Matei, C.; Nicolae, I.; Caruntu, A.; Tocut, S.M.; Popa, M.I.; Caruntu, C.; et al. Mediators of Inflammation—A Potential Source of Biomarkers in Oral Squamous Cell Carcinoma. *J. Immunol. Res.* **2018**, *2018*, 1061780. [CrossRef] [PubMed]
156. Madzharova, E.; Kastl, P.; Sabino, F.; Auf dem Keller, U. Post-Translational Modification-Dependent Activity of Matrix Metalloproteinases. *Int. J. Mol. Sci.* **2019**, *20*, 3077. [CrossRef]
157. Fanjul-Fernández, M.; Folgueras, A.R.; Cabrera, S.; López-Otín, C. Matrix Metalloproteinases: Evolution, Gene Regulation and Functional Analysis in Mouse Models. *Biochim. Biophys. Acta* **2010**, *1803*, 3–19. [CrossRef]

158. Jean, C.; Bogdanowicz, P.; Haure, M.-J.; Castex-Rizzi, N.; Fournié, J.-J.; Laurent, G. UVA activated Synthesis of Metalloproteinases 1, 3 and 9 Is Prevented by a Broadspectrum Sunscreen. *Photodermatol. Photoimmunol. Photomed.* **2011**, *27*, 318–324. [CrossRef] [PubMed]
159. Noubissi, F.K.; Yedjou, C.G.; Spiegelman, V.S.; Tchounwou, P.B. Cross-Talk between Wnt and Hh Signaling Pathways in the Pathology of Basal Cell Carcinoma. *Int. J. Environ. Res. Public Health* **2018**, *15*, 1442. [CrossRef] [PubMed]
160. Ciurea, M.E.; Cernea, D.; Georgescu, C.C.; Cotoi, O.S. Expression of CXCR4, MMP-13 and β-Catenin in Different Histological Subtypes of Facial Basal Cell Carcinoma. *Rom. J. Morphol. Embryol.* **2013**, *54*, 939–951. [PubMed]
161. Zhang, Y.; Dong, Z.; Bode, A.M.; Ma, W.-Y.; Chen, N.; Dong, Z. Induction of EGFR-Dependent and EGFR-Independent Signaling Pathways by Ultraviolet A Irradiation. *DNA Cell Biol.* **2001**, *20*, 769–779. [CrossRef]
162. Tampa, M.; Georgescu, S.R.; Mitran, C.I.; Mitran, M.I.; Matei, C.; Scheau, C.; Constantin, C.; Neagu, M. Recent Advances in Signaling Pathways Comprehension as Carcinogenesis Triggers in Basal Cell Carcinoma. *J. Clin. Med.* **2020**, *9*, 3010. [CrossRef] [PubMed]
163. Gonzalez-Avila, G.; Sommer, B.; García-Hernández, A.A.; Ramos, C. Matrix Metalloproteinases' Role in Tumor Microenvironment. In *Tumor Microenvironment*; Birbrair, A., Ed.; Springer International Publishing: Cham, Switzerland, 2020; Volume 1245, pp. 97–131. [CrossRef]
164. Yamaguchi, H.; Sakai, R. Direct Interaction between Carcinoma Cells and Cancer Associated Fibroblasts for the Regulation of Cancer Invasion. *Cancers* **2015**, *7*, 2054–2062. [CrossRef]
165. Löffek, S.; Schilling, O.; Franzke, C.-W. Series "Matrix Metalloproteinases in Lung Health and Disease": Biological Role of Matrix Metalloproteinases: A Critical Balance. *Eur. Respir. J.* **2011**, *38*, 191–208. [CrossRef]
166. Fallata, A.M.; Wyatt, R.A.; Levesque, J.M.; Dufour, A.; Overall, C.M.; Crawford, B.D. Intracellular Localization in Zebrafish Muscle and Conserved Sequence Features Suggest Roles for Gelatinase A Moonlighting in Sarcomere Maintenance. *Biomedicines* **2019**, *7*, 93. [CrossRef] [PubMed]
167. Limb, G.A.; Matter, K.; Murphy, G.; Cambrey, A.D.; Bishop, P.N.; Morris, G.E.; Khaw, P.T. Matrix Metalloproteinase-1 Associates with Intracellular Organelles and Confers Resistance to Lamin A/C Degradation during Apoptosis. *Am. J. Pathol.* **2005**, *166*, 1555–1563. [CrossRef]
168. Eguchi, T.; Kubota, S.; Kawata, K.; Mukudai, Y.; Uehara, J.; Ohgawara, T.; Ibaragi, S.; Sasaki, A.; Kuboki, T.; Takigawa, M. Novel Transcription-Factor-like Function of Human Matrix Metalloproteinase 3 Regulating the CTGF/CCN2 Gene. *Mol. Cell. Biol.* **2008**, *28*, 2391–2413. [CrossRef]
169. Kaplan, R.N.; Rafii, S.; Lyden, D. Preparing the "Soil": The Premetastatic Niche. *Cancer Res.* **2006**, *66*, 11089–11093. [CrossRef]
170. O'Grady, A.; Dunne, C.; O'Kelly, P.; Murphy, G.M.; Leader, M.; Kay, E. Differential Expression of Matrix Metalloproteinase (MMP)-2, MMP-9 and Tissue Inhibitor of Metalloproteinase (TIMP)-1 and TIMP-2 in Non-Melanoma Skin Cancer: Implications for Tumour Progression. *Histopathology* **2007**, *51*, 793–804. [CrossRef]
171. Didiasova, M.; Wujak, L.; Wygrecka, M.; Zakrzewicz, D. From Plasminogen to Plasmin: Role of Plasminogen Receptors in Human Cancer. *Int. J. Mol. Sci.* **2014**, *15*, 21229–21252. [CrossRef] [PubMed]
172. Georgescu, S.R.; Tampa, M.; Mitran, C.I.; Mitran, M.I.; Caruntu, C.; Caruntu, A.; Lupu, M.; Matei, C.; Constantin, C.; Neagu, M. Tumour Microenvironment in Skin Carcinogenesis. *Adv. Exp. Med. Biol.* **2020**, *1226*, 123–142. [CrossRef]
173. Ahmed, F.; Haass, N.K. Microenvironment-Driven Dynamic Heterogeneity and Phenotypic Plasticity as a Mechanism of Melanoma Therapy Resistance. *Front. Oncol.* **2018**, *8*, 173. [CrossRef] [PubMed]
174. Kaiser, U.; Loeffler, K.U.; Nadal, J.; Holz, F.G.; Herwig-Carl, M.C. Polarization and Distribution of Tumor-Associated Macrophages and COX-2 Expression in Basal Cell Carcinoma of the Ocular Adnexae. *Curr. Eye Res.* **2018**, *43*, 1126–1135. [CrossRef]
175. Tjiu, J.-W.; Chen, J.-S.; Shun, C.-T.; Lin, S.-J.; Liao, Y.-H.; Chu, C.-Y.; Tsai, T.-F.; Chiu, H.-C.; Dai, Y.-S.; Inoue, H.; et al. Tumor-Associated Macrophage-Induced Invasion and Angiogenesis of Human Basal Cell Carcinoma Cells by Cyclooxygenase-2 Induction. *J. Investig. Dermatol.* **2009**, *129*, 1016–1025. [CrossRef]
176. Padoveze, E.H.; Chiacchio, N.D.; Ocampo-Garza, J.; Cernea, S.S.; Belda, W.; Sotto, M.N. Macrophage Subtypes in Recurrent Nodular Basal Cell Carcinoma after Mohs Micrographic Surgery. *Int. J. Dermatol.* **2017**, *56*, 1366–1372. [CrossRef]
177. Gillette, M.A.; Satpathy, S.; Cao, S.; Dhanasekaran, S.M.; Vasaikar, S.V.; Krug, K.; Petralia, F.; Li, Y.; Liang, W.-W.; Reva, B.; et al. Clinical Proteomic Tumor Analysis Consortium. Proteogenomic Characterization Reveals Therapeutic Vulnerabilities in Lung Adenocarcinoma. *Cell* **2020**, *182*, 200–225.e35. [CrossRef]
178. Wu, M.; Li, X.; Liu, R.; Yuan, H.; Liu, W.; Liu, Z. Development and Validation of a Metastasis-Related Gene Signature for Predicting the Overall Survival in Patients with Pancreatic Ductal Adenocarcinoma. *J. Cancer* **2020**, *11*, 6299–6318. [CrossRef]
179. Omland, S.H. Local Immune Response in Cutaneous Basal Cell Carcinoma. *Dan. Med. J.* **2017**, *64*, B5412. [PubMed]
180. Pellegrini, C.; Orlandi, A.; Costanza, G.; Di Stefani, A.; Piccioni, A.; Di Cesare, A.; Chiricozzi, A.; Ferlosio, A.; Peris, K.; Fargnoli, M.C.; et al. Expression of IL-23/Th17-Related Cytokines in Basal Cell Carcinoma and in the Response to Medical Treatments. *PLoS ONE* **2017**, *12*, e0183415. [CrossRef] [PubMed]
181. Beksaç, B.; İlter, N.; Erdem, Ö.; Çakmak, P.; Çenetoğlu, S.; Yapar, D. Sparsity of Dendritic Cells and Cytotoxic T Cells in Tumor Microenvironment May Lead to Recurrence in Basal Cell Carcinoma. *Int. J. Dermatol.* **2020**, *59*, 1258–1263. [CrossRef]
182. Matei, C.; Tampa, M.; Caruntu, C.; Ion, R.-M.; Georgescu, S.-R.; Dumitrascu, G.R.; Constantin, C.; Neagu, M. Protein Microarray for Complex Apoptosis Monitoring of Dysplastic Oral Keratinocytes in Experimental Photodynamic Therapy. *Biol. Res.* **2014**, *47*, 33. [CrossRef] [PubMed]

183. Amano, T.; Fu, L.; Marshak, A.; Kwak, O.; Shi, Y.-B. Spatio-Temporal Regulation and Cleavage by Matrix Metalloproteinase Stromelysin-3 Implicate a Role for Laminin Receptor in Intestinal Remodeling during *Xenopus Laevis* Metamorphosis. *Dev. Dyn.* **2005**, *234*, 190–200. [CrossRef] [PubMed]

184. Kaporis, H.G.; Guttman-Yassky, E.; Lowes, M.A.; Haider, A.S.; Fuentes-Duculan, J.; Darabi, K.; Whynot-Ertelt, J.; Khatcherian, A.; Cardinale, I.; Novitskaya, I.; et al. Human Basal Cell Carcinoma Is Associated with Foxp3+ T Cells in a Th2 Dominant Microenvironment. *J. Investig. Dermatol.* **2007**, *127*, 2391–2398. [CrossRef] [PubMed]

185. Crawford, B.D.; Pilgrim, D.B. Ontogeny and Regulation of Matrix Metalloproteinase Activity in the Zebrafish Embryo by in Vitro and in Vivo Zymography. *Dev. Biol.* **2005**, *286*, 405–414. [CrossRef] [PubMed]

186. Keow, J.Y.; Pond, E.D.; Cisar, J.S.; Cravatt, B.F.; Crawford, B.D. Activity-Based Labeling of Matrix Metalloproteinases in Living Vertebrate Embryos. *PLoS ONE* **2012**, *7*, e43434. [CrossRef]

187. Jeffrey, E.J.; Crawford, B.D. The Epitope-Mediated MMP Activation Assay: Detection and Quantification of the Activation of Mmp2 in Vivo in the Zebrafish Embryo. *Histochem. Cell Biol.* **2018**, *149*, 277–286. [CrossRef]

188. Yucel, T.; Mutnal, A.; Fay, K.; Fligiel, S.E.G.; Wang, T.; Johnson, T.; Baker, S.R.; Varani, J. Matrix Metalloproteinase Expression in Basal Cell Carcinoma: Relationship between Enzyme Profile and Collagen Fragmentation Pattern. *Exp. Mol. Pathol.* **2005**, *79*, 151–160. [CrossRef]

189. Varani, J.; Hattori, Y.; Chi, Y.; Schmidt, T.; Perone, P.; Zeigler, M.E.; Fader, D.J.; Johnson, T.M. Collagenolytic and Gelatinolytic Matrix Metalloproteinases and Their Inhibitors in Basal Cell Carcinoma of Skin: Comparison with Normal Skin. *Br. J. Cancer* **2000**, *82*, 657–665. [CrossRef]

190. Son, K.D.; Kim, T.J.; Lee, Y.S.; Park, G.S.; Han, K.T.; Lim, J.S.; Kang, C.S. Comparative Analysis of Immunohistochemical Markers with Invasiveness and Histologic Differentiation in Squamous Cell Carcinoma and Basal Cell Carcinoma of the Skin. *J. Surg. Oncol.* **2008**, *97*. [CrossRef] [PubMed]

191. Chen, G.S.; Lu, M.P.; Wu, M.T. Differential expression of matrix metalloproteinase-2 by fibroblasts in co-cultures with keratinocytes, basal cell carcinoma and melanoma. *J. Dermatol.* **2006**, *33*, 609–615. [CrossRef]

192. Monhian, N.; Jewett, B.S.; Baker, S.R.; Varani, J. Matrix Metalloproteinase Expression in Normal Skin Associated with Basal Cell Carcinoma and in Distal Skin from the Same Patients. *Arch. Facial Plast. Surg.* **2005**, *7*, 238–243. [CrossRef]

193. Manola, I.; Mataic, A.; Drvar, D.L.; Pezelj, I.; Dzombeta, T.R.; Kruslin, B. Peritumoral Clefting and Expression of MMP-2 and MMP-9 in Basal Cell Carcinoma of the Skin. *In Vivo* **2020**, *34*, 1271–1275. [CrossRef]

194. Zlatarova, Z.I.; Softova, E.B.; Dokova, K.G.; Messmer, E.M. Expression of Matrix Metalloproteinase-1, -9, -13, and Tissue Inhibitor of Metalloproteinases-1 in Basal Cell Carcinomas of the Eyelid. *Graefes Arch. Clin. Exp. Ophthalmol.* **2012**, *250*, 425–431. [CrossRef]

195. Pettersen, J.S.; Fuentes-Duculan, J.; Suárez-Fariñas, M.; Pierson, K.C.; Pitts-Kiefer, A.; Fan, L.; Belkin, D.A.; Wang, C.Q.; Bhuvanendran, S. Tumor-Associated Macrophages in the Cutaneous SCC Microenvironment Are Heterogeneously Activated. *J. Investig. Dermatol.* **2011**, *131*, 1322–1330. [CrossRef]

196. Ciążyńska, M.; Bednarski, I.A.; Wódz, K.; Kolano, P.; Narbutt, J.; Sobjanek, M.; Woźniacka, A.; Lesiak, A. Proteins Involved in Cutaneous Basal Cell Carcinoma Development. *Oncol. Lett.* **2018**, *16*, 4064–4072. [CrossRef] [PubMed]

197. Hattori, Y.; Nerusu, K.C.; Bhagavathula, N.; Brennan, M.; Hattori, N.; Murphy, H.S.; Su, L.D.; Wang, T.S.; Johnson, T.M.; Varani, J. Vascular Expression of Matrix Metalloproteinase-13 (Collagenase-3) in Basal Cell Carcinoma. *Exp. Mol. Pathol.* **2003**, *74*, 230–237. [CrossRef]

198. El-Hawary, A.K.; Yassin, E.; Khater, A.; Abdelgaber, S. Expression of Matrix Metalloproteinase-13 and Ki-67 in Nonmelanoma Skin Cancer in Xeroderma Pigmentosum and Non-Xeroderma Pigmentosum. *Am. J. Dermatopathol.* **2013**, *35*, 45–49. [CrossRef] [PubMed]

199. Boyd, S.; Tolvanen, K.; Virolainen, S.; Kuivanen, T.; Kyllönen, L.; Saarialho-Kere, U. Differential Expression of Stromal MMP-1, MMP-9 and TIMP-1 in Basal Cell Carcinomas of Immunosuppressed Patients and Controls. *Virchows Arch.* **2008**, *452*, 83–90. [CrossRef]

200. Ghita, M.A.; Caruntu, C.; Rosca, A.E.; Kaleshi, H.; Caruntu, A.; Moraru, L.; Docea, A.O.; Zurac, S.; Boda, D.; Neagu, M.; et al. Reflectance Confocal Microscopy and Dermoscopy for in Vivo, Non-Invasive Skin Imaging of Superficial Basal Cell Carcinoma. *Oncol. Lett.* **2016**, *11*, 3019–3024. [CrossRef]

201. Poswar, F.O.; Fraga, C.A.C.; Farias, L.C.; Feltenberger, J.D.; Cruz, V.P.D.; Santos, S.H.S.; Silveira, C.M.; de Paula, A.M.B.; Guimarães, A.L.S. Immunohistochemical Analysis of TIMP-3 and MMP-9 in Actinic Keratosis, Squamous Cell Carcinoma of the Skin, and Basal Cell Carcinoma. *Pathol. Res. Pract.* **2013**, *209*, 705–709. [CrossRef] [PubMed]

202. Wollina, U.; Pabst, F.; Krönert, C.; Schorcht, J.; Haroske, G.; Klemm, E.; Kittner, T. High-Risk Basal Cell Carcinoma: An Update. *Expert Rev. of Dermatol.* **2010**, *5*, 357–368. [CrossRef]

203. Dimovska Nilsson, K.; Neittaanmäki, N.; Zaar, O.; Angerer, T.B.; Paoli, J.; Fletcher, J.S. TOF-SIMS Imaging Reveals Tumor Heterogeneity and Inflammatory Response Markers in the Microenvironment of Basal Cell Carcinoma. *Biointerphases* **2020**, *15*, 041012. [CrossRef]

204. Goździalska, A.; Wojas-Pelc, A.; Drąg, J.; Brzewski, P.; Jaśkiewicz, J.; Pastuszczak, M. Expression of Metalloproteinases (MMP-2 and MMP-9) in Basal-Cell Carcinoma. *Mol. Biol. Rep.* **2016**, *43*, 1027–1033. [CrossRef] [PubMed]

205. Matei, C.; Tampa, M.; Ion, R.; Neagu, M.; Constantin, C. Photodynamic properties of aluminium sulphonated phthalo-cyanines in human displazic oral keratinocytes experimental model. *Dig. J. Nanomater. Biostructures* **2012**, *7*, 1535–1547.

206. Orimoto, A.M.; Neto, C.F.; Pimentel, E.R.A.; Sanches, J.A.; Sotto, M.N.; Akaishi, E.; Ruiz, I.R.G. High Numbers of Human Skin Cancers Express MMP2 and Several Integrin Genes. *J. Cutan. Pathol.* **2008**, *35*, 285–291. [CrossRef] [PubMed]
207. Zhu, L.; Kohda, F.; Nakahara, T.; Chiba, T.; Tsuji, G.; Hachisuka, J.; Ito, T.; Tu, Y.; Moroi, Y.; Uchi, H.; et al. Aberrant Expression of S100A6 and Matrix Metalloproteinase 9, but Not S100A2, S100A4, and S100A7, Is Associated with Epidermal Carcinogenesis. *J. Dermatol. Sci.* **2013**, *72*, 311–319. [CrossRef] [PubMed]
208. Kadeh, H.; Saravani, S.; Heydari, F.; Shahraki, S. Differential Immunohistochemical Expression of Matrix Metalloproteinase-10 (MMP-10) in Non-Melanoma Skin Cancers of the Head and Neck. *Pathol. Res. Pract.* **2016**, *212*, 867–871. [CrossRef]
209. Cribier, B.; Noacco, G.; Peltre, B.; Grosshans, E. Expression of Stromelysin 3 in Basal Cell Carcinomas. *Eur. J. Dermatol.* **2001**, *11*, 530–533. [PubMed]
210. Greco, M.; Arcidiacono, B.; Chiefari, E.; Vitagliano, T.; Ciriaco, A.G.; Brunetti, F.S.; Cuda, G.; Brunetti, A. HMGA1 and MMP-11 Are Overexpressed in Human Non-Melanoma Skin Cancer. *Anticancer Res.* **2018**, *38*, 771–778. [CrossRef]
211. Oh, S.T.; Kim, H.S.; Yoo, N.J.; Lee, W.S.; Cho, B.K.; Reichrath, J. Increased Immunoreactivity of Membrane Type-1 Matrix Metalloproteinase (MT1-MMP) and β-Catenin in High-Risk Basal Cell Carcinoma. *Br. J. Dermatol.* **2011**, *165*, 1197–1204. [CrossRef]
212. Feinberg, T.Y.; Zheng, H.; Liu, R.; Wicha, M.S.; Yu, S.M.; Weiss, S.J. Divergent Matrix-Remodeling Strategies Distinguish Developmental from Neoplastic Mammary Epithelial Cell Invasion Programs. *Dev. Cell* **2018**, *47*, 145–160.e6. [CrossRef]
213. Philips, N.; Auler, S.; Hugo, R.; Gonzalez, S. Beneficial Regulation of Matrix Metalloproteinases for Skin Health. *Enzym. Res.* **2011**, *2011*, 427285. [CrossRef] [PubMed]
214. Tian, X.; Liu, Z.; Niu, B.; Zhang, J.; Tan, T.K.; Lee, S.R.; Zhao, Y.; Harris, D.C.H.; Zheng, G. E-Cadherin/β-Catenin Complex and the Epithelial Barrier. *J. Biomed. Biotechnol.* **2011**, *2011*, 567305. [CrossRef] [PubMed]
215. Vanjaka-Rogošić, L.; Puizina-Ivić, N.; Mirić, L.; Rogošić, V.; Kuzmić-Prusac, I.; Babić, M.S.; Vuković, D.; Mardešić, S. Matrix Metalloproteinases and E-Cadherin Immunoreactivity in Different Basal Cell Carcinoma Histological Types. *Acta Histochem.* **2014**, *116*, 688–693. [CrossRef] [PubMed]
216. Chu, C.Y.; Cha, S.T.; Chang, C.C.; Hsiao, C.H.; Tan, C.T.; Lu, Y.C.; Jee, S.H.; Kuo, M.L. Involvement of matrix metalloproteinase-13 in stromal-cell-derived factor 1 alpha-directed invasion of human basal cell carcinoma cells. *Oncogene* **2007**, *26*, 2491–2501. [CrossRef]
217. El-Khalawany, M.A.; Abou-Bakr, A.A. Role of Cyclooxygenase-2, Ezrin and Matrix Metalloproteinase-9 as Predictive Markers for Recurrence of Basal Cell Carcinoma. *J. Cancer Res. Ther.* **2013**, *9*, 613–617. [CrossRef]
218. Karahan, N.; Baspinar, S.; Bozkurt, K.K.; Caloglu, E.; Ciris, I.M.; Kapucuoglu, N. Increased Expression of COX-2 in Recurrent Basal Cell Carcinoma of the Skin: A Pilot Study. *Indian J. Pathol. Microbiol.* **2011**, *54*, 526–531. [CrossRef]
219. Napoli, S.; Scuderi, C.; Gattuso, G.; Di Bella, V.; Candido, S.; Basile, M.S.; Libra, M.; Falzone, L. Functional Roles of Matrix Metalloproteinases and Their Inhibitors in Melanoma. *Cells* **2020**, *9*, 1151. [CrossRef] [PubMed]
220. Zhong, Y.; Lu, Y.-T.; Sun, Y.; Shi, Z.-H.; Li, N.-G.; Tang, Y.-P.; Duan, J.-A. Recent Opportunities in Matrix Metalloproteinase Inhibitor Drug Design for Cancer. *Expert Opin. Drug Discov.* **2018**, *13*, 75–87. [CrossRef] [PubMed]
221. Murphy, G. Tissue inhibitors of metalloproteinases. *Genome Biol.* **2011**, *12*, 233. [CrossRef]
222. Brew, K.; Nagase, H. The Tissue Inhibitors of Metalloproteinases (TIMPs): An Ancient Family with Structural and Functional Diversity. *Biochim. Biophys. Acta BBA Mol. Cell Res.* **2010**, *1803*, 55–71. [CrossRef]
223. Raeeszadeh-Sarmazdeh, M.; Do, L.D.; Hritz, B.G. Metalloproteinases and Their Inhibitors: Potential for the Development of New Therapeutics. *Cells* **2020**, *34*, 1313. [CrossRef] [PubMed]
224. Fu, X.; Zhang, C.; Chen, C.; Zhang, L.; Wang, L.; Wang, B.; Liu, X.; Zhang, M. Expression and significance of matrix metalloproteinase and tissue inhibitor of matrix metalloproteinase in non-melanoma skin cancer. *Zhonghua Zhong Liu Za Zhi* **2012**, *34*, 369–373. [CrossRef] [PubMed]
225. Lee, M.-H.; Rapti, M.; Murphy, G. Unveiling the Surface Epitopes That Render Tissue Inhibitor of Metalloproteinase-1 Inactive against Membrane Type 1-Matrix Metalloproteinase. *J. Biol. Chem.* **2003**, *278*, 40224–40230. [CrossRef] [PubMed]
226. Lee, M.H.; Atkinson, S.; Rapti, M.; Handsley, M.; Curry, V.; Edwards, D.; Murphy, G. The Activity of a Designer Tissue Inhibitor of Metalloproteinases (TIMP)-1 against Native Membrane Type 1 Matrix Metalloproteinase (MT1-MMP) in a Cell-Based Environment. *Cancer Lett.* **2010**, *290*, 114–122. [CrossRef] [PubMed]

biomolecules

Article

AG-9, an Elastin-Derived Peptide, Increases In Vitro Oral Tongue Carcinoma Cell Invasion, through an Increase in MMP-2 Secretion and MT1-MMP Expression, in a RPSA-Dependent Manner

Clara Bretaudeau [1,2,3], Stéphanie Baud [1,2,4], Aurélie Dupont-Deshorgue [1,2], Rémi Cousin [1,2], Bertrand Brassart [1,2,†] and Sylvie Brassart-Pasco [1,2,*,†]

[1] Université de Reims Champagne-Ardenne (URCA), 51100 Reims, France; clara.bretaudeau@etudiant.univ-reims.fr (C.B.); stephanie.baud@univ-reims.fr (S.B.); aurelie.dupont@univ-reims.fr (A.D.-D.); remicousin70@gmail.com (R.C.); bertrand.brassart@univ-reims.fr (B.B.)
[2] CNRS, UMR 7369, Matrice Extracellulaire et Dynamique Cellulaire (MEDyC), 51100 Reims, France
[3] CHU Reims, Service d'Odontologie, 51100 Reims, France
[4] Plateau de Modélisation Moléculaire Multi-échelle, URCA, 51100 Reims, France
* Correspondence: sylvie.brassart-pasco@univ-reims.fr
† These authors contributed equally to this work.

Citation: Bretaudeau, C.; Baud, S.; Dupont-Deshorgue, A.; Cousin, R.; Brassart, B.; Brassart-Pasco, S. AG-9, an Elastin-Derived Peptide, Increases In Vitro Oral Tongue Carcinoma Cell Invasion, through an Increase in MMP-2 Secretion and MT1-MMP Expression, in a RPSA-Dependent Manner. *Biomolecules* **2021**, *11*, 39. https://doi.org/10.3390/biom11010039

Received: 12 November 2020
Accepted: 26 December 2020
Published: 30 December 2020

Abstract: Oral tongue squamous cell carcinoma is one of the most prevalent head and neck cancers. During tumor progression, elastin fragments are released in the tumor microenvironment. Among them, we previously identified a nonapeptide, AG-9, that stimulates melanoma progression in vivo in a mouse melanoma model. In the present paper, we studied AG-9 effect on tongue squamous cell carcinoma invasive properties. We demonstrated that AG-9 stimulates cell invasion in vitro in a modified Boyen chamber model. It increases MMP-2 secretion, analyzed by zymography and MT1-MMP expression, studied by Western blot. The stimulatory effect was mediated through Ribosomal Protein SA (RPSA) receptor binding as demonstrated by SiRNA experiments. The green tea-derived polyphenol, (−)-epigallocatechin-3-gallate (EGCG), was previously shown to bind RPSA. Molecular docking experiments were performed to compare the preferred areas of interaction of AG-9 and EGCG with RPSA and suggested overlapping areas. This was confirmed by competition assays. EGCG abolished AG-9-induced invasion, MMP-2 secretion, and MT1-MMP expression.

Keywords: elastin; ribosomal protein SA; tongue carcinoma; MMP-2; EGCG

1. Introduction

Head and neck tumors are a heterogeneous group of cancers occurring in the mouth, nose, pharynx, larynx, salivary glands, and sinuses. Approximately 90% of those cancers are squamous cell carcinomas [1]. They are typically characterized by a peak of incidence in the elderly and a strong correlation with chronic exposure to risk factors such as smoking and alcohol abuse [2]. Among head and neck squamous cell carcinomas (HNSCC), cancers of the oral cavity and pharynx display an increasing incidence since 1999, despite a decrease in tobacco use [2,3]. Many subsites have shown a rise in the number of cases, especially the anterior tongue [2]. Indeed, its incidence raised +1.8% per year on average between 2007 and 2016, particularly among young adults and especially in females [3–7]. Thus, in 2007, the prevalence in the US was 4422 cases, rising to 6155 cases in 2016 [3]. Despite advances in diagnosis and management of oral cancer in recent decades, the long term prognosis of patients with advanced-stage squamous cell carcinoma of the tongue is generally poor, with 5-year survival rates around 50% [8].

Elastic fibers are components of all mammalian connective tissues. Elastin, the major protein component of elastic fibers, has unique elastomeric properties which provide reversible deformability crucial for arterial vessels, lungs, and skin [9]. Elastic fibers are also stable components of tongue, providing the elasticity necessary to mechanical resistance during mastication [10].

During tumor progression, tumor and stromal cells secrete a wide variety of proteolytic enzymes including matrix metalloproteinases (MMPs), urokinase-type plasminogen activator (uPA), a disintegrin and metalloproteinases (ADAMs), a disintegrin-like and metalloproteinase with thrombospondin type 1 motif (ADAMTS), and cathepsins that are activated to degrade extracellular matrix (ECM) components and to favor tumor growth and cell dissemination. ECM-degrading enzymes also liberate active fragments from matrix components called matrikines [11]. Proteolytic degradation of elastin results in the liberation of bioactive elastin peptides, termed elastokines [12]. Elastases that degrade elastin belong to the classes of matrix metalloproteinases (MMPs), aspartic proteases, serine proteases, and cysteine proteases [13]. Elastokines containing the GxxPG motif such as VGVAPG (VG-6) promote tumor progression by stimulating cell invasion through the activation of proteolytic cascades, especially MMP-2 and uPA [14,15]. They also promote angiogenesis by stimulating the migration of endothelial cells through MMP-14 increase [16]. We have also demonstrated that an elastin nonapeptide of consensus sequence xGxPGxGxG, AGVPGLGVG (AG-9), favors the migration and invasion of tumor cells through MMP-2 and uPA increases [17,18]. These effects are mediated through the RPSA (ribosomal protein SA) receptor [19].

The 37/67-kDa laminin receptor was reported to bind elastin by Mecham et al. in 1989 [20]. The 37/67-kDa laminin receptor, RPSA, also referred as 67LR, LAMBR, LAMR1, lamR, LRP/LR, 37LRP, LBP, LBP/p40, NEM/1CHD4, SA, ICAS, and p40 is ubiquitously expressed. It allows cell adhesion to the basement membrane. The 37-, 53-, and 67 kDa forms are the major forms reported but additional high-molecular-weight (HMW) forms of 32, 37, 45, 53, 55, 67, 80, and >110 kDa were also reported. The conversion of the 37 kDa form to higher molecular weight species remains unclear [21]. RPSA not only localized on the cell surface but also in the nucleus, in association with nucleolar pre-40S ribosomes, small nucleolar ribonucleoproteins (snoRNPs), chromatin, histones, and in the cytosol as a ribosomal component or as actin and cytoskeletal stress fibers partner. It was reported to mediate cell proliferation, adhesion, differentiation, invasion, and angiogenesis. RPSA prevents cell apoptotic escape, allowing tumor progression [22].

The green tea-derived polyphenol, (−)-epigallocatechin-3-gallate (EGCG), is a small molecule with anti-cancer effects [23–25]. It inhibits the invasion of human oral cancer cells, decreases the production of MMPs and urokinase-plasminogen activator [26,27], prevents epithelial-mesenchymal transition [28], suppresses cell proliferation, promotes apoptosis and autophagy [29], and inhibits tumor growth [28]. Li et al. demonstrated that the inhibitory effect on cell proliferation, apoptosis, migration, and invasion of tongue squamous cell carcinoma was mediated through the hippo TAZ signaling pathway [30]. EGCG inhibitory effect is mediated through RPSA receptor. RPSA antibodies block EGCG anti-cancer activity but do not trigger the same effects, indicating that the polyphenol may act agonistically or allosterically [21]. EGCG was reported to exert its anti-cancer activity through the 10 amino acid sequence of RPSA, IPCNNKGAHS [31].

2. Materials and Methods

2.1. Peptide Synthesis

AG-9 was purchased from Proteogenix® (Schiltigheim, France). It was obtained by solid-phase synthesis using a FMOC (N-(9-fluorenyl) methoxy-carbonyl) derivative procedure. It was then purified by reverse phase high performance liquid chromatography using a C18 column, eluted by a gradient of acetonitrile in trifluoroacetic acid and lyophilized. Its purity (>98%) was assessed by HPLC and mass spectroscopy.

2.2. Cell Culture

Human tongue squamous cell carcinoma CAL 27 (ATCC® CRL-2095™) were purchased from ATCC. Cells were grown in DMEM (Dulbecco's modified Eagle medium) with 4.5 g/L glucose supplemented with 10% fetal bovine serum (FBS) at 37 °C in a humid atmosphere with 5% CO_2 in air. At 70–90% confluency, cells were subcultured according to ATCC protocol.

2.3. Cytotoxicity Assay

A total of 50,000 CAL 27 cells were seeded in 96-well plates in DMEM supplemented with 10% FBS. After cell adhesion, culture medium was removed and replaced by FBS-free medium. Cells were incubated for 24 h with or without effectors. Cells were then fixed with 1.1% glutaraldehyde for 15 min and stained with crystal violet for another 15 min. Dye was eluted with a 10% acetic acid solution. Absorbance was read at 560 nm using a Biochrom Asys UVM 340 microplate reader (Biochrom, Yvelines, France).

2.4. Proliferation Assay

A total of 2000 CAL 27 cells were seeded in 96-well plates in DMEM supplemented with 10% FBS. After cell adhesion, medium was removed and cells were cultivated in DMEM supplemented with 2.5% FBS with or without AG-9 for 24, 48, 72, or 96 h. Cells were then fixed with 1.1% glutaraldehyde for 15 min and stained with crystal violet for another 15 min. Dye was eluted with a 10% acetic acid solution. Absorbance was read at 560 nm using a Biochrom Asys UVM 340 microplate reader (Biochrom, Yvelines, France).

2.5. In Vitro Invasion Assays

Invasion was assessed in modified Boyden chambers (tissue culture treated, 6.5 mm diameter, 8 μm pore, Greiner-One, Courtaboeuf, France). Further, 5×10^4 cells were suspended in DMEM containing 10% FBS and seeded onto membranes coated with Matrigel® (10 μg/well). After cell adhesion, culture medium was removed from the upper compartment and replaced by DMEM containing 0.2% BSA ± effectors. DMEM supplemented with 10% FBS and 2% BSA was used as a chemoattractant. After a 72 h incubation period, cells were fixed with methanol and stained with crystal violet for 15 min. Cells remaining on the upper face of the membranes were scrapped. Crystal violet was eluted using 10% acetic acid and absorbance was read at 560 nm using a Biochrom Asys UVM 340 microplate reader (Biochrom, Yvelines, France).

2.6. Zymography Analyses

2.6.1. Cell Incubation with Effectors

At subconfluence, cells were washed twice with phosphate-buffered saline to remove residual FBS and incubated for 48 h in DMEM, with or without effectors. Conditioned media were harvested and centrifuged at $10,000 \times g$ for 10 min at 4 °C.

2.6.2. Gelatin Zymography

To study MMP-2 secretion by CAL 27 cells, conditioned media, diluted in $2 \times$ non-reducing Laemmli buffer, were electrophoresed in a 10% polyacrylamide SDS gel containing 0.1% (*w/v*) gelatin. The gels were washed twice for 30 min at room temperature in a 2.5% (*v/v*) Triton X-100 solution to remove SDS, then incubated at 37 °C for 24 h in 50 mM Tris-HCl (pH 7.6), 200 mM NaCl, 5 mM $CaCl_2$, stained for 30 min with 0.1% (*w/v*) Coomassie blue (G 250) in 45% (*v/v*) methanol/10% (*v/v*) acetic acid and destained in the same solution without dye.

2.6.3. Gelatin-Plasminogen Zymography

To study uPA secretion, CAL 27-conditioned media were electrophoresed in SDS-poly-acrylamide gels containing 1 mg/mL gelatin and 10 μg/mL plasminogen. The gels were washed twice for 30 min at room temperature in a 2.5% (*v/v*) Triton X-100 solution to remove SDS, then incubated at 37 °C for 24 h in 100 mM glycine, 10 mM EDTA (pH 8.3), stained for 30 min with 0.1% (*w/v*) Coomassie blue (G 250) in 45% (*v/v*) methanol/10% (*v/v*) acetic acid and destained in the same solution without dye.

2.7. Western-Blot Analyses

Samples were reduced by 10 mM of dithiothreitol and subjected to SDS-PAGE (0.1% SDS, 10% polyacrylamide gel) (50 µg total protein per lane), then transferred onto Immobilon-P membranes (Millipore, St Quentin-en-Yvelines, France). Membranes were blocked by incubation with 5% non-fat dry milk, 0.1% Tween-20 in 50 mM Tris-HCl buffer, 150 mM NaCl, pH 7.5 (TBS-T) for 2 h at room temperature. They were incubated overnight with the first antibody (anti-RPSA polyclonal antibody, Abcam Ab99484, diluted 1/3000; anti-MT1-MMP Ab38971 diluted 1/3000; anti-actin polyclonal antibody, Sigma Aldrich Biotechnology A2066, diluted 1/2000) and then for 1 h with the 1/10,000 diluted peroxidase-conjugated goat anti-rabbit secondary antibody (GE Healthcare, NA931V) at room temperature. Immune complexes were visualized using the ECL prime chemiluminescence detection kit (GE Healthcare, Orsay, France).

2.8. Immunocytochemistry

Cells were seeded on glass slides and incubated in 10% serum-containing medium for 16 h. Cells were incubated with an anti-RPSA antibody (Abcam, 137388) diluted 1/400 in culture medium supplemented with 1% BSA for 1 h on ice, washed and incubated for 30 min with the Alexa-488-conjugated secondary antibodies diluted 1/1000 in culture medium with 1% BSA, and then fixed for 10 min with 4% paraformaldehyde at room temperature. Immunofluorescence-labeled cell preparations were studied using a Zeiss LSM 710® NLO confocal laser scanning microscope (Carl ZEISS SAS, Marly-le-Roi, France) with the 63x oil-immersion objective (ON 1.4) coupled with CHAMELEON femtosecond Titanium-Sapphire Laser (Coherent, Santa Clara, CA, USA). Alexa 488 was excited by 488 nm line of Argon. Emitted signals were collected with 493–560 nm bandpass filter. Image acquisitions were performed with ZEN Software (Carl ZEISS SAS, Marly-le-Roi, France) and all acquisition settings were constant between specimens.

2.9. RNA Isolation and Real-Time PCR Analysis

Total RNA was isolated from cells using a RNeasy Plus Mini kit (Qiagen, Courtaboeuf, France) according to the manufacturer's instruction. The amount and integrity of isolated RNA were analyzed using the Bioanalyzer RNA 6000 nano assay (Agilent Technologies, Les Ulis, France) as recommended by the manufacturer. Total RNA was reverse transcribed using the first strand cDNA synthesis kit (Thermo Scientific, Illkirch, France) following the manufacturer's instructions. Real-time PCR analysis was conducted in 20 µL reaction mixture, using Thermo Scientific Maxima SYBR Green qPCR Master Mix, following the manufacturer's instructions. Relative expression of different gene transcripts was calculated by the $\Delta\Delta$Ct method. The Ct of the gene of interest was normalized to the Ct of the normalizer (EEF1A1). Fold changes (arbitrary units) were determined as $2-\Delta\Delta$Ct.

RT-qPCR primers were designed according to sequence of RPSA (NM_002295). The forward primer for RPSA was 5′-CCA-TTG-AAA-ACC-CTG-CTG-AT-3′ and the reverse primer was 5′-CTG-CCT-GGA-TCT-GGT-TAG-TGA-3′ with a 144 bp product. The forward primer for EEF1a1 was 5′-CTG-GAG-CCA-AGT-GCT-AAC-ATG-CC-3′ and the reverse primer was 5′-CCG-GGT-TTG-AGA-ACA-CCA-GTC-3′ with a 221 bp product. All primers were synthesized by Eurofins (Les Ulis, France).

2.10. SiRNA Transfection

SiRNA were transfected as previously described [19]. SiRNA specific to human RPSA and negative control siRNA (non-targeting pool), which do not target any gene, were purchased from Qiagen. The siRNA targets different regions of the RPSA mRNA: 1st siRNA target sequence (5′-AGG-CTC-TTA-AGC-AGC-ATG-GAA-3′), 2nd siRNA target sequence (5′-TAC-CTG-GGA-TTG-CAT-ATC-AAA-3′), 3rd siRNA target sequence (5′-TTG-CAT-ATC-AAA-GCA-TAA-TAA-3′), and 4th siRNA target sequence (5′-TCG-ACA-TGA-GTT-GTA-CTT-CTA-3′). Expression of RPSA mRNA and protein was confirmed by real-time PCR and Western blot.

2.11. Docking Experiments

Dockings of the AG-9 peptide and EGCG onto RPSA (RCSB Protein Data Bank 3BCH) were performed using Autodock software (version 4.2) [32]. We performed preliminary docking experiments to determine the relevant set of docking parameters. The software was used with a fixed RPSA and flexible AG-9 and EGCG ligands. Since RPSA is a large molecule, we performed several independent dockings targeting subvolumes of the protein; we considered 80 overlapping boxes with a volume of 31.5 Å × 31.5 Å × 31.5 Å. The spacing parameter used to compute the 3D maps in each box was set to 0.25 Å. The selected search method was the Lamarckian genetic algorithm, and for each docking experiment, 50 solutions per box were generated with the default parameters of Autodock except for the population size (200), number of energy evaluations (2.5×10^6), and maximum number of generations (270,000), which were derived from the preliminary study. The identification of the contacts between RPSA and the poses of each ligand was carried out by the evaluation of the intermolecular atomic distances. Contacts were counted for distances lower than the 5 Å threshold. Molecular models were graphed with VMD software, which is available online.

2.12. Statistical Analysis

Results were expressed as means $+/-$ standard deviation. Statistical significance between groups was assessed using unpaired Student's *t* test.

3. Results

3.1. CAL 27 Cells Express RPSA Receptor

Elastin was reported to interact with invasive cancer cells through RPSA [33]. Previous experiments from our laboratory identified RPSA as AG-9 receptor on HT-1080 human fibrosarcoma cell surface [19], and on MIA PaCa-2 cell pancreatic adenocarcinoma cells [34]. We first checked for RPSA expression in CAL 27 cells, compared to MIA PaCa-2 cells. By qPCR and Western blot analyses, we proved that RPSA was expressed by CAL 27 at the mRNA and protein level (Figure 1A,B respectively). By immuno-histochemistry, we confirmed that RPSA was present at the cell surface of CAL 27 cells.

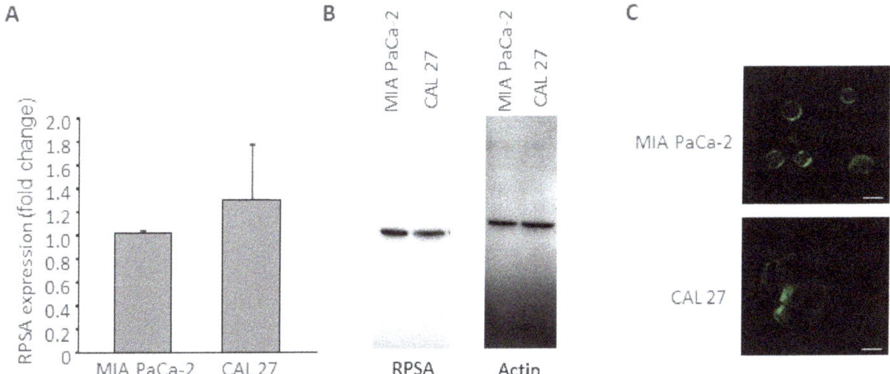

Figure 1. RPSA expression was studied at the mRNA level by qPCR (**A**) and at the protein level by Western blot (**B**). RPSA distribution at the cell surface was studied by immunocytochemistry (orthoslide) (**C**). Scale bar: 20 µM.

3.2. AG-9 Increases CAL 27 Invasion through Matrigel®, MMP-2 Secretion, and MT1-MMP Expression

Soluble kappa-elastin peptides were shown to regulate MT1-MMP and MMP-2 [14]. Both MMPs are largely involved in cell invasion. Recently, we reported that AG-9 peptide stimulates tumor cell invasion at lower concentrations than the well characterized VG-6 peptide [18]. CAL 27 invasion was studied in vitro in the presence of soluble kappa-elastin peptides or AG-9 peptide using the transwell invasion assay. After 72 h of incubation, kappa-elastin peptides and AG-9 peptide increased cell invasion by +22 and +25%, respectively (Figure 2A). The increase was independent of cell proliferation since the peptides did not significantly modified cell proliferation as demonstrated using the crystal violet assay (Figure 2B). MMP-2 secretion was studied by zymography. After 24 h of incubation, kappa-elastin peptides and AG-9 peptide increased MMP-2 cell secretion by +39 and +139%, respectively (Figure 2C). Kappa-elastin peptides and AG-9 peptide also induced MT1-MMP expression, studied by Western blot (Figure 2D).

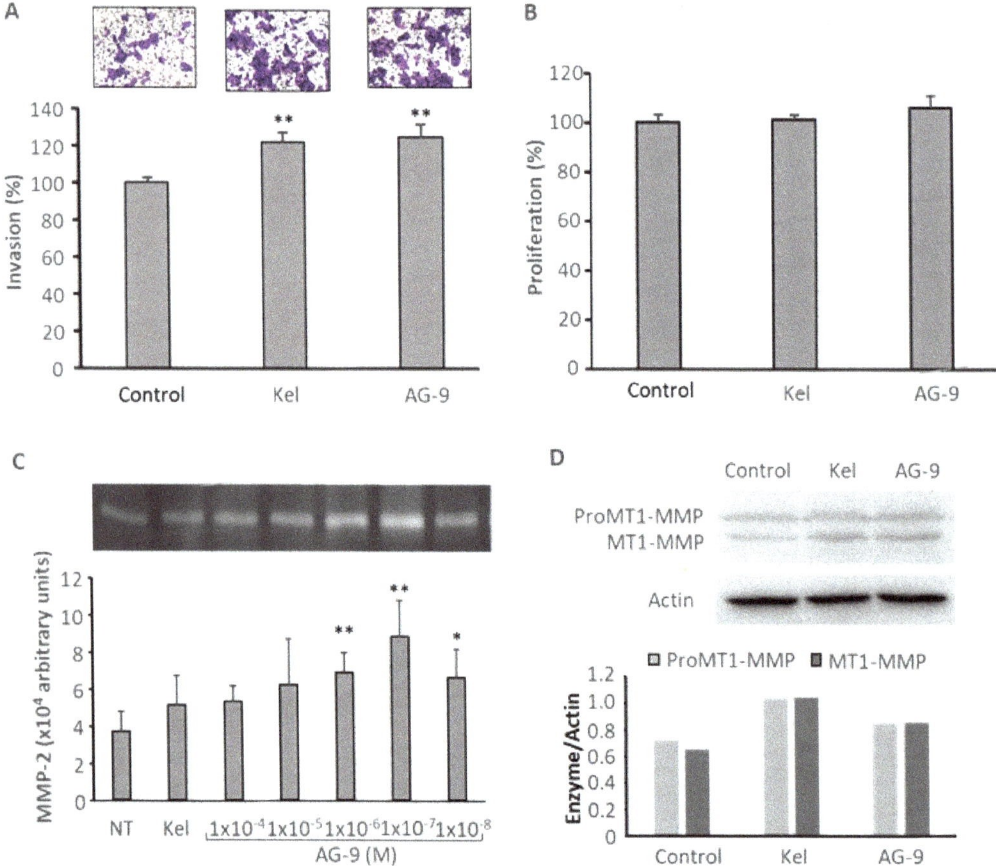

Figure 2. Cell invasion was studied using the transwell assay system coated with Matrigel® as described in the Materials and Methods section after 72 h of incubation with the different effectors: Kappa-elastin peptides 50 μg/mL, AG-9 peptide 10^{-7} M. **: $p < 0.01$ (**A**). Cell proliferation was assessed using crystal violet staining after 72 h of incubation with the effectors: kappa-elastin peptides 50 μg/mL, AG-9 peptide 10^{-7} M (**B**). MMP-2 secretion was studied by zymography after 24 h of incubation with kappa-elastin peptides 50 μg/mL or various amount of AG-9 ranging from 10^{-4} M to 10^{-8} M. *: $p < 0.05$; **: $p < 0.01$ (**C**). MT1-MMP expression was studied by Western blot after 24 h of incubation with the effectors: kappa-elastin peptides 50 μGg/mL, AG-9 peptide 10^{-7} M (**D**).

3.3. AG-9 Proinvasive Effects Are Mediated through RPSA Receptor

CAL 27 cells were tranfected with control (non-targeting) or RPSA siRNA. RPSA gene expression was measured by qPCR 48 h after transfection. RPSA gene expression was decreased by 62% after RPSA siRNA transfection compared to control (Figure 3A). RPSA protein expression was evaluated 72 h after transfection by Western blot. RPSA was decreased by 36% (Figure 3B). We performed invasion assays, zymography, and Western blot experiments. Even partial RPSA invalidation abolished the AG-9-induced effects on MMP-2 secretion (Figure 3C) and MT1-MMP expression (Figure 3D). The results confirm the involvement of RPSA receptor in AG-9-mediated effects.

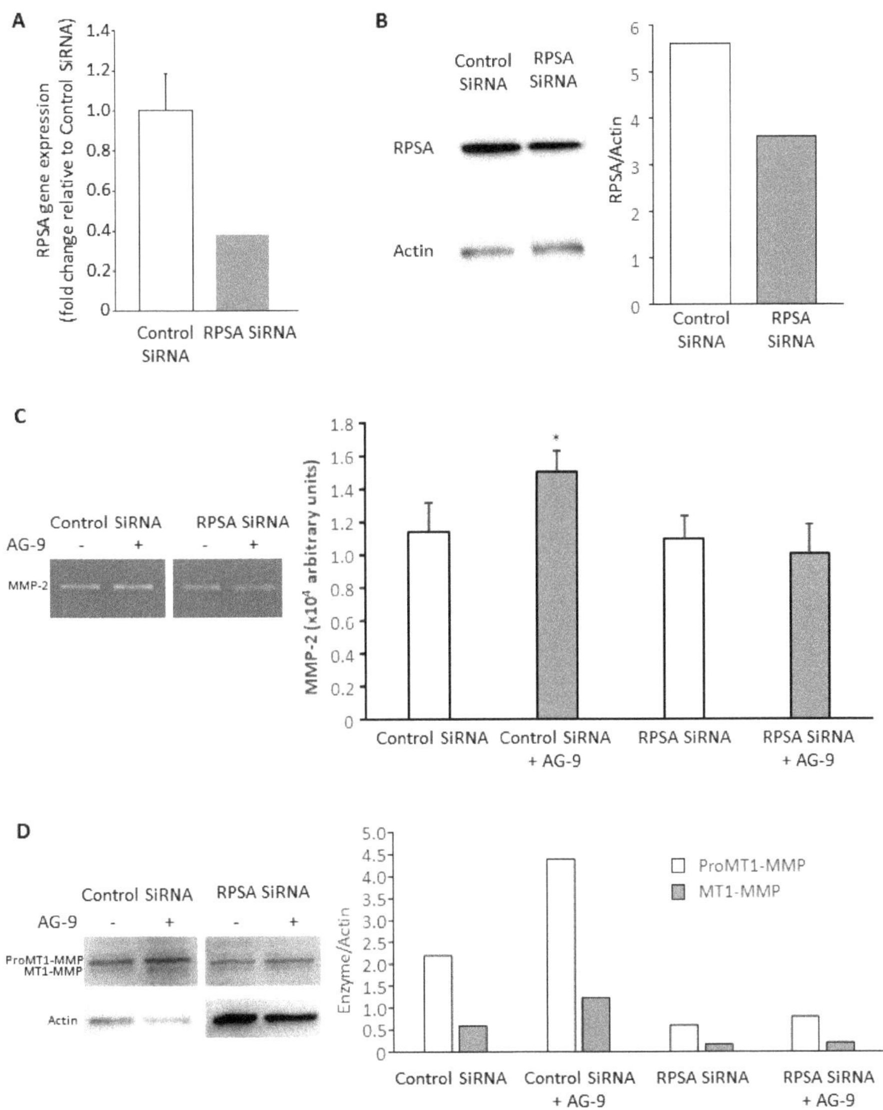

Figure 3. Cells were transfected with control or RPSA siRNA. RPSA gene expression was measured 48 h post-transfection (**A**) and protein expression was measured 72 h post-transfection (**B**). For MMP-2 secretion (**C**) and MT1-MMP expression (**D**) studies, cells were incubated for another 24 h with or without 10^{-7} M AG-9 peptide. *: $p < 0.05$.

3.4. In Silico Study of AG-9 and EGCG Binding on RPSA

EGCG was previously reported to bind RPSA [18,22,35]. EGCG may prevent AG-9 fixation on RPSA and represent a molecule of choice to limit AG-9 pro-tumoral effects on oral tongue squamous cell carcinoma. Clustering the 50 best results for each ligand, we first identified the preferred areas of interaction (PAI) with EGCG and then with AG-9 and found overlapping areas (Figure 4A,B). The comparison of the associated distribution profile of the free energy of binding demonstrated lower free energy of binding for EGCG onto RPSA than for AG-9 onto RPSA, corresponding to a stronger binding of the polyphenol onto the receptor (Figure 4C,D). The localization of the lowest free energy of binding pose of each ligand highlighted a colocalization onto RPSA (Figure 4E,F). Finally, the analysis of the RPSA residues making contact with the ligands evidenced common interactions with R^{117}, $^{120}RL^{121}$, and the region $^{140}VNLP^{143}$ (Figure 3G).

3.5. In Vitro EGCG Cytotoxicity on CAL 27 Cells

The aim of this part was to determine EGCG maximal concentration that could be used to counteract AG-9 stimulation without affecting cell viability. For this purpose, CAL 27 cells were incubated for 24 h with increasing amount of EGCG and cell viability was measured (Figure 5). Cell viability was 97.4% for 10 μM EGCG. Comparable results were obtained by Weisburgh et al. [36]. The same authors reported that cell viability of normal oral fibroblasts was also around 95% at this concentration.

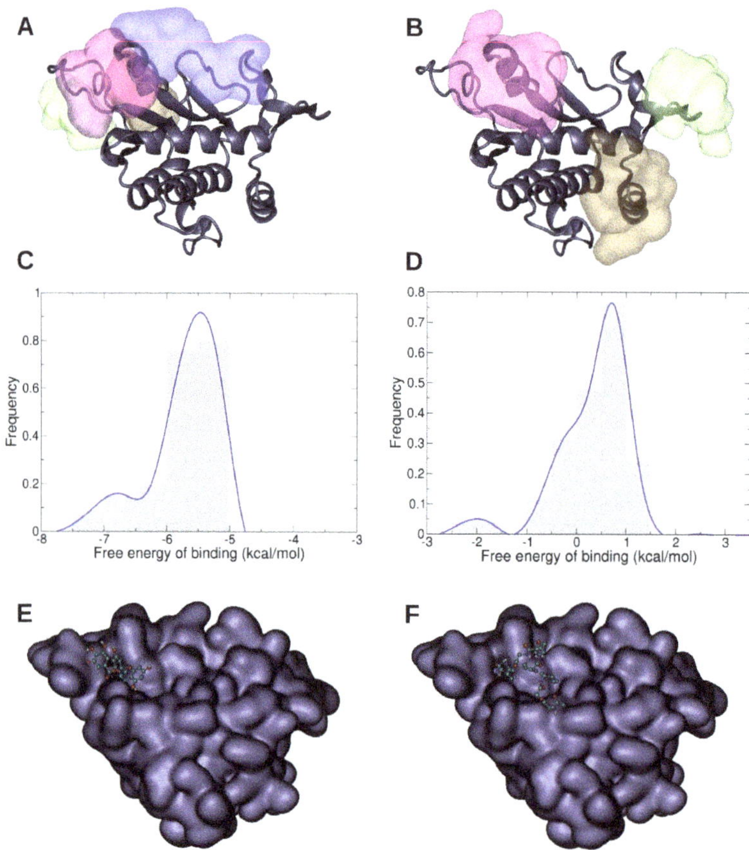

Figure 4. *Cont.*

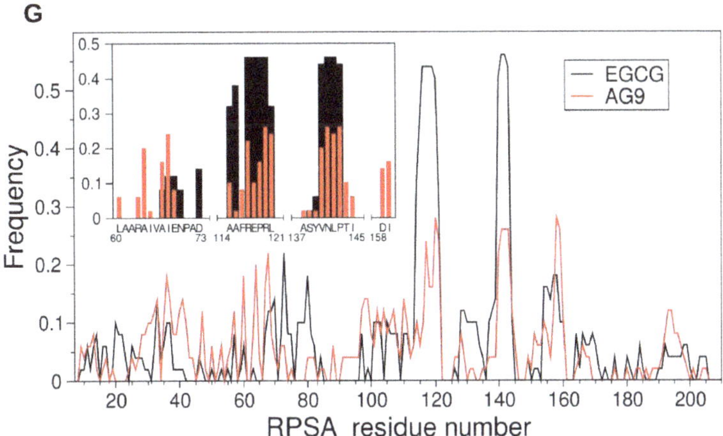

Figure 4. Docking experiments of green tea-derived polyphenol EGCG and AG-9 peptide onto RPSA were performed using Autodock software and evidenced the existence of 4 preferred area of interaction (PAI) for EGCG (**A**) and 3 PAI for AG-9 (**B**). The color code used for the representation of PAI is linked to their population: the most populated one is represented in pink, the second in green, the third in brown, and the fourth in blue. Comparison of the 50 best results of EGCG and AG-9 docking experiments displays an overlapping area (pink PAI). Frequence distribution diagrams of the free energy of binding of the 50 best poses of EGCG onto RPSA demonstrated that those free energies of binding are comprised between −7.75 and −4.75 kcal/mol. The most frequent poses have a free energy of binding of −5.5 kcal/mol (**C**). For AG-9 peptide, the free energy of binding of the 50 best results of docking onto RPSA is comprised between −2.75 and +1.75 kcal/mol and the energy of binding of the most frequent position of AG-9 onto RPSA is around 0.75 kcal/mol (**D**). The best docking poses of EGCG (**E**) and AG-9 (**F**) are located on the same area at the surface of RPSA. Comparison of the frequency of contacts made by RPSA residues with the 50 best results of EGCG (black line) and AG-9 (red line) reveals two favored regions (frequency above the threshold of 0.2) of the protein (**G**). A focus on the poses associated with the first cluster of EGCG and AG-9 (G/close-up) allows the identification of the following hot spots (frequency above the threshold of 0.2 for the two ligands): R^{117}, $^{120}RL^{121}$, and the sequence $^{140}VNLP^{143}$.

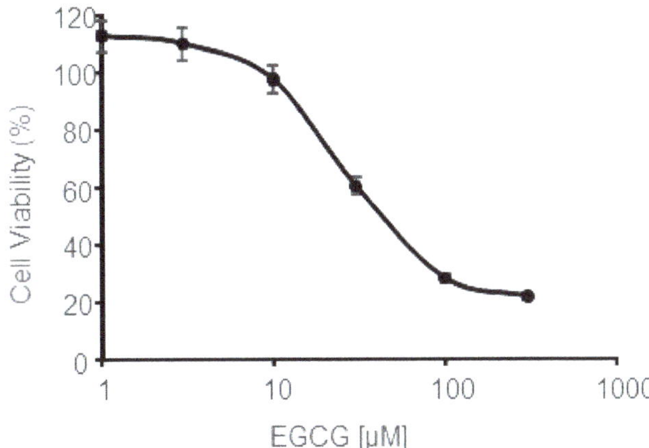

Figure 5. Cell viability was assessed using the crystal violet assay after 24 h of incubation with increasing concentrations of EGCG ranging from 1 to 300 μM.

We thus decided to use this concentration in the following experiments to try to block AG-9-stimulatory effects without affecting cell viability.

3.6. EGCG Prevents AG-9 Stimulatory Effect on CAL 27 Migration, Invasion, MMP-2 Secretion, and MT1-MMP Expression

Incubation with 10µM EGCG slightly decreased CAL 27 migration (-11%; Figure 6A) and significantly decreased cell invasion (-23%; Figure 6B) and it abolished AG-9 stimulatory effect (Figure 6). It also slightly decreased MMP-2 secretion and MT1-MMP expression (Figure 6C,D). This is in accordance with previously published papers [37]. Incubation with EGCG also prevented AG-9 stimulatory effect on cell migration (Figure 6A), invasion (Figure 6B), as well as on MMP-2 secretion (Figure 6C) and MT1-MMP expression (Figure 6D).

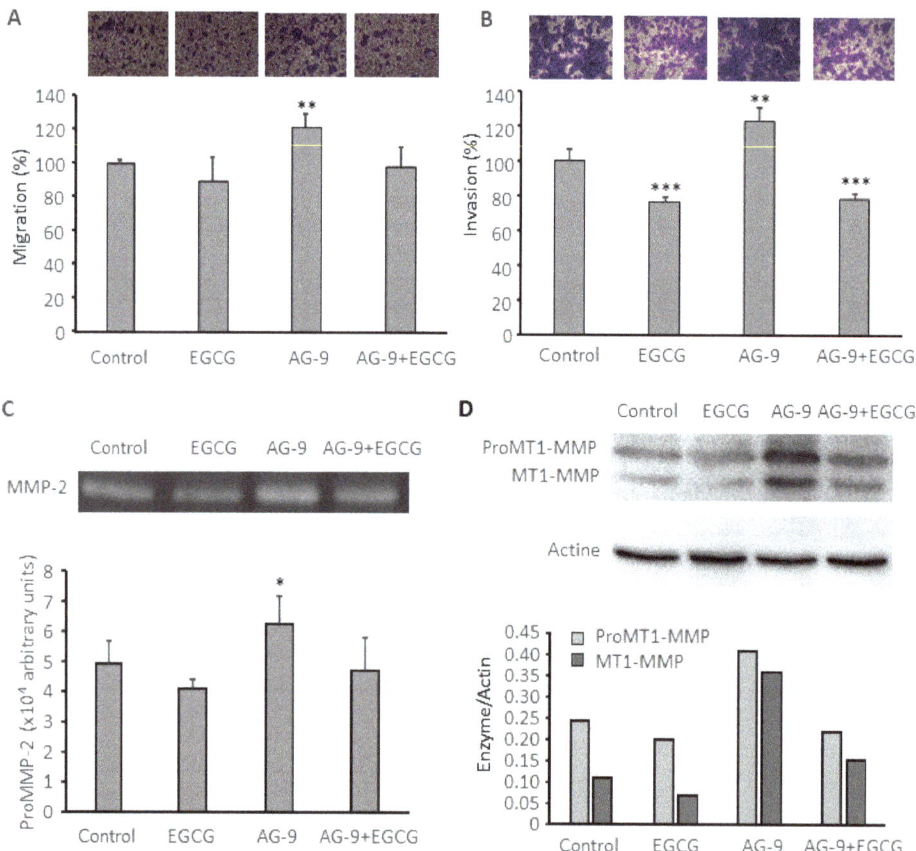

Figure 6. Cell migration was studied using the transwell assay system after 48 h of incubation with the different effectors: AG-9 10^{-7} M and EGCG 10 µM. **: $p < 0.01$ (**A**). Cell invasion was studied in transwells previously coated with Matrigel®. ***: $p < 0.001$, **: $p < 0.01$ (**B**). MMP-2 secretion was studied by zymography after 24 h of incubation with the different effectors: with the different effectors: kappa-elastin peptides 50 µg/mL, AG-9 peptide 10^{-7} M. *: $p < 0.05$ (**C**). MT1-MMP expression was studied by Western blot after 24 h of incubation with the different effectors: kappa-elastin peptides 50 µg/mL, AG-9 peptide 10^{-7} M (**D**).

4. Discussion

Oral squamous cell carcinoma (OSCC) is the most common malignant epithelial neoplasm affecting the oral cavity. The treatment of choice for OSCC is surgical resection. Adjuvant radiotherapy with or without chemotherapy is offered when there is a high risk of recurrence and after taking into consideration multiple factors, including patient's age and comorbidities, pathologic staging, margin status, the extent of nodal involvement, and other histopathologic characteristics of the primary tumor [38]. Despite the advances of therapeutic approaches, percentages of morbidity and mortality of OSCC have not improved significantly during the last 30 years. Developing new therapeutic approaches is thus challenging.

Cancer development leads to ECM degradation by tumor and stromal cells. Fragments with biological activities are released during this process, named matrikines. Elastin is the major component of elastic fibers. Its cleavage by elastase-proteinases such as metalloproteinases or leucocyte elastase is known to unmask cryptic sites within the macromolecule and to release matrikines, called elastin derived peptides (EDPs) or elastokines. These EDPs exert a wide range of biological activities. They influence cell survival, differentiation [39], proliferation, chemotaxis [18,40], migration [18,41], tumor progression [15,17,19,42], angiogenesis [16], atherogenesis, and aneurysm formation. Among all the EDPs described in the literature, two major consensus sequence were reported: the xGxxPG consensus sequence including the VGVAPG, VAPG, VGVPG, VGAPG, (VGVAPG)$_n$, and PGAIPG peptides and the xGxPGxGxG consensus sequence with the AGVPGLGVG, AGVPGFGVG, GLGVGVAPG, and GFGVGAGVP peptides. In vivo study showed that AG-9 peptide promotes melanoma progression even more than the well described VG-6 peptide. These results were confirmed by in vitro studies in proliferation assays, migration assays, adhesion assays, proteinase secretion studies, and pseudotube formation assays to investigate angiogenesis [18,34].

In the present paper, we report for the first time EDP-effect on oral tongue SCC. CAL 27 is one of the most frequently used cell lines in the field of OSCC studying [43]. Soluble elastin peptides were obtained by partial hydrolysis of elastin in 1 M KOH in 80 per cent aqueous ethanol (kappa-elastin). Additionally, 50 mM kappa-elastin increases CAL 27 invasion through transwell previously coated with Matrigel® which mimicks basement membrane. AG-9 also significantly increased CAL 27 invasion, as previously reported for MIA PaCa-2 pancreatic ductal adenocarcinoma cells [34]. EDP were previously reported to increase MMP secretion, especially MMP-2 [44,45]. These results were confirmed by zymography analysis of conditioned media from CAL 27. AG-9 increased MMP-2 secretion, as reported for MIA PaCa-2 cells; this effect was biphasic, with an optimal effect obtained for 1.10^{-7} M AG-9. Moreover, AG-9 increases MT1-MMP expression. MT1-MMP is able to degrade extracellular matrix by itself or to activate proMMP-2 at the cell surface.

The pro-tumor biological effects of AG-9 was previously reported to involve a lactose-insensitive receptor, the ribosomal protein SA (RPSA) [19]. The 37/67-kDa laminin receptor was reported to bind elastin by Mecham et al. in 1989 [20]. It allows cell adhesion to basement membrane. RPSA does not only localize on the cell surface but also in the nucleus, in association with nucleolar pre-40S ribosomes, small nucleolar ribonucleoproteins (snoRNPs), chromatin, histones, and in the cytosol as a ribosomal component or as actin and cytoskeletal stress fibers partner. It was reported to mediate cell proliferation, adhesion, differentiation, invasion, and angiogenesis. RPSA also prevents cell apoptotic escape, allowing tumor progression [22].

The green tea-derived polyphenol (−)-epigallocatechin-3-gallate (EGCG), was reported to affect cell behavior through RPSA binding [31]. We performed molecular docking experiments to determine its potential preferred areas of interaction with RPSA and especially the area with the lowest energy of binding that means the highest probability of binding. We did the same for the AG-9 peptide. The preferred area of interaction with the lowest energy for EGCG and AG-9 was broadly overlapping. We then formulated the hypothesis that EGCG could counteract the effects induced by the peptide AG-9. To test this hypothesis, we performed the invasion experiments in presence or absence of EGCG. We first observed that EGCG alone at 10 µM was able to decrease CAL 27 invasion. This was previously reported by Chang et al. with a concentration of 25 µM EGCG [37]. EGCG was also able to inhibit SCC-4, SCC-9, and SCC-15 cell [28–30]. other oral SCC and nasopharyngeal carcinoma cell invasion [46,47]. In addition, 10 µM EGCG also inhibits MMP-2 secretion as previously described for CAL 27 cells [37], nasopharyngeal carcinoma [47], and buccal mucosa cancer cells [26]. This was also observed in other oral SCC. Further, 10 µM EGCG prevents AG-9 induced invasion, MMP-2 secretion, and MT1-MMP expression. As suggested by molecular docking experiments, this may be due to steric hindrance or conformation modification as the PAIs of lowest binding energy of both molecules are very close. Binding energy of EGCG for RPSA are lower than those found with AG-9, suggesting that RPSA affinity for EGCG may be greater than for AG-9.

5. Conclusions

During cancer progression, tumor cells induce the expression of MMPs, such as MT1-MMP and MMP-2, which degrade ECM macromolecules like elastin and release EDPs, such as AG-9, that in turns stimulate MMP secretion, leading to an auto-amplification loop. EGCG prevents AG-9 stimulation and represents a molecule of choice to limit cancer progression in elastin-rich tissues.

Author Contributions: Conceptualization, S.B., B.B. and S.B.-P.; Formal analysis, S.B., B.B. and S.B.-P.; Investigation, C.B., S.B., A.D.-D., R.C., B.B. and S.B.-P.; Project administration, B.B. and S.B.-P.; Supervision, B.B. and S.B.-P.; Validation, S.B., B.B. and S.B.-P.; Visualization, S.B. and S.B.-P.; Writing—original draft, S.B., B.B. and S.B.-P. All authors have read and agreed to the published version of the manuscript.

Funding: This work was supported by the Centre National de la Recherche Scientifique (UMR 7369), the University of Reims Champagne-Ardenne.

Institutional Review Board Statement: Not applicable.

Informed Consent Statement: Not applicable.

Data Availability Statement: Data is contained within the article.

Acknowledgments: The authors thank the HPC-Regional Center ROMEO, the Multiscale Molecular Modeling Platform (P3M) and the PICT-IBiSA Platform of the University of Reims Champagne-Ardenne (France) for providing time and support.

Conflicts of Interest: The authors declare no conflict of interest.

References

1. Chi, A.C.; Day, T.A.; Neville, B.W. Oral cavity and oropharyngeal squamous cell carcinoma-an update. CA. *Cancer J. Clin.* **2015**. [CrossRef] [PubMed]
2. Kim, Y.J.; Kim, J.H. Increasing incidence and improving survival of oral tongue squamous cell carcinoma. *Sci. Rep.* **2020**. [CrossRef]
3. Ellington, T.D.; Henley, S.J.; Senkomago, V.; O'Neil, M.E.; Wilson, R.J.; Singh, S.; Thomas, C.C.; Wu, M.; Richardson, L.C. Trends in Incidence of Cancers of the Oral Cavity and Pharynx—United States 2007–2016. *MMWR Morb. Mortal. Wkly. Rep.* **2020**. [CrossRef] [PubMed]
4. Paderno, A.; Morello, R.; Piazza, C. Tongue carcinoma in young adults: A review of the literature. *Acta Otorhinolaryngol. Ital.* **2018**, *38*, 175–180. [PubMed]
5. Lenze, N.R.; Farquhar, D.R.; Dorismond, C.; Sheth, S.; Zevallos, J.P.; Blumberg, J.; Lumley, C.; Patel, S.; Hackman, T.; Weissler, M.C.; et al. Age and risk of recurrence in oral tongue squamous cell carcinoma: Systematic review. *Head Neck* **2020**, *42*, 3755–3768. [CrossRef] [PubMed]
6. Mukdad, L.; Heineman, T.E.; Alonso, J.; Badran, K.W.; Kuan, E.C.; St. John, M.A. Oral tongue squamous cell carcinoma survival as stratified by age and sex: A surveillance, epidemiology, and end results analysis. *Laryngoscope* **2019**, *129*. [CrossRef] [PubMed]
7. Tota, J.E.; Anderson, W.F.; Coffey, C.; Califano, J.; Cozen, W.; Ferris, R.L.; St. John, M.; Cohen, E.E.W.; Chaturvedi, A.K. Rising incidence of oral tongue cancer among white men and women in the United States, 1973–2012. *Oral Oncol.* **2017**. [CrossRef]
8. Gonzalez, M.; Riera, A. Tongue Cancer. Available online: https://www.ncbi.nlm.nih.gov/books/NBK562324/ (accessed on 30 December 2020).
9. Wagenseil, J.E.; Mecham, R.P. Vascular extracellular matrix and arterial mechanics. *Physiol. Rev.* **2009**, *89*, 957–989. [CrossRef]
10. Marettova, E.; Maretta, M. Distribution of elastic fibers in the cat tongue. *Rev. Med. Vet. (Toulouse).* **2012**, *163*, 577–580.
11. Maquart, F.X.; Pasco, S.; Ramont, L.; Hornebeck, W.; Monboisse, J.C. An introduction to matrikines: Extracellular matrix-derived peptides which regulate cell activity - Implication in tumor invasion. *Crit. Rev. Oncol. Hematol.* **2004**, *49*, 199–202. [CrossRef]

12. Duca, L.; Floquet, N.; Alix, A.J.P.; Haye, B.; Debelle, L. Elastin as a matrikine. *Crit. Rev. Oncol. Hematol.* **2004**, *49*, 235–244. [CrossRef]

13. Heinz, A. Elastases and elastokines: Elastin degradation and its significance in health and disease. *Crit. Rev. Biochem. Mol. Biol.* **2020**, *55*, 252–273. [CrossRef]

14. Brassart, B.; Randoux, A.; Hornebeck, W.; Emonard, H. Regulation of matrix metalloproteinase-2 (gelatinase A, MMP-2), membrane-type matrix metalloproteinase-1 (MT1-MMP) and tissue inhibitor of metalloproteinases-2 (TIMP-2) expression by elastin-derived peptides in human HT-1080 fibrosarcoma cell line. *Clin. Exp. Metastasis* **1998**, *16*, 489–500. [CrossRef]

15. Huet, E.; Brassart, B.; Cauchard, J.-H.; Debelle, L.; Birembaut, P.; Wallach, J.; Emonard, H.; Polette, M.; Hornebeck, W. Cumulative influence of elastin peptides and plasminogen on matrix metalloproteinase activation and type I collagen invasion by HT-1080 fibrosarcoma cells. *Clin. Exp. Metastasis* **2002**, *19*. [CrossRef]

16. Robinet, A.; Fahem, A.; Cauchard, J.-H.; Huet, E.; Vincent, L.; Lorimier, S.; Antonicelli, F.; Soria, C.; Crepin, M.; Hornebeck, W.; et al. Elastin-derived peptides enhance angiogenesis by promoting endothelial cell migration and tubulogenesis through upregulation of MT1-MMP. *J. Cell Sci.* **2005**, *118*, 343–356. [CrossRef]

17. Donet, M.; Brassart-Pasco, S.; Salesse, S.; Maquart, F.-X.; Brassart, B. Elastin peptides regulate HT-1080 fibrosarcoma cell migration and invasion through an Hsp90-dependent mechanism. *Br. J. Cancer* **2014**, *111*. [CrossRef]

18. Da Silva, J.; Lameiras, P.; Beljebbar, A.; Berquand, A.; Villemin, M.; Ramont, L.; Dukic, S.; Nuzillard, J.-M.; Molinari, M.; Gautier, M.; et al. Structural characterization and in vivo pro-tumor properties of a highly conserved matrikine. *Oncotarget* **2018**, *9*. [CrossRef]

19. Brassart, B.; Da Silva, J.; Donet, M.; Seurat, E.; Hague, F.; Terryn, C.; Velard, F.; Michel, J.; Ouadid-Ahidouch, H.; Monboisse, J.C.; et al. Tumour cell blebbing and extracellular vesicle shedding: Key role of matrikines and ribosomal protein SA. *Br. J. Cancer* **2019**, *120*, 453–465. [CrossRef]

20. Mecham, R.P.; Hinek, A.; Griffin, G.L.; Senior, R.M.; Liotta, L.A. The elastin receptor shows structural and functional similarities to the 67-kDa tumor cell laminin receptor. *J. Biol. Chem.* **1989**, *264*, 16652–16657.

21. Digiacomo, V.; Meruelo, D. Looking into laminin receptor: Critical discussion regarding the non-integrin 37/67-kDa laminin receptor/RPSA protein. *Biol. Rev.* **2016**, *91*, 288–310. [CrossRef]

22. Vania, L.; Morris, G.; Otgaar, T.C.; Bignoux, M.J.; Bernert, M.; Burns, J.; Gabathuse, A.; Singh, E.; Ferreira, E.; Weiss, S.F.T. Patented therapeutic approaches targeting LRP/LR for cancer treatment. *Expert Opin Ther Pat.* **2019**, *29*, 987–1009. [CrossRef]

23. Gan, R.Y.; Li, H.B.; Sui, Z.Q.; Corke, H. Absorption, metabolism, anti-cancer effect and molecular targets of epigallocatechin gallate (EGCG): An updated review. *Crit. Rev. Food Sci. Nutr.* **2018**, *58*, 924–941. [CrossRef]

24. Almatrood, S.A.; Almatroudi, A.; Khan, A.A.; Alhumaydh, F.A.; Alsahl, M.A.; Rahmani, A.H. Potential therapeutic targets of epigallocatechin gallate (EGCG), the most abundant catechin in green tea, and its role in the therapy of various types of cancer. *Molecules* **2020**, *25*, 3146. [CrossRef]

25. Aggarwal, V.; Tuli, H.S.; Tania, M.; Srivastava, S.; Ritzer, E.E.; Pandey, A.; Aggarwal, D.; Barwal, T.S.; Jain, A.; Kaur, G.; et al. Molecular mechanisms of action of epigallocatechin gallate in cancer: Recent trends and advancement. *Semin. Cancer Biol.* **2020**, 32461153. [CrossRef]

26. Ho, Y.C.; Yang, S.F.; Peng, C.Y.; Chou, M.Y.; Chang, Y.C. Epigallocatechin-3-gallate inhibits the invasion of human oral cancer cells and decreases the productions of matrix metalloproteinases and urokinase-plasminogen activator. *J. Oral Pathol. Med.* **2007**. [CrossRef] [PubMed]

27. Chiang, W.C.; Wong, Y.K.; Lin, S.C.; Chang, K.W.; Liu, C.J. Increase of MMP-13 expression in multi-stage oral carcinogenesis and epigallocatechin-3-gallate suppress MMP-13 expression. *Oral Dis.* **2006**. [CrossRef]

28. Chen, P.N.; Chu, S.C.; Kuo, W.H.; Chou, M.Y.; Lin, J.K.; Hsieh, Y.S. Epigallocatechin-3 gallate inhibits invasion, epithelial-mesenchymal transition, and tumor growth in oral cancer cells. *J. Agric. Food Chem.* **2011**. [CrossRef]

29. Irimie, A.I.; Braicu, C.; Zanoaga, O.; Pileczki, V.; Gherman, C.; Berindan-Neagoe, I.; Campian, R.S. Epigallocatechin-3-gallate suppresses cell proliferation and promotes apoptosis and autophagy in oral cancer SSC-4 cells. *Onco. Targets. Ther.* **2015**. [CrossRef]

30. Li, A.; Gu, K.; Wang, Q.; Chen, X.; Fu, X.; Wang, Y.; Wen, Y. Epigallocatechin-3-gallate affects the proliferation, apoptosis, migration and invasion of tongue squamous cell carcinoma through the Hippo-TAZ signaling pathway. *Int. J. Mol. Med.* **2018**. [CrossRef]

31. Fujimura, Y.; Sumida, M.; Sugihara, K.; Tsukamoto, S.; Yamada, K.; Tachibana, H. Green tea polyphenol EGCG sensing motif on the 67-kDa laminin receptor. *PLoS ONE* **2012**. [CrossRef]

32. Morris, G.M.; Huey, R.; Lindstrom, W.; Sanner, M.F.; Belew, R.K.; Goodsell, D.S.; Olson, A.J. AutoDock4 and AutoDockTools4: Automated docking with selective receptor flexibility. *J. Comput. Chem.* **2009**, *30*, 2785–2791. [CrossRef]

33. Fülöp, T.; Larbi, A. Putative role of 67 kDa elastin-laminin receptor in tumor invasion. *Semin. Cancer Biol.* **2002**. [CrossRef]

34. Lefebvre, T.; Rybarczyk, P.; Bretaudeau, C.; Vanlaeys, A.; Cousin, R.; Brassart-Pasco, S.; Chatelain, D.; Dhennin-Duthille, I.; Ouadid-Ahidouch, H.; Brassart, B.; et al. TRPM7/RPSA Complex Regulates Pancreatic Cancer Cell Migration. *Front. Cell Dev. Biol.* **2020**. [CrossRef] [PubMed]

35. Tachibana, H.; Koga, K.; Fujimura, Y.; Yamada, K. A receptor for green tea polyphenol EGCG. *Nat. Struct. Mol. Biol.* **2004**. [CrossRef] [PubMed]

36. Weisburg, J.H.; Weissman, D.B.; Sedaghat, T.; Babich, H. In vitro cytotoxicity of epigallocatechin gallate and tea extracts to cancerous and normal cells from the human oral cavity. *Basic Clin. Pharmacol. Toxicol.* **2004**. [CrossRef] [PubMed]

37. Chang, C.M.; Chang, P.Y.; Tu, M.G.; Lu, C.C.; Kuo, S.C.; Amagaya, S.; Lee, C.Y.; Jao, H.Y.; Chen, M.Y.; Yang, J.S. Epigallocatechin gallate sensitizes CAL-27 human oral squamous cell carcinoma cells to the anti-metastatic effects of gefitinib (Iressa) via synergistic suppression of epidermal growth factor receptor and matrix metalloproteinase-2. *Oncol. Rep.* **2012**. [CrossRef]

38. Zanoni, D.K.; Montero, P.H.; Migliacci, J.C.; Shah, J.P.; Wong, R.J.; Ganly, I.; Patel, S.G. Survival outcomes after treatment of cancer of the oral cavity (1985–2015). *Oral Oncol.* **2019**. [CrossRef]

39. Betre, H.; Ong, S.R.; Guilak, F.; Chilkoti, A.; Fermor, B.; Setton, L.A. Chondrocytic differentiation of human adipose-derived adult stem cells in elastin-like polypeptide. *Biomaterials* **2006**, *27*, 91–99. [CrossRef]

40. Long, M.M.; King, V.J.; Prasad, K.U.; Freeman, B.A.; Urry, D.W. Elastin repeat peptides as chemoattractants for bovine aortic endothelial cells. *J. Cell. Physiol.* **1989**, *140*, 512–518. [CrossRef] [PubMed]

41. Senior, R.M.; Griffin, G.L.; Mecham, R.P. Chemotactic activity of elastin-derived peptides. *J. Clin. Invest.* **1980**, *66*, 859–862. [CrossRef]

42. Toupance, S.; Brassart, B.; Rabenoelina, F.; Ghoneim, C.; Vallar, L.; Polette, M.; Debelle, L.; Birembaut, P. Elastin-derived peptides increase invasive capacities of lung cancer cells by post-transcriptional regulation of MMP-2 and uPA. *Clin. Exp. Metastasis* **2012**, *29*. [CrossRef]

43. Jiang, L.; Ji, N.; Zhou, Y.; Li, J.; Liu, X.; Wang, Z.; Chen, Q.; Zeng, X. CAL 27 is an oral adenosquamous carcinoma cell line. *Oral Oncol.* **2009**. [CrossRef]

44. Ntayi, C.; Labrousse, A.L.; Debret, R.; Birembaut, P.; Bellon, G.; Antonicelli, F.; Hornebeck, W.; Bernard, P. Elastin-Derived Peptides Upregulate Matrix Metalloproteinase-2-ediated Melanoma Cell Invasion Through Elastin-Binding Protein. *J. Invest. Dermatol.* **2004**. [CrossRef] [PubMed]

45. Szychowski, K.A.; Wójtowicz, A.K.; Gmiński, J. Impact of Elastin-Derived Peptide VGVAPG on Matrix Metalloprotease-2 and -9 and the Tissue Inhibitor of Metalloproteinase-1, -2, -3 and -4 mRNA Expression in Mouse Cortical Glial Cells In Vitro. *Neurotox. Res.* **2019**. [CrossRef] [PubMed]

46. Fang, C.Y.; Wu, C.C.; Hsu, H.Y.; Chuang, H.Y.; Huang, S.Y.; Tsai, C.H.; Chang, Y.; Tsao, G.S.W.; Chen, C.L.; Chen, J.Y. EGCG inhibits proliferation, invasiveness and tumor growth by up-regulation of adhesion molecules, suppression of gelatinases activity, and induction of apoptosis in nasopharyngeal carcinoma cells. *Int. J. Mol. Sci.* **2015**, *16*, 2530–2558. [CrossRef] [PubMed]

47. Ho, H.C.; Huang, C.C.; Lu, Y.T.; Yeh, C.M.; Ho, Y.T.; Yang, S.F.; Hsin, C.H.; Lin, C.W. Epigallocatechin-3-gallate inhibits migration of human nasopharyngeal carcinoma cells by repressing MMP-2 expression. *J. Cell. Physiol.* **2019**. [CrossRef]

 *biomolecules*

Review

Syndecan-4 as a Pathogenesis Factor and Therapeutic Target in Cancer

Jessica Oyie Sousa Onyeisi [1,2,*]**, Carla Cristina Lopes** [2,3] **and Martin Götte** [1,*]

[1] Department of Gynecology and Obstetrics, University Hospital Münster, Albert-Schweitzer-Campus 1, D11, 48149 Münster, Germany

[2] Disciplina de Biologia Molecular, Departamento de Bioquímica, Universidade Federal de São Paulo, São Paulo 04039-032, Brazil; cclazevedo@gmail.com

[3] Instituto de Ciências Ambientais, Químicas e Farmacêuticas, Universidade Federal de São Paulo, Diadema 09913-030, Brazil

* Correspondence: jessicaoyie@hotmail.com (J.O.S.O.); martingotte@uni-muenster.de (M.G.)

Abstract: Cancer is an important cause of morbidity and mortality worldwide. Advances in research on the biology of cancer revealed alterations in several key pathways underlying tumorigenesis and provided molecular targets for developing new and improved existing therapies. Syndecan-4, a transmembrane heparan sulfate proteoglycan, is a central mediator of cell adhesion, migration and proliferation. Although several studies have demonstrated important roles of syndecan-4 in cell behavior and its interactions with growth factors, extracellular matrix (ECM) molecules and cytoskeletal signaling proteins, less is known about its role and expression in multiple cancer. The data summarized in this review demonstrate that high expression of syndecan-4 is an unfavorable biomarker for estrogen receptor-negative breast cancer, glioma, liver cancer, melanoma, osteosarcoma, papillary thyroid carcinoma and testicular, kidney and bladder cancer. In contrast, in neuroblastoma and colorectal cancer, syndecan-4 is downregulated. Interestingly, syndecan-4 expression is modulated by anticancer drugs. It is upregulated upon treatment with zoledronate and this effect reduces invasion of breast cancer cells. In our recent work, we demonstrated that the syndecan-4 level was reduced after trastuzumab treatment. Similarly, syndecan-4 levels are also reduced after panitumumab treatment. Together, the data found suggest that syndecan-4 level is crucial for understanding the changes involving in malignant transformation, and also demonstrate that syndecan-4 emerges as an important target for cancer therapy and diagnosis.

Keywords: syndecan-4; heparan sulfate; cancer; prognosis; biomarker; signal transduction; proteoglycan; metastasis

Citation: Onyeisi, J.O.S.; Lopes, C.C.; Götte, M. Syndecan-4 as a Pathogenesis Factor and Therapeutic Target in Cancer. *Biomolecules* **2021**, *11*, 503. https://doi.org/10.3390/biom11040503

Academic Editor: George Tzanakakis

Received: 4 March 2021
Accepted: 24 March 2021
Published: 26 March 2021

Publisher's Note: MDPI stays neutral with regard to jurisdictional claims in published maps and institutional affiliations.

1. The Syndecan Family of Cell Surface Heparan Sulfate Proteoglycans

Syndecans are a family of four transmembrane heparan sulfate proteoglycans (syndecan-1, -2, -3 and -4) in mammals. Each syndecan has an extracellular ectodomain that displays low sequence homology, except for the consensus sequences of the attachment sites for carbohydrate chains of the glycosaminoglycan (GAG) type. Syndecans have predominantly heparan sulfate-GAG (HS-GAG) chains attached to the extracellular domain and in the case of Sdc1 and Sdc3, additional chondroitin sulfate GAG chains. HS is a long, unbranched carbohydrate composed of repetitive disaccharide units of N-acetylglucosamine-α-L-iduronic acid/β-D-glucuronic acid, which can be substituted with sulfate residues, thus generating a highly negatively charged biomolecule capable of interacting with numerous ligands relevant to tumor progression [1]. HS chains are synthesized via O-glycosylation in the Golgi apparatus in a series of acetylation and deacetylation and sulfation steps which result in a high degree of structural complexity [2]. The HS chains are covalently linked to serine residues of the core protein via a defined tetrasaccharide linker consisting of glucuronic acid, two galactose residues, and a xylose. The relevance of glycosaminoglycan attachment has been documented in rare human inherited diseases where this GAG attachment site cannot be formed efficiently, resulting in severe developmental defects [3,4]. It has a conserved transmembrane domain, and a short cytoplasmic domain that has two conserved regions, C1 and C2, proximal and distal, respectively, to the membrane, common to all syndecans. The C1 and C2 regions are separated by a variable (V) region unique to each syndecan. Syndecans-1–3 have a restricted tissue distribution, whereas syndecan-4 is expressed ubiquitously [5–7].

2. Syndecan-4

The syndecan-4 has abundant expression in liver, kidney, brain, lung, breast heart, skeletal muscle, skin and small intestine [2] (Gene Expression database (https://www.ncbi. nlm.nih.gov/gene/6385—accessed on 8 January 2021). As outlined in detail below, the broad effects of this molecule are exemplified by its ability to form a connection between the extracellular matrix (ECM) and intracellular signaling cascades and to affect the growth and differentiation of a number of tissues and organs.

2.1. Syndecan-4 Membrane Localization, Trafficking and Signaling

Syndecan-4 is important for the interplay between extracellular matrix and cytoplasmatic signaling molecules and scaffolding proteins. It contributes to several outside-in and inside-out signaling events, such as the sequestration and concentration of matrix components, as well as effects on cell–matrix adhesion, endocytosis, exosome biogenesis or cytokinesis [8]. Syndecan-4 is localized to the plasma membrane and also localizes in endocytic compartments such as early endosomes and multivesicular bodies, indicating internalization and trafficking along the endosomal/lysosomal degradation route during muscle cell differentiation. Furthermore, the syndecan-4/syntenin complex is essential for exosome biosynthesis and multivesicular bodies, which give rise to exosomes [9,10].

2.2. Syndecan-4 as an Extracellular Signaling Interface

Through heparan sulfate chains on its extracellular domain, syndecan-4 can bind to various heparin-binding growth factors, chemokines and morphogens [11] (Figure 1). Syndecan-4 is a powerful regulator of FGF-2 signaling and can modulate growth factor responses in multiple cell types. In addition, syndecan-4 is capable of signaling in response to FGF independently of FGF receptor interactions [12–14]. Elfenbein and collaborators demonstrated that syndecan-4-mediated modulation of FGF2-induced FGFR1 endocytosis and MAPK signaling represents a previously unappreciated mechanism of crosstalk between the two receptors binding the same ligand [15].

Apart from FGFR signaling, syndecan-4 also affects epidermal growth factor (EGF)-mediated signaling, albeit through a different mechanism: It has been shown that EGF-dependent cancer cell migration is mediated through a complex of human epidermal growth factor receptor-1 (EGFR), α6β4 integrin and Sdc4. In this context, syndecan-4 modulates signaling via interactions with the extreme C terminus of the β4 integrin cytoplasmic domain, thereby affecting epithelial cancer cell migration [16]. Moreover, an extracellular site comprising amino acids 87–131 in the ectodomain of syndecan-4 captures EGFR, thus affecting signaling in epithelial cancer cells [17]. Apart from receptor tyrosine kinases, syndecan-4 has also been shown to affect G-protein coupled receptor signaling. For example, syndecan-4 affects hepatoma and HeLa cell motility and invasion by facilitating signaling via chemokines such as RANTES/CCL5 and SDF-1 [18,19], which is in accordance with the role of heparan sulfate in chemokine signaling [20]. The importance of syndecan-4 for the communication between tumor cells and immune cells has also been highlighted in vivo, as syndecan-4-deficient mice show reduced Lewis lung carcinoma growth, less dendritic cell recruitment, and increased recruitment of natural killer cells [21]. Finally, syndecan-4 is also involved in mediating signaling via morphogens such as Wnt. This has been demonstrated in model organisms such as the frog *Xenopus*, where the Wnt modulator R-spondin 3 induces syndecan-4-dependent clathrin-mediated endocytosis of Wnt–receptor complexes, thus affecting morphogenesis [22]. In turn, noncanonical Wnt signaling induces ubiquitination and degradation of syndecan-4 in *Xenopus*, suggesting complex regulatory mechanisms [10]. In a cancer context, silencing of syndecan-4 expression was shown to exhibit an antitumoral effect on human papillary thyroid carcinoma cells by affecting apoptosis and epithelial-to-mesenchymal transition via the Wnt/beta-catenin pathway [23]. Moreover, invasive growth of melanoma cells can be inhibited by syndecan-4 knockdown and rescued by addition of Wnt5a, suggesting an impact of syndecan-4 on this signaling pathway in melanoma [24]. Besides regulating cellular signaling via soluble growth factors and morphogens, syndecan-4 also act as a receptor for ECM molecules. Syndecan-4 facilitates $\alpha5\beta1$ integrin binding to its substrate fibronectin, allowing maturation of focal adhesions [25,26]. The engagement of syndecan-4 by fibronectin triggers rapid endocytosis of $\alpha_5\beta_1$-integrin, due to activation of RhoG [27]. Furthermore, syndecan-4 phosphorylation is a control point for integrin recycling [28]. In addition, through heparan sulphate side chains, syndecan-4 interacts with transglutaminase type 2 (TG2), an extracellular matrix crosslinking enzyme, affecting fibrosis [29].

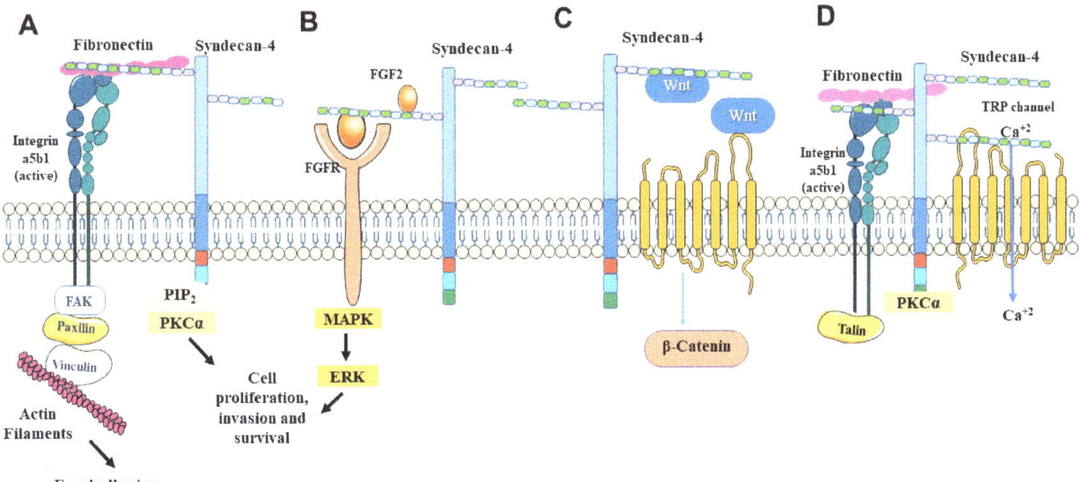

Figure 1. Overview of signaling pathways activated by syndecan-4. (**A**) Syndecan-4 and integrin signaling. (**B**) Syndecan-4 and growth factors. (**C**) Syndecan-4 and Wnt signaling. (**D**) Syndecan-4 and TRPC channels. See text for details.

Shedding of the Extracellular Domain

One mechanism by which syndecan-4 regulates its extracellular signaling is the proteolytic cleavage of its intact extracellular domain, in a process called shedding [30]. This cleavage is highly regulated by matrix metalloproteinases (MMPs) and can be accelerated under certain physiological conditions [31].

MMP9 has been shown to cleave syndecan-4 from HeLa cells, human primary macrophages and endothelial cells [32,33]. Our previous work demonstrated that syndecan-4 depletion decreased the expression of MMP3 in endometriosis, resulting in decreased invasive growth [34], and suggesting possible feedback loops. ADAMTS-1, a disintegrin and metalloproteinase with thrombospondin motifs, promotes syndecan-4 shedding, and this shedding disrupts cell adhesion and promotes cell migration [35,36].

Syndecan ectodomains can be cleaved by thrombin to produce bioactive fragments. For example, the recombinant ectodomains of human syndecan-3 and syndecan-4 induce significant decreases in endothelial barrier resistance and this involves Rho kinase pathway-mediated F-actin stress fiber formation and VE-cadherin junction disorganization [37]. Modification of the extracellular domain of syndecan-4 with highly flexible glycosaminoglycan side chains makes the receptor ideally suited to the detection of ligands that are dilute or distant from the membrane [38].

2.3. Intracellullar Signal Transduction Mechanisms

The cytoplasmic domain of syndecan-4 is distinct from the other syndecans in its capacity to bind phosphatidylinositol 4, 5-bisphosphate (PIP$_2$) and to activate protein kinase C-alpha (PKC-alpha) [39–41]. Syndecan-4 also provokes protein kinase Cα (PKCα) to phosphorylate the transient receptor potential canonical 7 cell membrane channel (TRPC7) that is involved in the regulation of cytosolic calcium levels to control myofibroblast differentiation [42,43]. (Figure 1). Syndecan-4 regulates downstream signaling pathways and the activity of the small GTPase Rac1 which orchestrates actin polymerization in migrating cells [38,44]. In addition, syndecan-4 can regulate the intracellular calcium distribution [43]. Syndecan-4 is known to regulate the organization of cytoskeleton, including focal adhesion and stress fiber formation, and Carvalheiro and coworkers demonstrated that the coupling of vinculin to F-actin demands syndecan-4. The authors showed that syndecan-4 acts as a central mediator that bridges fibronectin, integrin and intracellular components (actin and vinculin) and once silenced, the cytoskeleton protein network is disrupted [45]. Overall, these mechanisms expand the role of syndecan-4 beyond its classical function as a coreceptor for growth factor-mediated receptor tyrosine kinase signaling [12–14] and chemokine-mediated signaling via G-protein coupled heptahelical receptors [15,16].

3. Syndecan-4 and Cancer

3.1. Syndecan-4 Expression in Cancers

Given the multitude of syndecan-4-mediated cellular functions with relevance to tumor progression, it is not surprising that the expression of syndecan-4 is dysregulated in a number of malignant diseases, highlighting its importance as a pathogenesis factor and diagnostic marker (Table 1). In the following section, we will provide an overview of its clinicopathological relevance in a number of tumor entities.

Table 1. Syndecan-4 expression in different types of cancer.

Cancer Type	Syndecan-4 Expression	References
Breast Cancer (Estrogen receptor-negative)	Overexpressed	[46]
Colorectal	Reduced	[47]
Glioma	Overexpressed	[48]
Liver	Overexpressed	[49]
Melanoma	Overexpressed	[24]
Neuroblastoma	Reduced	[50]
Osteosarcoma	Overexpressed	[51]
Testicular	Overexpressed	[52]
Papillary Thyroid Carcinoma	Overexpressed	[23]
Kidney	Overexpressed	[53]
ladder	Overexpressed	[54]

3.1.1. Breast Cancer

Breast cancer (BC) is a complex heterogeneous form of cancer with numerous genetic alterations and distinct molecular subtypes [55]. According to the American Cancer Society, breast cancer is the most common cancer among women, accounting for nearly one in three cancers diagnosed in women [56]. Several studies have shown that heparan sulfate proteoglycans and specific genes involved in the synthesis and editing of heparan sulfate proteoglycans show altered expression in breast cancer [57–61]. Syndecan-4 is expressed in normal human mammary epithelium, and was initially described as being overexpressed in an estrogen receptor-negative, highly proliferative breast carcinoma subtype [46]. However, in a study on duplicate samples of benign and malignant breast cancer cases, syndecan-4 expression was found to be correlated with positive estrogen and progesterone receptor status, and found to exhibit an expression pattern distinct from syndecan-1, suggesting divergent pathobiological roles for these proteoglycans [62]. Besides being the second most abundant HSPG produced by most breast carcinoma cell lines [62], syndecan-4 is involved in membrane fixation of LL-37 and its pro-migratory effect in breast cancer cells [63]. Moreover, in vivo, targeting of syndecan-4 in murine 4T1 breast cancer cells inhibited the formation of early bone metastases [64]. Interestingly, a study demonstrated that estrogen receptor beta (ERβ) silencing in MDA-MB-231 breast cancer cells induces the expression of syndecan-4 [65], suggesting endocrine regulation of syndecan-4 in this tumor entity. This view is supported by investigations of menstrual cycle-dependent expression changes in healthy breast tissue, where syndecan-4 mRNA expression was significantly lower among parous women in the progesterone-dominated luteal phase compared to the estrogen-dominated follicular phase [66]. Apart from steroidal mechanisms, inhibition of the receptor tyrosine kinase IGFR was shown to downregulate syndecan-4 levels in estrogen receptor-positive breast cancer cells via an endocytic mechanism [67].

3.1.2. Colon Cancer

Heparan sulfate proteoglycans, as well as heparan sulfate remodeling enzymes, are molecules involved in colorectal cancer tumorigenesis [68]. In normal epithelial cells and tissues, the expression level of syndecan-4 is high, however, syndecan-4 is significantly reduced in highly metastatic colon carcinoma cells (KM1214) [47,69]. Hypoxia is one of the factors regulating syndecan-4 expression in human colon cancer cells, as it can induce its expression, along with alpha 5 integrin [70]. Interestingly, hetero-oligomerization with syndecan-2 reduces both syndecan-4-dependent PKCα activation and cell adhesion and syndecan-2-mediated migration and anchorage-independent growth in colon cancer cells, suggesting a functional interplay of syndecans in tumor progression [71]. Moreover, the recombinant syndecan-4 ectodomain is capable of inducing the expression of the epidermal growth factors erb-b2 and erb-b3 in colon cancer cells, suggesting a regulatory crosstalk between these receptor tyrosine kinases and the proteoglycan [68]. In turn, experimental lung metastasis of the murine colon cancer cell line MC-38 resulted in an induction of syndecan-4 expression in blood vessels at the metastatic site, suggesting a possible role for this proteoglycan in the metastatic niche [72].

3.1.3. Glioma

Human glioma is the most common type of primary brain tumor worldwide. Using proteomics analysis, a recent study showed the pull-down of multiple cancer-related proteoglycans with key roles in the pathogenesis of glioma [73]. All malignant glioma cell lines and glioblastoma specimens expressed all types of syndecans at the mRNA level. Syndecan-4 is highly expressed on the surface of glioma cells [48]. Interestingly, syndecan-4 mRNA expression has been indicated as a novel marker for the prediction of glioblastoma multiforme patients' response to treatment with the WT1 peptide vaccine [74], and its expression is altered in pediatric astrocytoma [75].

3.1.4. Liver Cancer

Liver cancer is the most frequent cause of cancer deaths across the globe [76]. Alterations in proteoglycan expression interfere with the physiologic function of the liver on several levels and, in addition, this affects cancer cell signaling pathways, facilitating tumorigenesis [77,78]. Syndecan-4 is expressed in human normal liver [79]. In both hepatocellular carcinoma (HCC) and cholangiocarcinoma, increased levels of syndecan-4 were found [49]. Notably, studies in a susceptible mouse model of Moloney murine leukemia, virus infection demonstrated that provirus integration at a site upstream of the first exon of the syndecan-4 gene resulted in particularly fast-growing hepatocellular carcinomas [80]. Moreover, syndecan-4 plays an important coreceptor role in the effects of the chemokine SDF-1 on human hepatoma cell growth, migration, and invasion [81].

3.1.5. Melanoma

Melanoma is a highly aggressive skin cancer [82]. Along with beta 3 integrin and WNT5A, syndecan-4 is part of a gene signature characteristic for metastatic disease in melanoma [83]. Indeed, syndecan-4 is an important component of the Wnt5A autocrine signaling loop and its overexpression is correlated to increased metastatic potential in melanoma patients. In addition, the knockdown of syndecan-4 caused decreases in cell invasion of metastatic melanoma cells [24]. Moreover, a study has shown that inhibition of syndecan-binding protein syntenin-1 (SDCBP) expression by siRNA impaired the ability of uveal melanoma cells to migrate in a wound-healing assay [84].

In contrast, a different study demonstrated that reduction in syndecan-4 expression in melanoma cells resulted in downregulation of FGF-2 signaling, leading to an increase in tumor cell motility and decreased adhesion to fibronectin, demonstrating a regulatory role of syndecan-4 on these cell functions [85]. Of note, syndecan-4 is required for the activating function of latent heparanase in the activation of VLA4 integrin in melanoma cells [86]. Finally, syndecan-4 overexpression significantly reduces the migration of A375 melanoma cells, whereas its siRNA knockdown enhanced their migration, consistent with the observation that syndecan-4 overexpression reduced lung and popliteal lymph node metastasis of B16F10 melanoma cells in mice. Notably, syntenin overexpression could compensate for the effect of syndecan-4 depletion, suggesting functional interactions [87].

3.1.6. Neuroblastoma

Neuroblastoma is a pediatric malignancy that originates from the neural crest. Previous works have shown that extracellular matrix components contribute to tumor progression in neuroblastoma [88,89]. Using microarray dataset analysis, Knelson and collaborators demonstrated that syndecan-4 expression is reduced in neuroblastoma in comparison with benign neuroblastic tumors and is high in the Schwannian stroma [50].

3.1.7. Osteosarcoma

Osteosarcoma is the most common malignant bone tumor in young adults and children [90,91]. Pathogenesis of osteosarcoma implicates qualitative and quantitative changes in the proteoglycans [92,93]. A study has shown that syndecan-4 expression is upregulated in high-grade osteosarcoma when compared to other tissues. In addition, its overexpression was significantly associated with a larger tumor size, distant metastasis and poor overall survival [51]. The expression of syndecan-4 on osteosarcoma cell lines in vitro can be induced by the cytokines IL-1b and IL-6, but not by the osteotropic hormones parathyroid hormone (PTH(1–34)), and 1,25(OH)2-vitamin D3 [94]. Mechanistically, syndecan-4 mediates tumorigenic properties of osteosarcoma cells via cell surface interactions with autotaxin-β [64].

3.1.8. Testicular Germ Cell Tumors

Testicular germ cell tumors are the most common malignancy of young adult males [95,96]. They are classified into seminomatous germ cell tumors (testicular germ cell tumors, TGCTs) and nonseminomatous germ cell tumors (NSGCTs), the latter being either undifferentiated (embryonal carcinoma) or differentiated (teratoma, yolk sac tumor and choriocarcinoma) [97]. In both seminomatous testicular germ cell tumors (TGCTs) and nonseminomatous germ cell tumors (NSGCTs), significantly increased expression of syndecan-4 was detected in tumor cells. Syndecan-4 is differentially expressed in seminomas and NSGCTs and might be a useful marker [52]. Studies on rat Sertoli cell development have demonstrated that syndecan-4 expression in healthy testes is regulated by protein kinase C, follicle-stimulating hormone and the second messenger cAMP, providing possible avenues for pharmacological intervention in the context of malignant disease [98,99].

3.1.9. Papillary Thyroid Cancer

Papillary thyroid cancer is the most common type of thyroid cancer [100,101]. To obtain proteomic profiles from various thyroid cancer cell lines that represent the range of thyroid cancers of follicular cell origin, a study used a proteomics strategy targeting cell surface and secreted proteins and identified syndecan-1 and syndecan-4 as glycoproteins uniquely expressed by the various thyroid cancer cell lines [102]. Using a microarray, two recent studies have shown that syndecan-4 expression levels among the papillary thyroid cancer tissues are higher than that in normal thyroid tissues [103,104]. Interestingly, syndecan-4 gene silencing represses EMT, and enhances cell apoptosis by suppressing the activation of the Wnt/β-catenin signaling pathway in human papillary thyroid carcinoma [23].

3.1.10. Kidney Cancer

Renal cell carcinoma, also known as hypernephroma, renal adenocarcinoma or Grawitz tumor, is the most common malignant type of kidney cancer [105]. Renal cell carcinoma is characterized by profound changes in cellular metabolism such as glucose and glutamine utilization, lipid metabolism and mitochondrial function [106]. A study demonstrated that metastatic Caki-1 and ACHN cells (human renal adenocarcinoma) expressed higher levels of syndecan-4 mRNA than primary renal cell carcinoma cell lines. The authors concluded that upregulation of syndecan-4 mRNA plays an important role in the development of renal cell carcinoma and advanced forms of the disease with metastasis [53]. In contrast, a study utilizing data from the Human Protein Atlas dataset assigned a positive prognostic value to high syndecan-4 protein expression in renal cell carcinoma [107]. Different methodological approaches, such as the assessment of syndecan-4 mRNA vs. protein levels, may account for the discordant results of these studies.

3.1.11. Bladder Cancer

Bladder cancer is the most common malignancy of the urinary tract and is common in women and the fourth most common malignancy in men [108,109]. The expression of syndecan-1, -2 and -3 is decreased while syndecan-4 is increased in bladder cancers compared to normal tissues [54,110]. Functionally, interactions of syndecan-4 and angiomodulin have been found to be responsible for the formation of cord-like structures in the human bladder carcinoma cell line ECV-304 [111].

3.2. Syndecan-4 in Cancer Biology

The dysregulated expression of syndecan-4 in numerous tumor entities (see Table 1) suggests a possible mechanistic contribution to cancer progression. Indeed, proteoglycans are capable of modulating virtually all hallmarks of cancer [112–114]. In the following section, we will highlight the role of syndecan-4 in selected tumor-associated cellular functions.

3.2.1. Survival

Disruption of cell–matrix attachment results in a loss of prosurvival signals and culminates in programmed cell death, referred to as anoikis [115]. Tumor cells often resist anoikis, survive and grow in the absence of anchorage to the extracellular matrix (ECM). In our previous work, we have demonstrated that the acquisition of anoikis resistance by blocking adhesion to the substrate upregulates syndecan-4 expression in endothelial cells and syndecan-4 gene silencing reverses the transformed phenotype of anoikis-resistant endothelial cells [116,117]. The Ras/Raf/MAPK (MEK)/ERK pathway plays a crucial role in the survival and development of tumor cells [118,119]. Neel and collaborators demonstrated that SDC4-ROS1 and SLC34A2-ROS1 fusion oncoproteins reside on endosomes and activate the MAPK pathway. Moreover, they showed that knockdown of these fusion proteins resulted in suppression of the RAS/MAPK pathway [120]. In addition, knockdown of syndecan-4 in human papillary thyroid carcinoma cells promoted apoptosis via the Wnt/beta catenin pathway [23]. Moreover, knockdown of syndecan-4 reduces macrophage cell surface TG2 activity and apoptotic cell clearance [121]. Finally, association of the chemokine SDF-1 with syndecan-4 increases the resistance of hepatoma cells to TNF-alpha-induced apoptosis [80].

3.2.2. Proliferation

Proliferation is an important part of cancer development and progression [122]. The Ras–Raf–MEK–ERK signaling cascade is crucial for controlling this process [123]. Several studies have shown that syndecan-4 signaling can lead to ERK activation and induce cell proliferation [124,125]. Syndecan-4 promotes cytokinesis in a phosphorylation-dependent manner in MCF-7 breast adenocarcinoma cells, which shed the ectodomain of syndecan-4 periodically in a cell cycle-dependent way, reaching the maximum at the G2/M phase [126]. Several works have demonstrated that syndecan-4 gene silencing suppresses the cell cycle progression, decreasing the transition from G1 to S phase and decreasing the levels of cyclin D1 and cyclin E in different cancer cell lines [117,127,128]. Recently, it was demonstrated that the prometastatic integrin-interacting factor autotaxin-beta promotes osteosarcoma cell proliferation via a mechanism that requires a physical interaction with syndecan-4 [64], expanding the range of mechanisms by which syndecan-4 regulates tumor cell growth.

3.2.3. Adhesion

Tumor cells often show a decrease in cell–cell and/or cell–matrix adhesion [129]. Besides that, changes in cell adhesion molecules play a causal role in tumor dissemination [130]. Syndecan-4 is an important regulator of cell adhesion [11]. $\alpha_V\beta_1$ integrin and syndecan-4 are key players of the interaction with vitronectin in bladder cancer cells. Although these surface receptors share a similar role, the energy landscapes of single molecular complexes reveal higher (integrins) and lower (syndecans) energy barriers. The shape of the energy landscape agrees with the binding site structures of both complexes [131]. Notably, epithelial cell spreading depends on the interaction of syndecan-4 with integrins, as $\beta4$ integrin mutants deficient in syndecan-4 recognition act in a dominant negative manner to block EGFR-dependent cell spreading [16]. Recent works identified PAR-3 as a syndecan-4-binding protein and the syndecan-4/PAR-3 signaling complex participates in Thy-1/CD90-induced focal adhesion disassembly in mesenchymal cells [132].

3.2.4. Cell Migration

There are several mechanisms by which syndecan-4 contributes to tumor cell migration. Syndecan-4 promotes cell migration in a variety of cells. Syndecan-4 is involved in membrane fixation of cathelicidin LL-37 and its promigratory effect in breast cancer cells [63]. It is also involved in RANTES/CCL5-induced migration and invasion of human hepatoma cells [19]. Ochieng and collaborators have demonstrated that knockdown of syndecan-4 significantly attenuated the invasive capacity and uptake of labeled exosomes and FNH (fetuin-A and histones) nanoparticles of LN229, a highly aggressive glioblastoma cell line [133]. Similar results were found in anoikis-resistant endothelial cells, and syndecan-4 silencing led to downregulation of the invasive capacity of anoikis-resistant endothelial cells [117]. Syndecan-4 is also capable of modulating the effect of other matrix constituents in tumor cell migration: In breast cancer, the protease ADAMTS-15 reduces cell migration on fibronectin and laminin matrices. Notably, the inhibitory effect could be rescued by knockdown of syndecan-4, suggesting a mechanistic role in this context which is worthy of further exploration [134].

3.2.5. Tumor Angiogenesis

Angiogenesis is an important step for cancer growth and progression [135]. The ability of syndecan-4 to regulate angiogenesis, the formation of new blood vessels from an existing vasculature, is suggested by the observation that the codelivery of proteoliposomes with FGF-2 increased the cellular uptake, trafficking and nuclear localization of the growth factor. These alterations in cellular signaling, trafficking and nuclear localization led to increased proliferation, migration and angiogenic differentiation in response to FGF-2 treatment [13]. Moreover, the ability of endothelial tube formation in matrigel is reduced upon syndecan-4 silencing, which was partially attributed to the role of syndecan-4 in coupling viculin to F-actin, and to connecting actin filopodial protrusions to vascular endothelial cadherin-rich junctions [45].

In endometriosis, a disease characterized by invasive growth of endometrial tissue at ectopic sites, syndecan-4 is upregulated along with constituents of TGF-beta signaling, and modulates invasion via regulation of MMP and RAC1 expression [34,136]. Moreover, in a mouse model of pathological lymphangiogenesis, syndecan-4 was the predominant heparan sulfate proteoglycan in mouse lymphatic endothelia, and showed a VEGF-C-induced association with VEGF receptor-3 at the lymphatic cell surface. Notably, syndecan-4-deficient mice showed an impaired pathological lymphangiogenesis in this model, suggesting a possible coreceptor function for VEGF-C [21]. Recent findings document not only a role for syndecan-4 in lymphangiogenesis, but also in classical VEGFA-mediated angiogenesis: Syndecan-4-deficient mice show reduced angiogenesis not only in a model of diabetic retinopathy, but also in a melanoma model of tumor angiogenesis. The impact of syndecan-4 on these processes was shown to involve a mechanism by which syndecan-4 localized at endothelial cell junctions interacts with vascular endothelial cadherin, and participates in its internalization in response to VEGFA [137].

3.3. Syndecan-4 as a Target for Anticancer Drugs

The dysregulation of syndecan-4 in various cancers, and its mechanistic contribution to multiple steps of tumor progression, mark it as an attractive potential target for cancer therapy. Several studies have demonstrated that therapeutics currently used in clinics have an impact on syndecan-4 expression and its functions (Table 2).

Table 2. Effects of anticancer drugs on syndecan-4 expression in different cell models.

Anticancer Drug	Cell Type	Biological Effects	References
Trastuzumab (Herceptin®)	Anoikis-resistant endothelial cells	Decreases syndecan-4 expression	[138]
Panitumumab (Vectibix®)	Colon cancer		[139]
Bisphosphonate Zoledronic acid (ASCO)	Breast cancer	Syndecan-4 upregulation	[140]

3.3.1. Trastuzumab

Trastuzumab (Herceptin®) is a humanized recombinant monoclonal antibody (mAb) of the immunoglobulin G1 type, approved by the FDA for treatment of breast and gastric cancer with overexpression of ErbB2 (HER2) [141,142]. Recently, we demonstrated that trastuzumab reduces syndecan-4 expression in anoikis-resistant endothelial cells and this interaction controls cellular events, such as proliferation, adhesion and angiogenesis, in these cells [138].

3.3.2. Panitumumab

Panitumumab is a human monoclonal antibody (pMAb) approved by the FDA in 2007, which inhibits epidermal growth factor receptor (EGFR). Panitumumab is used for the treatment of patients with metastatic colorectal cancer. A first study reported that panitumumab (pmAb) significantly decreases the expression of syndecan-4 [139].

3.3.3. Bisphosphonate Zoledronic Acid (ASCO)

Bisphosphonate zoledronic acid (ASCO) inhibits osteoclast-mediated bone resorption. It has been approved for treatment of patients with advanced lung cancer, renal cancer and other solid tumors with bone metastases or multiple myeloma and for the management of tumor-induced hypercalcemia [143,144]. Interestingly, Dedes and collaborators demonstrated that syndecan-4 expression is upregulated upon treatment with zoledronate [140].

Overall, these data suggest that syndecan-4 expression and function is already modulated by existing anticancer drugs, and therefore part of the therapeutic response. Syndecan-4 has been targeted or utilized as part of therapeutic approaches in nonmalignant diseases, and this knowledge could be utilized in a cancer context in the future: For example, antibody-mediated inhibition of syndecan-4 has been proposed as a treatment for osteoarthritis [145,146]. Moreover, syndecan-4 enhances uptake of liposomes for therapeutic gene delivery [147]. Notably, binding of the cell-penetrating peptide Xentry to syndecan-4 has been utilized to target therapeutics to melanoma cells in vitro [148]. While not specific to syndecan-4 alone, a targeting of the heparan sulfate moiety may expand the range of approaches hampering syndecan-4 function in cancer [149,150].

4. Conclusions

In this review, we demonstrated that changes in the expression of syndecan-4 contribute to the development and progression of cancer, and have a diagnostic and prognostic value in numerous tumor entities. Besides that, we showed that syndecan-4 has an important role in the hallmarks of cancer, modulating multiple steps of tumor progression, including unlimited cell proliferation, resistance to apoptosis, invasive growth and metastasis, tumor angiogenesis and tumor-associated inflammation. Syndecan-4 mediates these processes as a signaling interface at the cell surface, acting as a classical heparan sulfate coreceptor for soluble ligands such as growth factors and chemokines, but also via interactions of its protein moiety with growth factor receptors and integrins. The data reviewed in this article support that a targeting of syndecan-4, or a modulation of its expression using already available drugs, may be promising strategy for the treatment of different types of cancers.

Author Contributions: J.O.S.O.: performed literature searches, designed tables and Figure 1, wrote the first manuscript draft; C.C.L.: supervised J.O.S.O., revised manuscript draft; M.G.: conceived the review, revised manuscript draft, performed additional literature searches, revised Figure 1, cosupervised J.O.S.O. All authors have read and agreed to the published version of the manuscript.

Funding: We acknowledge funding by German Academic Exchange Service DAAD (Co-funded Research Grants–Short-Term Grants-no.: 91782193) (to J.O.S.O.). The authors acknowledge funding by the Open Access Fund of the University of Muenster WWU.

Institutional Review Board Statement: Not applicable.

Informed Consent Statement: Not applicable.

Data Availability Statement: Not applicable.

Conflicts of Interest: The authors declare no conflict of interest.

References

1. Hassan, N.; Greve, B.; Espinoza-Sánchez, N.A.; Götte, M. Cell-surface heparan sulfate proteoglycans as multifunctional integrators of signaling in cancer. *Cell Signal* **2021**, *77*, 109822. [CrossRef]
2. Bernfield, M.; Götte, M.; Park, P.W.; Reizes, O.; Fitzgerald, M.L.; Lincecum, J.; Zako, M. Functions of cell surface heparan sulfate proteoglycans. *Annu. Rev. Biochem.* **1999**, *68*, 729–777. [CrossRef]

3. Götte, M.; Spillmann, D.; Yip, G.W.; Versteeg, E.; Echtermeyer, F.G.; van Kuppevelt, T.H.; Kiesel, L. Changes in heparan sulfate are associated with delayed wound repair, altered cell migration, adhesion and contractility in the galactosyltransferase I (ß4GalT-7) deficient form of Ehlers–Danlos syndrome. *Hum. Mol. Genet.* **2008**, *17*, 996–1009. [CrossRef] [PubMed]
4. Malfait, F.; Castori, M.; Francomano, C.A.; Giunta, C.; Kosho, T.; Byers, P.H. The Ehlers–Danlos syndromes. *Nat. Rev. Dis. Primers* **2020**, *6*, 1–25. [CrossRef]
5. Fears, C.Y.; Woods, A. The role of syndecans in disease and wound healing. *Matrix Biol.* **2006**, *25*, 443–456. [CrossRef] [PubMed]
6. Zimmermann, P.; David, G. The syndecans, tuners of transmembrane signaling. *FASEB J.* **1999**, *13*, S91–S100. [CrossRef] [PubMed]
7. Gondelaud, F.; Ricard-Blum, S. Structures and interactions of syndecans. *FEBS J.* **2019**, *286*, 2994–3007. [CrossRef]
8. Elfenbein, A.; Simons, M. Syndecan-4 signaling at a glance. *J. Cell Sci.* **2013**, *126*, 3799–3804. [CrossRef] [PubMed]
9. Rønning, S.B.; Carlson, C.R.; Stang, E.; Kolset, S.O.; Hollung, K.; Pedersen, M.E. Syndecan-4 regulates muscle differentiation and is internalized from the plasma membrane during myogenesis. *PLoS ONE* **2015**, *10*, e0129288. [CrossRef]
10. Carvallo, L.; Muñoz, R.; Bustos, F.; Escobedo, N.; Carrasco, H.; Olivares, G.; Larraín, J. Non-canonical Wnt signaling induces ubiquitination and degradation of Syndecan-4. *J. Biol. Chem.* **2010**, *285*, 29546–29555. [CrossRef]
11. Karamanos, N.K.; Piperigkou, Z.; Theocharis, A.D.; Watanabe, H.; Franchi, M.; Baud, S.; Brezillon, S.; Götte, M.; Passi, A.; Vigetti, D. Proteoglycan chemical diversity drives multifunctional cell regulation and therapeutics. *Chem. Rev.* **2018**, *118*, 9152–9232. [CrossRef] [PubMed]
12. Lopes, C.C.; Dietrich, C.P.; Nader, H.B. Specific structural features of syndecans and heparan sulfate chains are needed for cell signaling. *Braz. J. Med. Biol. Res.* **2006**, *39*, 157–167. [CrossRef]
13. Jang, E.; Albadawi, H.; Watkins, M.T.; Edelman, E.R.; Baker, A.B. Syndecan-4 proteoliposomes enhance fibroblast growth factor-2 (FGF-2)–induced proliferation, migration, and neovascularization of ischemic muscle. *Proc. Natl. Acad. Sci. USA* **2012**, *109*, 1679–1684. [CrossRef] [PubMed]
14. Horowitz, A.; Tkachenko, E.; Simons, M. Fibroblast growth factor–specific modulation of cellular response by syndecan-4. *J. Cell Biol.* **2002**, *157*, 715–725. [CrossRef]
15. Elfenbein, A.; Lanahan, A.; Zhou, T.X.; Yamasaki, A.; Tkachenko, E.; Matsuda, M.; Simons, M. Syndecan 4 regulates FGFR1 signaling in endothelial cells by directing macropinocytosis. *Sci. Signal.* **2012**, *5*, 36. [CrossRef]
16. Wang, H.; Jin, H.; Beauvais, D.M.; Rapraeger, A.C. Cytoplasmic domain interactions of syndecan-1 and syndecan-4 with α6β4 integrin mediate human epidermal growth factor receptor (HER1 and HER2)-dependent motility and survival. *J. Biol. Chem.* **2014**, *289*, 30318–30332. [CrossRef]
17. Wang, H.; Jin, H.; Rapraeger, A.C. Syndecan-1 and syndecan-4 capture epidermal growth factor receptor family members and the α3β1 integrin via binding sites in their ectodomains: Novel synstatins prevent kinase capture and inhibit α6β4-integrin-dependent epithelial cell motility. *J. Biol. Chem.* **2015**, *290*, 26103–26113. [CrossRef]
18. Brule, S.; Friand, V.; Sutton, A.; Baleux, F.; Gattegno, L.; Charnaux, N. Glycosaminoglycans and syndecan-4 are involved in SDF-1/CXCL12-mediated invasion of human epitheloid carcinoma HeLa cells. *Biochim. Biophys. Acta (BBA) Gen. Subj.* **2009**, *1790*, 1643–1650. [CrossRef] [PubMed]
19. Charni, F.; Friand, V.; Haddad, O.; Hlawaty, H.; Martin, L.; Vassy, R.; Oudar, O.; Gattegno, L.; Charnaux, N.; Sutton, A. Syndecan-1 and syndecan-4 are involved in RANTES/CCL5-induced migration and invasion of human hepatoma cells. *Biochim. Biophys. Acta (BBA) Gen. Subj.* **2009**, *1790*, 1314–1326. [CrossRef] [PubMed]
20. Kumar, A.V.; Katakam, S.K.; Urbanowitz, A.-K.; Gotte, M. Heparan sulphate as a regulator of leukocyte recruitment in inflammation. *Curr. Protein Pept. Sci.* **2015**, *16*, 77–86. [CrossRef] [PubMed]
21. Johns, S.C.; Yin, X.; Jeltsch, M.; Bishop, J.R.; Schuksz, M.; El Ghazal, R.; Wilcox-Adelman, S.A.; Alitalo, K.; Fuster, M.M. Functional importance of a proteoglycan coreceptor in pathologic lymphangiogenesis. *Circ. Res.* **2016**, *119*, 210–221. [CrossRef] [PubMed]
22. Ohkawara, B.; Glinka, A.; Niehrs, C. Rspo3 binds syndecan 4 and induces Wnt/PCP signaling via clathrin-mediated endocytosis to promote morphogenesis. *Dev. Cell* **2011**, *20*, 303–314. [CrossRef] [PubMed]
23. Chen, L.-L.; Gao, G.-X.; Shen, F.-X.; Chen, X.; Gong, X.-H.; Wu, W.-J. SDC4 gene silencing favors human papillary thyroid carcinoma cell apoptosis and inhibits epithelial mesenchymal transition via Wnt/β-catenin pathway. *Mol. Cells* **2018**, *41*, 853. [PubMed]
24. O'Connell, M.P.; Fiori, J.L.; Kershner, E.K.; Frank, B.P.; Indig, F.E.; Taub, D.D.; Hoek, K.S.; Weeraratna, A.T. Heparan sulfate proteoglycan modulation of Wnt5A signal transduction in metastatic melanoma cells. *J. Biol. Chem.* **2009**, *284*, 28704–28712. [CrossRef] [PubMed]
25. Couchman, J.R.; Woods, A. Syndecan-4 and integrins: Combinatorial signaling in cell adhesion. *J. Cell Sci.* **1999**, *112*, 3415–3420. [PubMed]
26. Bass, M.D.; Morgan, M.R.; Humphries, M.J. Integrins and syndecan-4 make distinct, but critical, contributions to adhesion contact formation. *Soft Matter* **2007**, *3*, 372–376. [CrossRef]
27. Bass, M.D.; Williamson, R.C.; Nunan, R.D.; Humphries, J.D.; Byron, A.; Morgan, M.R.; Martin, P.; Humphries, M.J. A syndecan-4 hair trigger initiates wound healing through caveolin-and RhoG-regulated integrin endocytosis. *Dev. Cell* **2011**, *21*, 681–693. [CrossRef]
28. Morgan, M.R.; Hamidi, H.; Bass, M.D.; Warwood, S.; Ballestrem, C.; Humphries, M.J. Syndecan-4 phosphorylation is a control point for integrin recycling. *Dev. Cell* **2013**, *24*, 472–485. [CrossRef] [PubMed]

29. Scarpellini, A.; Huang, L.; Burhan, I.; Schroeder, N.; Funck, M.; Johnson, T.S.; Verderio, E.A. Syndecan-4 knockout leads to reduced extracellular transglutaminase-2 and protects against tubulointerstitial fibrosis. *J. Am. Soc. Nephrol.* **2014**, *25*, 1013–1027. [CrossRef]

30. Piperigkou, Z.; Mohr, B.; Karamanos, N.; Götte, M. Shed proteoglycans in tumor stroma. *Cell Tissue Res.* **2016**, *365*, 643–655. [CrossRef] [PubMed]

31. Manon-Jensen, T.; Itoh, Y.; Couchman, J.R. Proteoglycans in health and disease: The multiple roles of syndecan shedding. *FEBS J.* **2010**, *277*, 3876–3889. [CrossRef] [PubMed]

32. Brule, S.; Charnaux, N.; Sutton, A.; Ledoux, D.; Chaigneau, T.; Saffar, L.; Gattegno, L. The shedding of syndecan-4 and syndecan-1 from HeLa cells and human primary macrophages is accelerated by SDF-1/CXCL12 and mediated by the matrix metalloproteinase-9. *Glycobiology* **2006**, *16*, 488–501. [CrossRef] [PubMed]

33. Ramnath, R.; Foster, R.R.; Qiu, Y.; Cope, G.; Butler, M.J.; Salmon, A.H.; Mathieson, P.W.; Coward, R.J.; Welsh, G.I.; Satchell, S.C. Matrix metalloproteinase 9-mediated shedding of syndecan 4 in response to tumor necrosis factor α: A contributor to endothelial cell glycocalyx dysfunction. *FASEB J.* **2014**, *28*, 4686–4699. [CrossRef] [PubMed]

34. Chelariu-Raicu, A.; Wilke, C.; Brand, M.; Starzinski-Powitz, A.; Kiesel, L.; Schüring, A.N.; Götte, M. Syndecan-4 expression is upregulated in endometriosis and contributes to an invasive phenotype. *Fertil. Steril.* **2016**, *106*, 378–385. [CrossRef] [PubMed]

35. Rodríguez-Manzaneque, J.C.; Carpizo, D.; del Carmen Plaza-Calonge, M.; Torres-Collado, A.X.; Thai, S.N.-M.; Simons, M.; Horowitz, A.; Iruela-Arispe, M.L. Cleavage of syndecan-4 by ADAMTS1 provokes defects in adhesion. *Int. J. Biochem. Cell Biol.* **2009**, *41*, 800–810. [CrossRef] [PubMed]

36. Lambert, J.; Makin, K.; Akbareian, S.; Johnson, R.; Alghamdi, A.A.; Robinson, S.D.; Edwards, D.R. ADAMTS-1 and syndecan-4 intersect in the regulation of cell migration and angiogenesis. *J. Cell Sci.* **2020**, *133*. [CrossRef]

37. Jannaway, M.; Yang, X.; Meegan, J.E.; Coleman, D.C.; Yuan, S.Y. Thrombin-cleaved syndecan-3/-4 ectodomain fragments mediate endothelial barrier dysfunction. *PLoS ONE* **2019**, *14*, e0214737. [CrossRef] [PubMed]

38. Bass, M.D.; Roach, K.A.; Morgan, M.R.; Mostafavi-Pour, Z.; Schoen, T.; Muramatsu, T.; Mayer, U.; Ballestrem, C.; Spatz, J.P.; Humphries, M.J. Syndecan-4–dependent Rac1 regulation determines directional migration in response to the extracellular matrix. *J. Cell Biol.* **2007**, *177*, 527–538. [CrossRef] [PubMed]

39. Oh, E.-S.; Woods, A.; Couchman, J.R. Syndecan-4 proteoglycan regulates the distribution and activity of protein kinase C. *J. Biol. Chem.* **1997**, *272*, 8133–8136. [CrossRef] [PubMed]

40. Suhovskih, A.V.; Mostovich, L.A.; Kunin, I.S.; Boboev, M.M.; Nepomnyashchikh, G.I.; Aidagulova, S.V.; Grigorieva, E.V. Proteoglycan expression in normal human prostate tissue and prostate cancer. *ISRN Oncol.* **2013**, *2013*, 680136. [CrossRef]

41. Oh, E.-S.; Woods, A.; Lim, S.-T.; Theibert, A.W.; Couchman, J.R. Syndecan-4 proteoglycan cytoplasmic domain and phosphatidylinositol 4, 5-bisphosphate coordinately regulate protein kinase C activity. *J. Biol. Chem.* **1998**, *273*, 10624–10629. [CrossRef]

42. Gopal, S.; Søgaard, P.; Multhaupt, H.A.; Pataki, C.; Okina, E.; Xian, X.; Pedersen, M.E.; Stevens, T.; Griesbeck, O.; Park, P.W. Transmembrane proteoglycans control stretch-activated channels to set cytosolic calcium levels. *J. Cell Biol.* **2015**, *210*, 1199–1211. [CrossRef]

43. Becsky, D.; Szabo, K.; Gyulai-Nagy, S.; Gajdos, T.; Bartos, Z.; Balind, A.; Dux, L.; Horvath, P.; Erdelyi, M.; Homolya, L. Syndecan-4 Modulates Cell Polarity and Migration by Influencing Centrosome Positioning and Intracellular Calcium Distribution. *Front. Cell Dev. Biol.* **2020**, *8*. [CrossRef] [PubMed]

44. Elfenbein, A.; Rhodes, J.M.; Meller, J.; Schwartz, M.A.; Matsuda, M.; Simons, M. Suppression of RhoG activity is mediated by a syndecan 4–synectin–RhoGDI1 complex and is reversed by PKCα in a Rac1 activation pathway. *J. Cell Biol.* **2009**, *186*, 75–83. [CrossRef] [PubMed]

45. Cavalheiro, R.; Lima, M.; Jarrouge-Bouças, T.; Viana, G.; Lopes, C.; Coulson-Thomas, V.; Dreyfuss, J.; Yates, E.; Tersariol, I.; Nader, H. Coupling of vinculin to F-actin demands Syndecan-4 proteoglycan. *Matrix Biol.* **2017**, *63*, 23–37. [CrossRef] [PubMed]

46. Baba, F.; Swartz, K.; Van Buren, R.; Eickhoff, J.; Zhang, Y.; Wolberg, W.; Friedl, A. Syndecan-1 and syndecan-4 are overexpressed in an estrogen receptor-negative, highly proliferative breast carcinoma subtype. *Breast Cancer Res. Treat.* **2006**, *98*, 91–98. [CrossRef] [PubMed]

47. Park, H.; Kim, Y.; Lim, Y.; Han, I.; Oh, E.-S. Syndecan-2 Mediates Adhesion and Proliferation of Colon Carcinoma Cells. *J. Biol. Chem.* **2002**, *277*, 29730–29736. [CrossRef] [PubMed]

48. Watanabe, A.; Mabuchi, T.; Satoh, E.; Furuya, K.; Zhang, L.; Maeda, S.; Naganuma, H. Expression of syndecans, a heparan sulfate proteoglycan, in malignant gliomas: Participation of nuclear factor-κB in upregulation of syndecan-1 expression. *J. Neuro-Oncol.* **2006**, *77*, 25–32. [CrossRef]

49. Roskams, T.; De Vos, R.; David, G.; Van Damme, B.; Desmet, V. Heparan sulphate proteoglycan expression in human primary liver tumours. *J. Pathol.* **1998**, *185*, 290–297. [CrossRef]

50. Knelson, E.H.; Gaviglio, A.L.; Nee, J.C.; Starr, M.D.; Nixon, A.B.; Marcus, S.G.; Blobe, G.C. Stromal heparan sulfate differentiates neuroblasts to suppress neuroblastoma growth. *J. Clin. Investig.* **2014**, *124*, 3016–3031. [CrossRef]

51. Na, K.Y.; Bacchini, P.; Bertoni, F.; Kim, Y.W.; Park, Y.-K. Syndecan-4 and fibronectin in osteosarcoma. *Pathology* **2012**, *44*, 325–330. [CrossRef] [PubMed]

52. Labropoulou, V.T.; Skandalis, S.S.; Ravazoula, P.; Perimenis, P.; Karamanos, N.K.; Kalofonos, H.P.; Theocharis, A.D. Expression of Syndecan-4 and Correlation with Metastatic Potential in Testicular Germ Cell Tumours. *BioMed Res. Int.* **2013**, *2013*, 1–10. [CrossRef] [PubMed]

53. Erdem, M.; Erdem, S.; Sanli, O.; Sak, H.; Kilicaslan, I.; Sahin, F.; Telci, D. Up-regulation of TGM2 with ITGB1 and SDC4 is important in the development and metastasis of renal cell carcinoma. *Urol. Oncol. Semin. Orig. Investig.* **2014**, *32*, 25.e13–25.e20. [CrossRef] [PubMed]

54. Marzioni, D.; Lorenzi, T.; Mazzucchelli, R.; Capparuccia, L.; Morroni, M.; Fiorini, R.; Bracalenti, C.; Catalano, A.; David, G.; Castellucci, M.; et al. Expression of Basic Fibroblast Growth Factor, its Receptors and Syndecans in Bladder Cancer. *Int. J. Immunopathol. Pharmacol.* **2009**, *22*, 627–638. [CrossRef] [PubMed]

55. Feng, Y.; Spezia, M.; Huang, S.; Yuan, C.; Zeng, Z.; Zhang, L.; Ji, X.; Liu, W.; Huang, B.; Luo, W. Breast cancer development and progression: Risk factors, cancer stem cells, signaling pathways, genomics, and molecular pathogenesis. *Genes Dis.* **2018**, *5*, 77–106. [CrossRef]

56. Siegel, R.L.; Miller, K.D.; Fuchs, H.E.; Jemal, A. Cancer Statistics, 2021. *CA Cancer J. Clin.* **2021**, *71*, 7–33. [CrossRef] [PubMed]

57. Koo, C.-Y.; Sen, Y.-P.; Bay, B.-H.; Yip, G.W. Targeting heparan sulfate proteoglycans in breast cancer treatment. *Recent Pat. Anti Cancer Drug Discov.* **2008**, *3*, 151–158. [CrossRef]

58. Fernández-Vega, I.; García, O.; Crespo, A.; Castañón, S.; Menéndez, P.; Astudillo, A.; Quirós, L.M. Specific genes involved in synthesis and editing of heparan sulfate proteoglycans show altered expression patterns in breast cancer. *BMC Cancer* **2013**, *13*, 1–16. [CrossRef]

59. Okolicsanyi, R.K.; Van Wijnen, A.J.; Cool, S.M.; Stein, G.S.; Griffiths, L.R.; Haupt, L.M. Heparan sulfate proteoglycans and human breast cancer epithelial cell tumorigenicity. *J. Cell. Biochem.* **2014**, *115*, 967–976. [CrossRef] [PubMed]

60. Kumar, A.V.; Brézillon, S.; Untereiner, V.; Sockalingum, G.D.; Katakam, S.K.; Mohamed, H.T.; Kemper, B.; Greve, B.; Mohr, B.; Ibrahim, S.A. HS2ST1-dependent signaling pathways determine breast cancer cell viability, matrix interactions, and invasive behavior. *Cancer Sci.* **2020**, *111*, 2907. [CrossRef] [PubMed]

61. Teixeira, F.C.; Vijaya Kumar, A.; Kumar Katakam, S.; Cocola, C.; Pelucchi, P.; Graf, M.; Kiesel, L.; Reinbold, R.; Pavão, M.S.; Greve, B. The heparan sulfate sulfotransferases HS2ST1 and HS3ST2 are novel regulators of breast cancer stem-cell properties. *Front. Cell Dev. Biol.* **2020**, *8*, 992. [CrossRef]

62. Lendorf, M.E.; Manon-Jensen, T.; Kronqvist, P.; Multhaupt, H.A.; Couchman, J.R. Syndecan-1 and syndecan-4 are independent indicators in breast carcinoma. *J. Histochem. Cytochem.* **2011**, *59*, 615–629. [CrossRef] [PubMed]

63. Habes, C.; Weber, G.; Goupille, C. Sulfated Glycoaminoglycans and Proteoglycan Syndecan-4 Are Involved in Membrane Fixation of LL-37 and Its Pro-Migratory Effect in Breast Cancer Cells. *Biomolecules* **2019**, *9*, 481. [CrossRef]

64. Leblanc, R.; Sahay, D.; Houssin, A.; Machuca-Gayet, I.; Peyruchaud, O. Autotaxin-β interaction with the cell surface via syndecan-4 impacts on cancer cell proliferation and metastasis. *Oncotarget* **2018**, *9*, 33170. [CrossRef]

65. Piperigkou, Z.; Bouris, P.; Onisto, M.; Franchi, M.; Kletsas, D.; Theocharis, A.D.; Karamanos, N.K. Estrogen receptor beta modulates breast cancer cells functional properties, signaling and expression of matrix molecules. *Matrix Biol.* **2016**, *56*, 4–23. [CrossRef] [PubMed]

66. Hallberg, G.; Andersson, E.; Naessén, T.; Ordeberg, G.E. The expression of syndecan-1, syndecan-4 and decorin in healthy human breast tissue during the menstrual cycle. *Reprod. Biol. Endocrinol.* **2010**, *8*, 35. [CrossRef]

67. Afratis, N.A.; Bouris, P.; Skandalis, S.S.; Multhaupt, H.A.; Couchman, J.R.; Theocharis, A.D.; Karamanos, N.K. IGF-IR cooperates with ERalpha to inhibit breast cancer cell aggressiveness by regulating the expression and localisation of ECM molecules. *Sci. Rep.* **2017**, *7*, 40138. [CrossRef] [PubMed]

68. Vicente, C.M.; da Silva, D.A.; Sartorio, P.V.; Silva, T.D.; Saad, S.S.; Nader, H.B.; Forones, N.M.; Toma, L. Heparan sulfate proteoglycans in human colorectal cancer. *Anal. Cell. Pathol.* **2018**, 8389595. [CrossRef]

69. Han, I.; Park, H.; Oh, E.-S. New insights into syndecan-2 expression and tumourigenic activity in colon carcinoma cells. *J. Mol. Histol.* **2004**, *35*, 319–326. [CrossRef] [PubMed]

70. Koike, T.; Kimura, N.; Miyazaki, K.; Yabuta, T.; Kumamoto, K.; Takenoshita, S.; Chen, J.; Kobayashi, M.; Hosokawa, M.; Taniguchi, A.; et al. Hypoxia induces adhesion molecules on cancer cells: A missing link between Warburg effect and induction of selectin-ligand carbohydrates. *Proc. Natl. Acad. Sci. USA* **2004**, *101*, 8132–8137. [CrossRef] [PubMed]

71. Choi, Y.; Kwon, M.-J.; Lim, Y.; Yun, J.-H.; Lee, W.; Oh, E.-S. Trans-regulation of Syndecan Functions by Hetero-oligomerization. *J. Biol. Chem.* **2015**, *290*, 16943–16953. [CrossRef]

72. Roblek, M.; Strutzmann, E.; Zankl, C.; Adage, T.; Heikenwalder, M.; Atlic, A.; Weis, R.; Kungl, A.; Borsig, L. Targeting of CCL2-CCR2-Glycosaminoglycan Axis Using a CCL2 Decoy Protein Attenuates Metastasis through Inhibition of Tumor Cell Seeding. *Neoplasia* **2016**, *18*, 49–59. [CrossRef] [PubMed]

73. Bang-Christensen, S.R.; Pedersen, R.S.; Pereira, M.A.; Clausen, T.M.; Løppke, C.; Sand, N.T.; Ahrens, T.D.; Jørgensen, A.M.; Lim, Y.C.; Goksøyr, L. Capture and detection of circulating glioma cells using the recombinant VAR2CSA malaria protein. *Cells* **2019**, *8*, 998. [CrossRef] [PubMed]

74. Takashima, S.; Oka, Y.; Fujiki, F.; Morimoto, S.; Nakajima, H.; Nakae, Y.; Nakata, J.; Nishida, S.; Hosen, N.; Tatsumi, N. Syndecan-4 as a biomarker to predict clinical outcome for glioblastoma multiforme treated with WT1 peptide vaccine. *Future Sci. OA* **2016**, *2*, FSO96. [CrossRef] [PubMed]

75. López-Aguilar, J.E.; Velázquez-Flores, M.A.; Simón-Martínez, L.A.; Ávila-Miranda, R.; Rodríguez-Florido, M.A.; Garrido, R.R.-E. Circulating microRNAs as biomarkers for pediatric astrocytomas. *Arch. Med. Res.* **2017**, *48*, 323–332. [CrossRef]

76. Anwanwan, D.; Singh, S.K.; Singh, S.; Saikam, V.; Singh, R. Challenges in liver cancer and possible treatment approaches. *Biochim. Biophys. Acta (BBA) Rev. Cancer* **2020**, *1873*, 188314. [CrossRef]

77. Baghy, K.; Tátrai, P.; Regős, E.; Kovalszky, I. Proteoglycans in liver cancer. *World J. Gastroenterol.* **2016**, *22*, 379. [CrossRef]
78. Tanaka, Y.; Tateishi, R.; Koike, K. Proteoglycans are attractive biomarkers and therapeutic targets in hepatocellular carcinoma. *Int. J. Mol. Sci.* **2018**, *19*, 3070. [CrossRef] [PubMed]
79. Roskams, T.; Moshage, H.; De Vos, R.; Guido, D.; Yap, P.; Desmet, V. Heparan sulfate proteoglycan expression in normal human liver. *Hepatology* **1995**, *21*, 950–958. [CrossRef]
80. Renard, C.-A.; Transy, C.; Tiollais, P.; Buendia, M. Infection of WHV/c-myc transgenic mice with Moloney murine leukaemia virus and proviral insertion near the syndecan-4 gene in an early liver tumour. *Res. Virol.* **1998**, *149*, 133–143. [CrossRef]
81. Sutton, A.; Friand, V.; Brulé-Donneger, S.; Chaigneau, T.; Ziol, M.; Sainte-Catherine, O.; Poiré, A.; Saffar, L.; Kraemer, M.; Vassy, J. Stromal cell-derived factor-1/chemokine (CXC motif) ligand 12 stimulates human hepatoma cell growth, migration, and invasion. *Mol. Cancer Res.* **2007**, *5*, 21–33. [CrossRef] [PubMed]
82. Shih, C.-Y.; Cheng, Y.-C.; Hsieh, C.; Tseng, T.; Jiang, S.; Lee, S.-C. Drug-selected population in melanoma A2058 cells as melanoma stem-like cells retained angiogenic features—The potential roles of heparan-sulfate binding ANGPTL4 protein. *Aging* **2020**, *12*, 22700. [CrossRef]
83. Tímár, J.; Mészáros, L.; Ladányi, A.; Puskás, L.; Rásó, E. Melanoma genomics reveals signatures of sensitivity to bio-and targeted therapies. *Cell. Immunol.* **2006**, *244*, 154–157. [CrossRef] [PubMed]
84. Gangemi, R.; Mirisola, V.; Barisione, G.; Fabbi, M.; Brizzolara, A.; Lanza, F.; Mosci, C.; Salvi, S.; Gualco, M.; Truini, M.; et al. Mda-9/syntenin is expressed in uveal melanoma and correlates with metastatic progression. *PLoS ONE* **2012**, *7*, e29989. [CrossRef]
85. Chalkiadaki, G.; Nikitovic, D.; Berdiaki, A.; Sifaki, M.; Krasagakis, K.; Katonis, P.; Karamanos, N.K.; Tzanakakis, G.N. Fibroblast growth factor-2 modulates melanoma adhesion and migration through a syndecan-4-dependent mechanism. *Int. J. Biochem. Cell Biol.* **2009**, *41*, 1323–1331. [CrossRef] [PubMed]
86. Gerber, U.; Hoß, S.G.; Shteingauz, A.; Jüngel, E.; Jakubzig, B.; Ilan, N.; Blaheta, R.; Schlesinger, M.; Vlodavsky, I.; Bendas, G. Latent Heparanase Facilitates VLA-4–Mediated Melanoma Cell Binding and Emerges as a Relevant Target of Heparin in the Interference with Metastatic Progression. *Semin. Thromb. Hemost.* **2015**, *41*, 244–254. [CrossRef] [PubMed]
87. Choi, Y.; Yun, J.-H.; Yoo, J.; Lee, I.; Kim, H.; Son, H.-N.; Kim, I.-S.; Yoon, H.S.; Zimmermann, P.; Couchman, J.R. New structural insight of C-terminal region of Syntenin-1, enhancing the molecular dimerization and inhibitory function related on Syndecan-4 signaling. *Sci. Rep.* **2016**, *6*, 1–16.
88. Lam, W.A.; Cao, L.; Umesh, V.; Keung, A.J.; Sen, S.; Kumar, S. Extracellular matrix rigidity modulates neuroblastoma cell differentiation and N-myc expression. *Mol. Cancer* **2010**, *9*, 1–7. [CrossRef]
89. Joshi, S. Targeting the tumor microenvironment in neuroblastoma: Recent advances and future directions. *Cancers* **2020**, *12*, 2057. [CrossRef]
90. Meazza, C.; Scanagatta, P. Metastatic osteosarcoma: A challenging multidisciplinary treatment. *Expert Rev. Anticancer Ther.* **2016**, *16*, 543–556. [CrossRef]
91. Yang, C.; Tian, Y.; Zhao, F.; Chen, Z.; Su, P.; Li, Y.; Qian, A. Bone Microenvironment and Osteosarcoma Metastasis. *Int. J. Mol. Sci.* **2020**, *21*, 6985. [CrossRef] [PubMed]
92. Benayahu, D.; Shur, I.; Marom, R.; Meller, I.; Issakov, J. Cellular and molecular properties associated with osteosarcoma cells. *J. Cell. Biochem.* **2002**, *84*, 108–114. [CrossRef]
93. Nagarajan, A.; Malvi, P.; Wajapeyee, N. Heparan sulfate and heparan sulfate proteoglycans in cancer initiation and progression. *Front. Endocrinol.* **2018**, *9*, 483. [CrossRef] [PubMed]
94. Birch, M.; Skerry, T. Differential regulation of syndecan expression by osteosarcoma cell lines in response to cytokines but not osteotropic hormones. *Bone* **1999**, *24*, 571–578. [CrossRef]
95. Chieffi, P.; De Martino, M.; Esposito, F. New anti-cancer strategies in testicular germ cell tumors. *Recent Pat. Anti Cancer Drug Discov.* **2019**, *14*, 53–59. [CrossRef]
96. Shen, H.; Shih, J.; Hollern, D.P.; Wang, L.; Bowlby, R.; Tickoo, S.K.; Thorsson, V.; Mungall, A.J.; Newton, Y.; Hegde, A.M. Integrated molecular characterization of testicular germ cell tumors. *Cell Rep.* **2018**, *23*, 3392–3406. [CrossRef] [PubMed]
97. Pedraza, A.M.; Stephenson, A.J. Prognostic markers in clinical stage I seminoma and nonseminomatous germ cell tumours. *Curr. Opin. Urol.* **2018**, *28*, 448–453. [CrossRef] [PubMed]
98. Brucato, S.; Villers, C. Protein kinase C regulation of glypican-1, syndecan-1 and syndecan-4 mRNAs expression during rat Sertoli cell development. *Biochimie* **2002**, *84*, 681–686. [CrossRef]
99. Brucato, S.; Bocquet, J.; Villers, C. Regulation of glypican-1, syndecan-1 and syndecan-4 mRNAs expression by follicle-stimulating hormone, cAMP increase and calcium influx during rat Sertoli cell development. *Eur. J. Biochem.* **2002**, *269*, 3461–3469. [CrossRef]
100. Carling, T.; Udelsman, R. Thyroid cancer. *Annu. Rev. Med.* **2014**, *65*, 125–137. [CrossRef]
101. Liang, W.; Sun, F. Identification of key genes of papillary thyroid cancer using integrated bioinformatics analysis. *J. Endocrinol. Investig.* **2018**, *41*, 1237–1245. [CrossRef]
102. Arcinas, A.; Yen, T.-Y.; Kebebew, E.; Macher, B.A. Cell surface and secreted protein profiles of human thyroid cancer cell lines reveal distinct glycoprotein patterns. *J. Proteome Res.* **2009**, *8*, 3958–3968. [CrossRef] [PubMed]
103. Reyes, I.; Reyes, N.; Suriano, R.; Iacob, C.; Suslina, N.; Policastro, A.; Moscatello, A.; Schantz, S.; Tiwari, R.K.; Geliebter, J. Gene expression profiling identifies potential molecular markers of papillary thyroid carcinoma. *Cancer Biomark.* **2019**, *24*, 71–83. [CrossRef]

104. Abdullah, M.I.; Junit, S.M.; Ng, K.L.; Jayapalan, J.J.; Karikalan, B.; Hashim, O.H. Papillary thyroid cancer: Genetic alterations and molecular biomarker investigations. *Int. J. Med. Sci.* **2019**, *16*, 450. [CrossRef] [PubMed]

105. Hancock, S.B.; Georgiades, C.S. Kidney cancer. *Cancer J.* **2016**, *22*, 387–392. [CrossRef] [PubMed]

106. Hoerner, C.R.; Miao, S.Y.; Hsieh, J.J.; Fan, A.C. Targeting Metabolic Pathways in Kidney Cancer: Rationale and Therapeutic Opportunities. *Cancer J.* **2020**, *26*, 407–418. [CrossRef] [PubMed]

107. Majo, S.; Courtois, S.; Souleyreau, W.; Bikfalvi, A.; Auguste, P. Impact of Extracellular Matrix Components to Renal Cell Carcinoma Behavior. *Front. Oncol.* **2020**, *10*, 625. [CrossRef] [PubMed]

108. Szarvas, T.; Reis, H.; Kramer, G.; Shariat, S.F.; Vom Dorp, F.; Tschirdewahn, S.; Schmid, K.W.; Kovalszky, I.; Rübben, H. Enhanced stromal syndecan-1 expression is an independent risk factor for poor survival in bladder cancer. *Hum. Pathol.* **2014**, *45*, 674–682. [CrossRef] [PubMed]

109. Lenis, A.T.; Lec, P.M.; Chamie, K. Bladder Cancer: A Review. *JAMA* **2020**, *324*, 1980–1991. [CrossRef]

110. Nord, H.; Segersten, U.; Sandgren, J.; Wester, K.; Busch, C.; Menzel, U.; Komorowski, J.; Dumanski, J.P.; Malmström, P.U.; de Ståhl, T.D. Focal amplifications are associated with high grade and recurrences in stage Ta bladder carcinoma. *Int. J. Cancer* **2010**, *126*, 1390–1402. [CrossRef]

111. Kishibe, J.; Yamada, S.; Okada, Y.; Sato, J.; Ito, A.; Miyazaki, K.; Sugahara, K. Structural requirements of heparan sulfate for the binding to the tumor-derived adhesion factor/angiomodulin that induces cord-like structures to ECV-304 human carcinoma cells. *J. Biol. Chem.* **2000**, *275*, 15321–15329. [CrossRef] [PubMed]

112. Hanahan, D.; Weinberg, R.A. The hallmarks of cancer. *Cell* **2000**, *100*, 57–70. [CrossRef]

113. Ibrahim, S.A.; Hassan, H.; Götte, M. Micro RNA regulation of proteoglycan function in cancer. *FEBS J.* **2014**, *281*, 5009–5022. [CrossRef] [PubMed]

114. Pickup, M.W.; Mouw, J.K.; Weaver, V.M. The extracellular matrix modulates the hallmarks of cancer. *EMBO Rep.* **2014**, *15*, 1243–1253. [CrossRef] [PubMed]

115. Gupta, P.; Gupta, N.; Fofaria, N.M.; Ranjan, A.; Srivastava, S.K. HER2-mediated GLI2 stabilization promotes anoikis resistance and metastasis of breast cancer cells. *Cancer Lett.* **2019**, *442*, 68–81. [CrossRef] [PubMed]

116. Carneiro, B.R.; Pernambuco Filho, P.C.A.; de Sousa Mesquita, A.P.; da Silva, D.S.; Pinhal, M.A.S.; Nader, H.B.; Lopes, C.C. Acquisition of anoikis resistance up-regulates syndecan-4 expression in endothelial cells. *PLoS ONE* **2014**, *9*, e116001. [CrossRef] [PubMed]

117. Onyeisi, J.O.S.; de Almeida Pernambuco Filho, P.C.; de Sousa Mesquita, A.P.; de Azevedo, L.C.; Nader, H.B.; Lopes, C.C. Effects of syndecan-4 gene silencing by micro RNA interference in anoikis resistant endothelial cells: Syndecan-4 silencing and anoikis resistance. *Int. J. Biochem. Cell Biol.* **2020**, *128*, 105848. [CrossRef] [PubMed]

118. Low, H.B.; Zhang, Y. Regulatory roles of MAPK phosphatases in cancer. *Immune Netw.* **2016**, *16*, 85. [CrossRef]

119. Guo, Y.J.; Pan, W.W.; Liu, S.B.; Shen, Z.F.; Xu, Y.; Hu, L.L. ERK/MAPK signalling pathway and tumorigenesis. *Exp. Ther. Med.* **2020**, *19*, 1997–2007. [CrossRef]

120. Neel, D.S.; Allegakoen, D.V.; Olivas, V.; Mayekar, M.K.; Hemmati, G.; Chatterjee, N.; Blakely, C.M.; McCoach, C.E.; Rotow, J.K.; Le, A. Differential subcellular localization regulates oncogenic signaling by ROS1 kinase fusion proteins. *Cancer Res.* **2019**, *79*, 546–556. [CrossRef] [PubMed]

121. Nadella, V.; Wang, Z.; Johnson, T.S.; Griffin, M.; Devitt, A. Transglutaminase 2 interacts with syndecan-4 and CD44 at the surface of human macrophages to promote removal of apoptotic cells. *Biochim. Biophys. Acta (BBA) Mol. Cell Res.* **2015**, *1853*, 201–212. [CrossRef] [PubMed]

122. Feitelson, M.A.; Arzumanyan, A.; Kulathinal, R.J.; Blain, S.W.; Holcombe, R.F.; Mahajna, J.; Marino, M.; Martinez-Chantar, M.L.; Nawroth, R.; Sanchez-Garcia, I. Sustained proliferation in cancer: Mechanisms and novel therapeutic targets. *Semin. Cancer Biol.* **2015**, *35*, S25–S54. [CrossRef]

123. Mebratu, Y.; Tesfaigzi, Y. How ERK1/2 activation controls cell proliferation and cell death: Is subcellular localization the answer? *Cell Cycle* **2009**, *8*, 1168–1175. [CrossRef] [PubMed]

124. Chua, C.C.; Rahimi, N.; Forsten-Williams, K.; Nugent, M.A. Heparan sulfate proteoglycans function as receptors for fibroblast growth factor-2 activation of extracellular signal–regulated kinases 1 and 2. *Circ. Res.* **2004**, *94*, 316–323. [CrossRef]

125. Corti, F.; Finetti, F.; Ziche, M.; Simons, M. The syndecan-4/protein kinase Cα pathway mediates prostaglandin E2-induced extracellular regulated kinase (ERK) activation in endothelial cells and angiogenesis in vivo. *J. Biol. Chem.* **2013**, *288*, 12712–12721. [CrossRef] [PubMed]

126. Keller-Pinter, A.; Bottka, S.; Timar, J.; Kulka, J.; Katona, R.; Dux, L.; Deak, F.; Szilak, L. Syndecan-4 promotes cytokinesis in a phosphorylation-dependent manner. *Cell. Mol. Life Sci.* **2010**, *67*, 1881–1894. [CrossRef] [PubMed]

127. Qin, Y.; Zhu, Y.; Luo, F.; Chen, C.; Chen, X.; Wu, M. Killing two birds with one stone: Dual blockade of integrin and FGF signaling through targeting syndecan-4 in postoperative capsular opacification. *Cell Death Dis.* **2017**, *8*, e2920. [CrossRef] [PubMed]

128. Keller-Pinter, A.; Szabo, K.; Kocsis, T.; Deak, F.; Ocsovszki, I.; Zvara, A.; Puskas, L.; Szilak, L.; Dux, L. Syndecan-4 influences mammalian myoblast proliferation by modulating myostatin signalling and G1/S transition. *FEBS Lett.* **2018**, *592*, 3139–3151. [CrossRef] [PubMed]

129. Cavallaro, U.; Christofori, G. Cell adhesion in tumor invasion and metastasis: Loss of the glue is not enough. *Biochim. Biophys. Acta (BBA) Rev. Cancer* **2001**, *1552*, 39–45. [CrossRef]

130. Janiszewska, M.; Primi, M.C.; Izard, T. Cell adhesion in cancer: Beyond the migration of single cells. *J. Biol. Chem.* **2020**, *295*, 2495–2505. [CrossRef] [PubMed]

131. Lekka, M.; Herman, K.; Zemła, J.; Bodek, Ł.; Pyka-Fościak, G.; Gil, D.; Dulińska-Litewka, J.; Ptak, A.; Laidler, P. Probing the recognition specificity of αVβ1 integrin and syndecan-4 using force spectroscopy. *Micron* **2020**, *137*, 102888. [CrossRef] [PubMed]

132. Valdivia, A.; Cárdenas, A.; Brenet, M.; Maldonado, H.; Kong, M.; Díaz, J.; Burridge, K.; Schneider, P.; San Martín, A.; García-Mata, R. Syndecan-4/PAR-3 signaling regulates focal adhesion dynamics in mesenchymal cells. *Cell Commun. Signal.* **2020**, *18*, 1–19. [CrossRef] [PubMed]

133. Ochieng, J.; Nangami, G.; Sakwe, A.; Rana, T.; Ingram, S.; Goodwin, J.S.; Moye, C.; Lammers, P.; Adunyah, S.E. Extracellular histones are the ligands for the uptake of exosomes and hydroxyapatite-nanoparticles by tumor cells via syndecan-4. *FEBS Lett.* **2018**, *592*, 3274–3285. [CrossRef] [PubMed]

134. Kelwick, R.; Wagstaff, L.; Decock, J.; Roghi, C.; Cooley, L.S.; Robinson, S.D.; Arnold, H.; Gavrilović, J.; Jaworski, D.M.; Yamamoto, K. Metalloproteinase-dependent and-independent processes contribute to inhibition of breast cancer cell migration, angiogenesis and liver metastasis by a disintegrin and metalloproteinase with thrombospondin motifs-15. *Int. J. Cancer* **2015**, *136*, E14–E26. [CrossRef] [PubMed]

135. De Palma, M.; Biziato, D.; Petrova, T.V. Microenvironmental regulation of tumour angiogenesis. *Nat. Rev. Cancer* **2017**, *17*, 457. [CrossRef]

136. Ponandai-Srinivasan, S.; Saare, M.; Boggavarapu, N.R.; Frisendahl, C.; Ehrström, S.; Riethmüller, C.; García-Uribe, P.A.; Rettkowski, J.; Iyengar, A.; Salumets, A.; et al. Syndecan-1 modulates the invasive potential of endometrioma via TGF-β signalling in a subgroup of women with endometriosis. *Hum. Reprod.* **2020**, *35*, 2280–2293.

137. De Rossi, G.; Vähätupa, M.; Cristante, E.; Arokiasamy, S.; Liyanage, S.E.; May, U.; Pellinen, L.; Uusitalo-Järvinen, H.; Bainbridge, J.W.; Jarvinen, T.A. Pathological Angiogenesis Requires Syndecan-4 for Efficient VEGFA (Vascular Endothelial Growth Factor A)-Induced VE-Cadherin Internalization. *Arterioscler. Thromb. Vasc. Biol.* **2021**. [CrossRef] [PubMed]

138. Onyeisi, J.O.S.; Castanho de Almeida Pernambuco Filho, P.; de Araujo Lopes, S.; Nader, H.B.; Lopes, C.C. Heparan sulfate proteoglycans as trastuzumab targets in anoikis-resistant endothelial cells. *J. Cell. Biochem.* **2019**, *120*, 13826–13840. [CrossRef] [PubMed]

139. Gialeli, C.; Theocharis, A.; Kletsas, D.; Tzanakakis, G.; Karamanos, N. Expression of matrix macromolecules and functional properties of EGF-responsive colon cancer cells are inhibited by panitumumab. *Investig. New Drugs* **2013**, *31*, 516–524. [CrossRef] [PubMed]

140. Dedes, P.; Gialeli, C.; Tsonis, A.; Kanakis, I.; Theocharis, A.; Kletsas, D.; Tzanakakis, G.; Karamanos, N. Expression of matrix macromolecules and functional properties of breast cancer cells are modulated by the bisphosphonate zoledronic acid. *Biochim. Biophys. Acta (BBA) Gen. Subj.* **2012**, *1820*, 1926–1939. [CrossRef] [PubMed]

141. Cameron, D.; Piccart-Gebhart, M.J.; Gelber, R.D.; Procter, M.; Goldhirsch, A.; de Azambuja, E.; Castro, G., Jr.; Untch, M.; Smith, I.; Gianni, L. 11 years' follow-up of trastuzumab after adjuvant chemotherapy in HER2-positive early breast cancer: Final analysis of the HERceptin Adjuvant (HERA) trial. *Lancet* **2017**, *389*, 1195–1205. [CrossRef]

142. Cobleigh, M.A.; Vogel, C.L.; Tripathy, D.; Robert, N.J.; Scholl, S.; Fehrenbacher, L.; Wolter, J.M.; Paton, V.; Shak, S.; Lieberman, G. Multinational study of the efficacy and safety of humanized anti-HER2 monoclonal antibody in women who have HER2-overexpressing metastatic breast cancer that has progressed after chemotherapy for metastatic disease. *J. Clin. Oncol.* **1999**, *17*, 2639. [CrossRef] [PubMed]

143. Green, J.R.; Müller, K.; Jaeggi, K.A. Preclinical pharmacology of CGP 42′ 446, a new, potent, heterocyclic bisphosphonate compound. *J. Bone Miner. Res.* **1994**, *9*, 745–751. [CrossRef] [PubMed]

144. Van Acker, H.H.; Anguille, S.; Willemen, Y.; Smits, E.L.; Van Tendeloo, V.F. Bisphosphonates for cancer treatment: Mechanisms of action and lessons from clinical trials. *Pharmacol. Ther.* **2016**, *158*, 24–40. [CrossRef]

145. Echtermeyer, F.; Bertrand, J.; Dreier, R.; Meinecke, I.; Neugebauer, K.; Fuerst, M.; Lee, Y.J.; Song, Y.W.; Herzog, C.; Theilmeier, G. Syndecan-4 regulates ADAMTS-5 activation and cartilage breakdown in osteoarthritis. *Nat. Med.* **2009**, *15*, 1072–1076. [CrossRef] [PubMed]

146. Godmann, L.; Bollmann, M.; Korb-Pap, A.; König, U.; Sherwood, J.; Beckmann, D.; Mühlenberg, K.; Echtermeyer, F.; Whiteford, J.; De Rossi, G. Antibody-mediated inhibition of syndecan-4 dimerisation reduces interleukin (IL)-1 receptor trafficking and signalling. *Ann. Rheum. Dis.* **2020**, *79*, 481–489. [CrossRef] [PubMed]

147. Letoha, T.; Kolozsi, C.; Ékes, C.; Keller-Pintér, A.; Kusz, E.; Szakonyi, G.; Duda, E.; Szilák, L. Contribution of syndecans to lipoplex-mediated gene delivery. *Eur. J. Pharm. Sci.* **2013**, *49*, 550–555. [CrossRef]

148. Montrose, K.; Yang, Y.; Sun, X.; Wiles, S.; Krissansen, G.W. Xentry, a new class of cell-penetrating peptide uniquely equipped for delivery of drugs. *Sci. Rep.* **2013**, *3*, 1–7. [CrossRef]

149. Yip, G.W.; Smollich, M.; Götte, M. Therapeutic value of glycosaminoglycans in cancer. *Mol. Cancer Ther.* **2006**, *5*, 2139–2148. [CrossRef] [PubMed]

150. Espinoza-Sánchez, N.A.; Götte, M. Role of cell surface proteoglycans in cancer immunotherapy. *Semin. Cancer Biol.* **2020**, *62*, 48–67. [CrossRef]

biomolecules

MDPI

Review

Syndecans and Pancreatic Ductal Adenocarcinoma

Nausika Betriu †, Juan Bertran-Mas †, Anna Andreeva and Carlos E. Semino *

Tissue Engineering Research Laboratory, Department of Bioengineering, IQS-School of Engineering, Ramon Llull University, 08017 Barcelona, Spain; nausikabetriur@iqs.url.edu (N.B.); juanbertranm@iqs.url.edu (J.B.-M.); annaandreeva@iqs.url.edu (A.A.)
* Correspondence: carlos.semino@iqs.url.edu; Tel.: +34-93-267-2107
† These authors contributed equally to this work.

Abstract: Pancreatic Ductal Adenocarcinoma (PDAC) is a fatal disease with poor prognosis because patients rarely express symptoms in initial stages, which prevents early detection and diagnosis. Syndecans, a subfamily of proteoglycans, are involved in many physiological processes including cell proliferation, adhesion, and migration. Syndecans are physiologically found in many cell types and their interactions with other macromolecules enhance many pathways. In particular, extracellular matrix components, growth factors, and integrins collect the majority of syndecans associations acting as biochemical, physical, and mechanical transducers. Syndecans are transmembrane glycoproteins, but occasionally their extracellular domain can be released from the cell surface by the action of matrix metalloproteinases, converting them into soluble molecules that are capable of binding distant molecules such as extracellular matrix (ECM) components, growth factor receptors, and integrins from other cells. In this review, we explore the role of syndecans in tumorigenesis as well as their potential as therapeutic targets. Finally, this work reviews the contribution of syndecan-1 and syndecan-2 in PDAC progression and illustrates its potential to be targeted in future treatments for this devastating disease.

Keywords: pancreatic ductal adenocarcinoma; syndecans; proteoglycans; tumor progression; angiogenesis

Citation: Betriu, N.; Bertran-Mas, J.; Andreeva, A.; Semino, C.E. Syndecans and Pancreatic Ductal Adenocarcinoma. *Biomolecules* **2021**, *11*, 349. https://doi.org/10.3390/biom11030349

Academic Editors: George Tzanakakis and Dragana Nikitovic

Received: 25 January 2021
Accepted: 22 February 2021
Published: 25 February 2021

Publisher's Note: MDPI stays neutral with regard to jurisdictional claims in published maps and institutional affiliations.

1. Introduction

The pancreas is an organ that functions as part of the gastrointestinal system. Even though it is primarily an exocrine gland, it also has an endocrine function. The endocrine pancreas is constituted of pancreatic islets, which produce and secrete insulin and glucagon hormones (among others), to regulate blood glucose levels and glucose intake by cells. The exocrine pancreas instead is constituted by the pancreatic duct and acinar cells. It is in charge of producing enzymes such as proteases, lipases, and amylases that are released into the duodenum to support nutrient digestion. Pancreatic dysfunction can lead to digestion problems as well as dysregulation of blood glucose homeostasis due to different diseases including diabetes mellitus, chronic and acute pancreatitis, and hereditary pancreatitis. Moreover, the most relevant for patient's survival is, without doubt, pancreatic cancer in the form of pancreatic ductal adenocarcinoma (PDAC). PDAC accounts for more than 90% of all pancreatic malignancies and is usually diagnosed at very advanced stages due to the lack of efficient screening tests present for early detection as well as due to its asymptomatic nature at early stages. Moreover, some of its symptoms, like abdominal pain, jaundice, and dark urine are not specific to this disease, which makes it even more difficult to diagnose [1]. The cause of pancreatic cancer remains unknown, but a few authors have recently expressed that some of the risk factors of developing PDAC are cigarette smoking, a diet based on a high intake of fat and meat, diabetes, alcohol abuse, and family history [1,2]. There are three precursors from which PDAC originates [3], namely intraductal papillary mucinous neoplasm (IPMN), mucinous cystic neoplasm (MCN) and pancreatic intraepithelial neoplasm (PanIN), the last one being the most common precursor lesion of pancreatic cancer. PanINs are non-invasive epithelial neoplasms usually located at the head of the pancreas. PanINs can be divided into PanIN-1, PanIN-2, and PanIN-3, depending on its stage. These neoplasms are considered to be the first steps before PDAC development and each one is associated with specific mutations. For example, PanIN-1 is characterized by alterations in Epithelial Growth Factor Receptor (EGFR) signaling and KRAS mutations, while PanIN-2 and -3 are characterized by the inactivation of tumor suppressor genes like CDKN2A, SMAD4, and TP53 [4].

As mentioned above, KRAS mutation is one of the first events in PanIN progression into PDAC. Constitutive activation of mutant KRas promotes plasticity of neoplastic cells and tumor development. Moreover, KRas activates signaling events such as MAPK and PI3K/Akt pathways, which regulate genes involved in cell proliferation, migration, survival/apoptosis, and metastasis [4]. These signaling pathways are normally activated through different cellular receptors such as tyrosine kinase receptors (TKRs), a receptor family which participates in a wide range of cellular processes, like cell proliferation, growth, and invasion [5]. Most of these biochemical pathways, such as cell proliferation, survival and metastasis, are possible because of the extracellular matrix (ECM). The ECM is a three-dimensional network of fibrous proteins, glycoproteins, proteoglycans, and polysaccharides with different biochemical and physical properties synthesized and secreted by stromal cells, mainly fibroblasts. Furthermore, the ECM provides structural and biochemical support for organs and tissues, epithelial cell layers as basal membrane, and individual cells as a substrate for migration [6,7]. The ECM, as well as cell–cell contact and diverse signaling molecules, serves as an information exchanger between cells forming a tissue. These interactions provide the necessary information to preserve cellular differentiation and thus create complex tissue structures. During the early stages of tumorigenesis, the ECM suffers a remodeling that supports tumor initiation and proliferation. This remodeling is caused mainly by enzymes called matrix metalloproteinases (MMP), overexpressed in most human cancers, which hydrolyze the ECM proteins [8].

Pancreatic cancer is characterized by a desmoplastic reaction, in which the deposition of abundant amounts of ECM by stromal pancreatic stellate cells (PSCs) exert biochemical and mechanical effects on PDAC cells [9]. The degree of stromal reaction predicts an aggressive phenotype, and it has been related to chemotherapeutic resistance. In fact, the accumulation of ECM components in PDAC is so high that the stroma accounts for more than 90% of the total tumor mass [10]. For this reason, the pancreatic cancer stroma has attracted the attention of many researchers in terms of targeting the ECM for the development of new PDAC treatments. However, special care should be taken to avoid the depletion of the entire ECM, as this could have dramatic consequences [10]. The degradation of the ECM components and the basal membrane implies the destruction of a physical barrier, which would allow stromal invasion by cancer cells as well as the migration of endothelial cells into the matrix, resulting in neovascularization. Moreover, during the ECM degradation, different growth factors may be released, a fact that would support the proliferation of cancer cells and tumor growth.

Another important component of the ECM which also plays a significant role in cancer cells are the proteoglycans (PGs). PGs are not only restricted to extracellular locations but are also present in cell membranes, acting as receptors to transduce extracellular signals. They are involved in functions like tissue repair and the development and maintenance of homeostatic intracellular processes. The PGs are constituted by core proteins where glycosaminoglycans (GAGs) are covalently attached. There are different types of GAGs but the most important ones are heparan sulfate (HS) and hyaluronan or hyaluronic acid (HA).

Heparan sulfate may be the most complex GAG identified until now, because of modifications and sulfations. As mentioned above, HS can be attached to a proteoglycan, forming heparan sulfate proteoglycans (HSPGs) and both are crucial for cancer initiation and progression. Importantly, HSPGs are able to modulate the interaction between growth factors and TKRs as well as the intracellular transport of extracellular vesicles carrying proteins and nucleic acids implicated in the development of cancer [11,12]. Different studies sustain the importance of these HSPGs in the development and physiology of the organisms, affecting metabolism, transport, and information transfer [13]. The major HSPGs present on the cell surface are syndecans and glypicans. Syndecans are a group of four members classified as "full-time" HSPGs because their function is restricted to the HS chains attached to the PG core protein [14]. The aim of this work is to review the interaction of syndecans with cellular receptors, their role in PDAC progression, as well as their inhibition as a potential PDAC treatment. Also, we will examine the role of syndecan-2 in promoting angiogenesis and its shedding as a key factor for angiogenesis inhibition.

2. Syndecan Structure and Interactions

Syndecans are type I glycoproteins encoded by the genes SDC1, SDC2, SDC3, and SDC4, which are expressed throughout the body, participating in different functions and pathways (Table 1) [15]. Syndecan structure can be divided into three differentiated domains, the ectodomain (the N-terminal polypeptide where GAGs are attached), a single transmembrane domain, and the C-terminal cytoplasmic domain (Figure 1a).

Table 1. Syndecans location and principal functions as transmembrane receptors.

Syndecan	Main Location	Function
Syndecan-1	Epithelial and plasma cells [16]	Cooperates with several integrins ($\alpha v\beta 3$, $\alpha v\beta 5$, $\alpha 2\beta 1$, $\alpha 3\beta 1$, and $\alpha 6\beta 4$) through the core protein. Plays roles in matrix remodeling, cell adhesion and spreading, migration, and cytoskeleton rearrangements [17]. Present in breast [18,19], prostate [20], colorectal [21], and pancreatic [22,23] cancers
Syndecan-2	Mesenchymal cells [24]	Important regulator of angiogenesis [25,26]. Present in some cancers such as PDAC [27] and colon cancer [28]. Regulates actin cytoskeleton organization, especially in lung cancer [29]
Syndecan-3	Brain, nervous system, and cartilage [30,31]	Important for brain development as well as in feeding behaviors (is upregulated in the hypothalamus in response to food deprivation] [32]. Plays roles in rheumatoid arthritis disease [33], angiogenesis [34], and HIV-1 infection [35]
Syndecan-4	Most tissues [36,37]	Plays roles in cell mechanics [38] and induces focal adhesion formation [39] and cytoskeleton organization [40,41]

The ectodomain is important for cell–cell and cell–matrix interactions via attached GAGs, and introduces variances between syndecan types, unlike the other two domains that are highly conserved. In particular, heparan sulfate (HS) and chondroitin sulfate (CS) are the GAGs present in syndecan-1 and syndecan-3, while for syndecan-2 and syndecan-4, there are exclusively HS chains [42]. As GAGs are able to bind ECM molecules, syndecans, which interact with the cytoskeleton through its cytoplasmatic domain, may promote mechanical cellular responses [43]. Moreover, syndecans cooperate with specific integrins through the core protein to mediate cell adhesion [44,45]. For example, syndecan-1 cooperates with several integrins ($\alpha v\beta 3$, $\alpha v\beta 5$, $\alpha 2\beta 1$, $\alpha 3\beta 1$, and $\alpha 6\beta 4$) [46,47], while syndecan-2 interacts with $\beta 1$ integrins [48].

The transmembrane domain is necessary for the dimerization of core proteins into homodimers, which are essential for activating cascade signals' [49]. The cytoplasmic domain is directly linked to the cytoskeleton, interacting with several kinases, and promoting different cellular functions [50]. It is divided into three subdomains (Figure 2b), two constant regions (C1 and C2) separated by a variable region (V), which is unique for every syndecan. The C1 region, the membrane-proximal domain, is associated with cytoskeletal interactions and endocytosis [51]. For example, in syndecan-2, the C1 region interacts with ezrin, an actin-associated cytoskeletal protein. Since C1 is a conserved region, the possibility exists that other syndecans could also be able to generate this interaction [52]. In syndecan-3, C1 binds to Src kinase and one of its substrates, cortactin, initiating neurite outgrowth [53]. On the other hand, the C2 region, the membrane-distal domain, interacts with PDZ binding proteins (synbindin, synectin, and syntenin) [50]. In particular, the interaction between syndecans and their adaptor protein syntenin is crucial for the biogenesis and loading of exosomes, a type of secreted vesicle involved in physiopathological processes such as cardiovascular diseases, neurodegeneration, and tumor progression. Exosomes secreted by cancer cells can contribute to tumor progression by fostering angiogenesis and the migration of tumor cells [54].

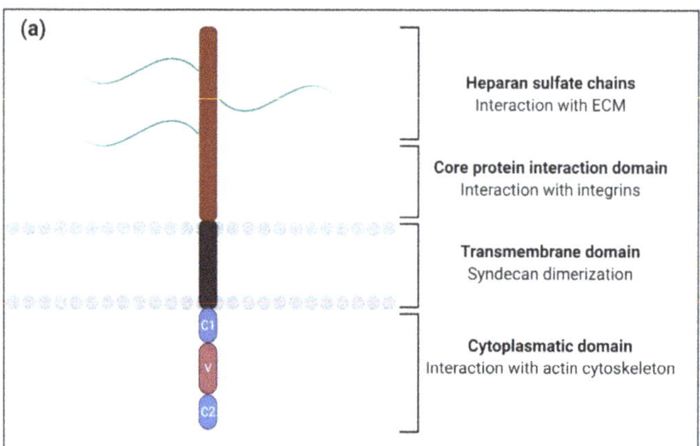

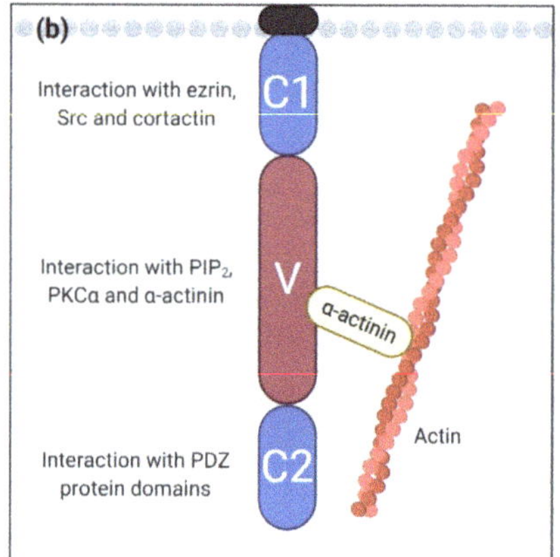

Figure 1. Syndecan structure. (**a**) Schematic representation of syndecan domains and their functions; (**b**) Close-up of the syndecan cytoplasmatic domain showing the constants (C1/C2) and the variable (V) regions and their potential interactions. The interaction between the variable region with α-actinin links syndecans to the cytoskeleton.

The V region is where most intracellular interactions between syndecans and other molecules take place. In syndecan-4, this region binds to phosphatidylinositol (4,5)-bisphosphate (PIP$_2$) and its binding promotes the activation of PKCα [55,56]. Moreover, recent studies demonstrated that the V region motif (KKXXXKK) acts as a scaffold for PKCα, regulating its localization, activity, and stability [40,57]. This binding also triggers the activation of RhoA and RhoGTPases, and through syndecan-4 cooperation, it promotes focal adhesion formation [58]. The syndecan-4-PKCα interaction might be regulated by the phosphorylation of Ser183 residue [59]. This phosphorylation decreases the affinity of syndecan-4 to PIP$_2$ and consequently to PKCα, which in turn decreases and prevents cell migration. However, this phosphorylation promotes an alternative binding of α-actinin to the V region [41,60] which creates a direct linkage between syndecan-4 and the actin cytoskeleton [38]. It has been demonstrated that interactions between the V region and α-actinin and/or PKCα are necessary for signal mechanotransduction and mechanical adaptation to force. Deletion of α-actinin and PKCα binding sites has been associated with defects in cytoskeletal organization, stress fiber assembly, and the inhibition of myofibroblast differentiation [38].

It is well known that syndecans interact with other membrane receptors, acting as co-receptors. They are able to associate with integrins, growth factor receptors (GFRs), as well as adhesion, invasion, angiogenic and migration promoters, and extracellular matrix glycoproteins and collagens [61]. An important aspect of this co-receptor behavior is the interaction with growth factors (GFs) and their respective GFRs. For example, Vascular Endothelial Growth Factor (VEGF) plays a key role in blood vessel formation and maintenance during development. Syndecans are believed to act as Vascular Endothelial Growth Factor Receptor (VEGFR) co-receptors by binding to its ligand and increasing its local concentration in the cell plasmatic membrane, facilitating their binding to VEGFR [62]. Indeed, in the absence of the HS chains that are present in syndecans, the VEGF would not be able to find some of its membrane receptors. Syndecans also promote ligand-receptor binding in the case of Fibroblast Growth Factor (FGF) and its receptor, where the O-sulfation pattern of the HS chains seems to be crucial for its binding capacity to FGF [63,64].

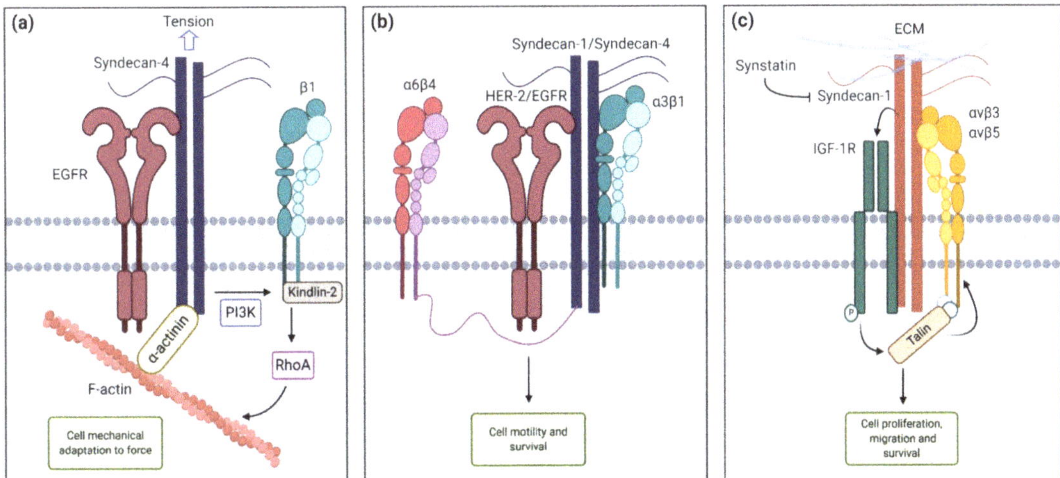

Figure 2. Syndecan functions and interactions. (**a**) Syndecan-4 cooperates with Epithelial Growth Factor Receptor (EGFR) and β1-integrin to tune cell mechanics in response to tension (**b**) Syndecan-1 and syndecan-4 interact with HER-2 and EGFR respectively, and with α6β4 and α3β1 to promote cell motility and survival (**c**) Syndecan-1 interaction with insulin-like growth factor-1 receptor (IGF1-R) activates talin, which in turn activates αvβ3 and αvβ5 integrins, leading to cell proliferation, migration, and survival. These interactions can be blocked with the synstatin peptide.

The Epithelial Growth Factor Receptor (EGFR) also interacts with the ectodomain of syndecans. In particular, Chronopoulos et al. [38] demonstrated that when inhibiting the EGFR with Gefitinib, pancreatic stellate cells (PSC) were unable to react to external tension applied to syndecan-4, preventing adaptive cell stiffening and also force-induced PI3K activation. They concluded that syndecan-4, in cooperation with EGFR and β-1 integrin, tunes cell mechanics in response to localized tension via kindlin-2 and RhoA in a PI3K-dependent manner (Figure 2a). In another work, Wang et al. [65] described, using HaCaT and MCF10A epithelial cell lines, that syndecan-1 and syndecan-4 directly interact with HER-2 and EGFR, respectively, and also capture α3β1 integrin. This ternary complex interacts with α6β4 integrin via its cytoplasmatic link with the syndecan, leading to activation and downstream signaling that promote motility and survival (Figure 2b) [65]. Other described interactions include the association between syndecan-1, the insulin-like growth factor-1 receptor (IGF1-R) and αvβ3 or αvβ5 in both human mammary carcinoma cells and endothelial cells undergoing angiogenesis. This mechanism requires syndecan-1 clustering or engaging with matrix ligands. When captured by syndecan-1, IGF1-R suffers autophosphorylation and activation. This initiates an inside-out signaling that activates talin, which in turn promotes integrin activation. This pathway can be competitively blocked by the Synstatin (SSTN$_{92-119}$), a short peptide that displaces IGF1-R and integrins from syndecan-1, preventing its activation [66–68] (Figure 2c).

Other examples include the cooperation between neuronal Thy-1, syndecan-4, and αvβ3 integrin in astrocyte cells to activate PKCα and in consequence, the RhoA pathway. This response was inhibited when SDC4 expression was silenced or a syndecan-4 mutant lacking the intracellular domain was overexpressed [69]. Syndecan-4 and α5β1 integrin also cooperates to activate PKCα in melanoma cells [70]. Moreover, the Thy-1-integrin linkage is relevant in melanoma invasion, myocyte transmigration through endothelial cells, and host defense mechanisms. In fact, the triple cooperation between Thy-1, syndecan-4, and α5β1 integrin is responsible for mediating the contractility response to mechanosignals in melanoma cells [71].

Finally, the interaction of syndecans with integrins can also be in an indirect way via intermediate receptors. For example, syndecan-2 has been described to interact with the protein tyrosine phosphatase receptor CD148 to promote Src and PI3K signaling, which in turn regulate β1 integrin-mediated adhesion processes, like angiogenesis and inflammation [48].

3. Syndecan Shedding

Syndecans are usually cleaved near the membrane by a process called shedding. The extracellular domain of the syndecans is released to the extracellular space and is converted into a soluble effector that can bind to growth factors, extracellular ligands, membrane receptors, etc. During this event, the GAGs are even more important because they interact with a great percentage of molecules in this shed-syndecan form. This highly regulated process is carried out by matrix metalloproteinases (MMPs), also known as sheddases of syndecans, which usually cleave before a hydrophobic residue. In humans, 23 different MMPs are known to exist, 14 of them expressed in vasculature. MMPs are produced by multiple tissues and cells: connective tissue, pro-inflammatory, and uteroplacental cells, fibroblasts, osteoblasts, endothelial cells, vascular smooth muscle cells, macrophages, neutrophils, lymphocytes, and cytotrophoblasts. MMPs are regulated at many levels, including at the mRNA and protein level (this last one by activation of the proenzyme to its active form), but also by growth factors. For example, VEGF-A downregulation promotes a decrease in MMP-2 expression [72], while the presence of Epithelial Growth Factor (EGF) upregulates MMP-1 and MMP-9 transcripts [73]. However, the most important regulators of MMPs are the tissue inhibitors of matrix metalloproteinases (TIMPs), which coexist in balance with MMPs. When MMPs are upregulated or TIMPs are downregulated (compared to homeostatic values), an imbalance that can lead to various pathological conditions such as heart failure, osteoarthritis, and cancer is produced [74]. MMPs activity is difficult to detect, except for during tissue repair processes such as wound healing and menstruation [75]. Some of the MMPs are excreted as proMMPs and then activated on the cell surface by various factors such as heat, low pH, chaotropic agents, and thiol-modifying agents, like N-ethylamine [74]. MMP expression or activity is influenced by hormones, GFs, and cytokines [76]. For example, in the case of menstruation, as a clear example of MMP action, the ovarian sex hormones regulate the expression of these MMPs during endometrium tissue remodeling. MMPs are important in many biological processes, such as cell proliferation, migration and invasion, ECM remodeling, and vascularization. These processes are constantly occurring in the whole body, but if not controlled, they can lead to different diseases such as cancer. MMPs participate in these pathological processes by cleaving GFs and PGs and by playing a role in ECM degradation.

Syndecans are a target for shedding by MMPs and this process has multiple effects on cell signaling (Figure 3). As extracellular soluble molecules, shed syndecans can bind ligands, preventing them from finding their transmembrane receptor. Moreover, as soluble effectors, they can start new signaling pathways through binding to other transmembrane receptors [77]. The syndecan shedding through MMPs is known to be induced by GFs [78], chemokines [79], bacterial virulence factors [80], trypsin [81], insulin (especially for patients with diabetes mellitus) [82], heparinase [83], and cell stress [74]. Intracellular mechanisms can also lead to syndecan shedding [84], as the tyrosine residues of the cytoplasmic domain are an important site of phosphorylation to induce this process [78,84]. But why and when does this shedding happen? It is known that syndecan shedding happens mainly in three main processes: wound healing, cancer, and bacterial pathogenesis.

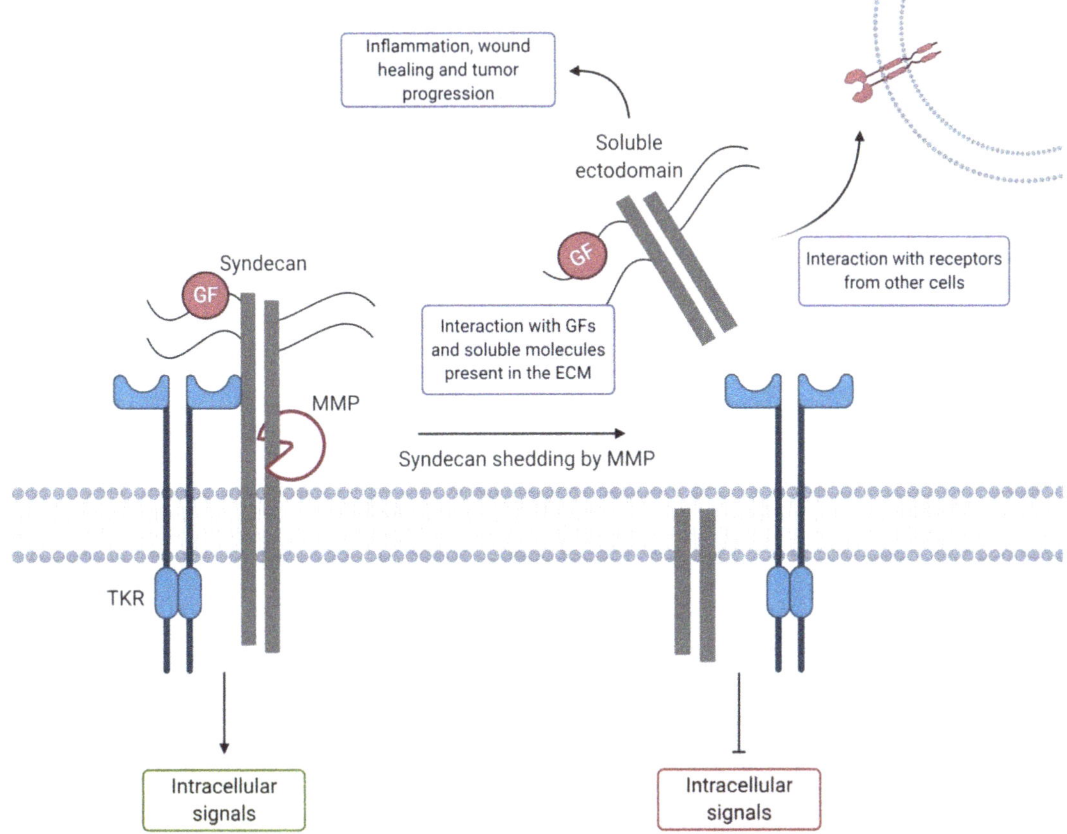

Figure 3. Syndecan shedding. Matrix metalloproteinases (MMPs] cleave syndecans near the transmembrane domain. Syndecan shedding prevents the transduction of intracellular signals that otherwise would be activated by the syndecan ectodomain. The soluble shed syndecan can bind to growth factors, extracellular matrix (ECM) substrates, and also receptors from other cells, which generate new signals or modify others. Soluble syndecans have implications in inflammation, cancer progression, angiogenesis, and wound healing.

3.1. Syndecan Shedding during Wound Healing

Wound healing is a biological process divided into three steps, inflammation, proliferation, and regeneration. Syndecans, as a major source of GAGs, control a large number of cytokines through GAG binding. For example, they are involved in leukocyte recruitment [85] and during inflammation, syndecan expression is upregulated [86]. In a recent study, the authors reported an upregulation of syndecan-2 in response to acute-induced inflammation in mice colons. This study also reports high expression levels of syndecan-2 in other inflammation-associated diseases, like colitis and sepsis, as this high-level expression is a good marker for acute inflammation and acute inflammation-associated diseases [87]. Another study using human umbilical vascular endothelial cells (HUVECs) demonstrated that syndecan-4 expression increased after inducing inflammation using lipopolysaccharides (LPS) and IL-β1 (which together mimic a bacterial infection and a general inflammatory condition). The authors also generated SDC4 knock-down endothelial cells, in which the absence of syndecan-4 in the membrane prevented the MMP from cleaving its ectodomain, delaying wound healing and tube formation [88]. This was described before using a mice model in which SDC4 knock-out (KO) induced wound healing deficiencies and impaired angiogenesis (Table 2) [89]. In another study, Chen et al. [90] reported that in MMP-7 null mice, the epithelial cells were unable to repair the wound after an injury, demonstrating the importance of syndecan-1 shedding by MMP-7 in re-epithelization after injury [90].

3.2. Syndecan Shedding during Tumor Development

Tumor development implies the participation of oncogenic genes, but also its growth, invasion, and spreading through metastasis is the result of the cooperation between GFs, integrins, TKRs, and syndecans. Syndecans can promote cell proliferation, survival, and invasion through different signaling pathways, for instance, by the activation of the KRas/MAPK pathway by syndecan-2 or the Wnt pathway by syndecan-1. These examples represent syndecans acting as membrane receptors, but as mentioned above, this type of proteoglycans can also act as soluble molecules when shed from the surface of the cell membrane. Syndecan shedding has implications in tumor progression, especially in metastasis. Soluble syndecans have attracted the attention of many researchers with the aim of establishing correlations between them and tumor progression. For example, in heparinase-induced shedding by myeloma cells, the soluble syndecan-1 promotes endothelial cell invasion and angiogenesis. This is achieved by the binding of syndecan-1 HS chains to VEGF [91]. A recent study involving breast cancer patients demonstrates that syndecan-1 levels in its soluble form increased compared to healthy patients [19]. They also report that these high levels of soluble syndecan-1 correlate with tumor size. This correlation has unknown implications, but it could serve as a serum biomarker for breast cancer. Another example is the work done by Szarvas et al. [20], in which they analyzed the levels of soluble syndecan-1 in the serum of prostate cancer patients and found out it was higher compared to the control group [20]. Although they could not explain the reason for these increased levels, they maintain that the syndecan-1 ectodomain present in the serum could be used as a biomarker to improve the prognosis of prostate cancer. Indeed, syndecan-1 can be used as a biomarker for a lot of tumors, such as lung cancer, hepatocellular carcinoma, and bladder cancer, among others [92].

3.3. Syndecan Shedding during Bacterial Pathogenesis

Some bacterial pathogens take advantage of syndecan shedding in order to inhibit the immune response of the host and enhance their own pathogenicity. For example, *Staphylococcus aureus*, a Gram-positive bacterium responsible for causing different diseases like pneumonia, toxic shock syndrome, osteomyelitis, and sepsis, generates syndecan-1 shedding through the α- and β-toxin, the latter one also being able to induce syndecan-4 shedding. Even though these toxins do not promote syndecan shedding directly, they stimulate signaling pathways that induce it, such as the activation of tyrosine kinase receptors [93]. Another example would be *Pseudomonas aeruginosa*, an opportunistic Gram-negative bacterium that induces syndecan-1 shedding through the protease LasA [94]. Therefore, considering the role of syndecan shedding in promoting bacterial pathogenesis, the authors suggest the use of shedding inhibitors to treat bacterial infections.

4. Syndecans in Cancer

In addition to playing many roles in development and signaling under physiological conditions, syndecans are also important in the progression of some malignancies. Their expression levels increase during cancer progression and therefore they can be considered good candidates for possible prognostic markers. For example, syndecan-1, also known as CD138, has been described to be present in multiple myeloma [95], breast [18,19], colorectal [21], and pancreatic carcinomas [22,23], among others. In highly metastatic breast cancer, syndecan-1 levels are higher than in those with low metastatic character, suggesting that syndecan-1 could be a remarkable biomarker for metastatic breast cancers. In particular, syndecan-1 promotes tumorigenesis via the Wnt pathway. Generation of

SDC1-null transgenic mice, crossed with Wnt1-expressing mice in the mammary gland, showed that Wnt1-induced hyperplasia was reduced by 70% and that the Wnt pathway was inhibited (Table 2) [18]. Syndecan-1 also promotes cell proliferation, adhesion, and angiogenesis in mouse embryonic fibroblasts (MEFs). When co-cultured with highly invasive carcinoma cells, those MEFs that were $SDC1^{+/+}$ enhanced the growth of carcinoma cells by 40% compared with $SDC1^{-/-}$ MEFs [96]. Syndecan-1 has also been used as a target for multiple myeloma treatment. In particular, Indatuximab ravtansine (BT062), an antibody-drug conjugate that binds to syndecan-1, has recently been successfully used in clinical trials in patients with relapsed multiple myeloma, stabilizing or improving disease in almost 80% of patients [97].

Syndecan-2 is associated with breast cancer's metastatic ability (angiogenesis and neo-vascularization), morphology, and invasion index, in part by regulating RhoGTPases [98]. It is a survival predictor for head and neck cancer [99] and is also associated with colorectal cancer, in which it has implications in the migratory behavior of highly metastatic tumor cells [28].

Syndecan-3 enhances epithelial-to-mesenchymal transition (EMT) in metastatic prostate cancer, suggesting that its attenuation, and consequently its signaling pathways, could lead to a better therapeutic outcome in prostate cancer [100]. As other syndecans, syndecan-3 has a role in angiogenesis, with a high expression in tumoral stromal vessels [101].

Finally, syndecan-4 is known to be expressed in breast cancer, regulating cell adhesion and spreading and also interacting with GFRs. For example, the estradiol–estrogen receptor complex initiates the growth and progression of hormone-dependent breast cancer, and it seems that syndecan-4 participates in this pathway as a mediator factor, being activated by the Insulin-like Growth Factor Receptor (IGFR) [102]. In addition, the expression levels of SDC4 in human breast cancer links with the FGF2R complex formation, indicating that syndecan-4 regulates this FGF2R complex formation in human breast cancer [103].

4.1. Syndecans in PDAC

4.1.1. Syndecan-1

Pancreatic ductal adenocarcinoma is one of the most lethal human pathologies, mainly because of its late detection and metastatic capacity [3,104]. It is the only gastrointestinal pathology showing syndecan-1 upregulation [22]. This overexpression correlates with cell proliferation, differentiation, and invasion for the development of PDAC. The stroma of PDAC has increased levels of syndecan-1 compared to a normal stroma. Indeed, the shift of syndecan-1 expression from the epithelium to the stroma is a poor prognostic factor in PDAC [105]. It has also been suggested that syndecan-1 performs different functions depending on its location: epithelial syndecan-1 promotes an epithelial morphology while stromal syndecan promotes tumorigenesis [105]. Heparanase (HPA) is an endoglycosidase able to specifically degrade the HS chains of syndecan-1. It promotes tumor progression and metastasis by enhancing the synthesis and shedding of syndecan-1 [83]. HPA is overexpressed in pancreatic cancer [106] and its expression has been correlated with cancer cell invasion and lymph node metastasis in PDAC patients [107,108]. Moreover, HPA modulates the response of pancreatic cancer to radiotherapy. In particular, ionizing radiation (IR) upregulates HPA expression, promoting the invasive ability of pancreatic cancer cells in vitro and in orthotopic tumors in vivo. This could be one explanation not only for tumor resistance to radiotherapy but also for its effect in enhancing tumor dissemination. Combined treatment with an HPA inhibitor and IR attenuated spreading in orthotopic pancreatic tumors in mice [109]. The HPA/syndecan-1 axis promotes the upregulation of FGF2, which in turn activates the PI3K/Akt pathway and EMT in cultured pancreatic cancer cell lines [110]. At the same time, FGF2 also promotes syndecan-1 shedding via MMP7 in Panc-1 cells [111]. Moreover, syndecan-1 shedding in this cell line is enhanced by treatment with the chemotherapeutic drugs bortezomib and doxorubicin [112]. Increased levels of shed syndecan-1 have been reported in some cancers such as breast [19] and prostate [20] and have been correlated with poor prognosis in patients with lung cancer [113] and myeloma [91]. To our knowledge, there are no in vivo studies reporting the shedding of syndecan-1 in pancreatic cancer. However, the fact that HPA, FGF2, and MMP7 are all upregulated in PDAC tissue samples [110,114,115] makes it reasonable to assume that syndecan-1 shedding could also be happening during pancreatic cancer progression. This is further supported by the fact that in pancreatic cancer, epithelial syndecan-1 is produced by the epithelial cancer cells [22], but the origin of the stromal syndecan is unknown. This stromal syndecan-1 could be produced by mesenchymal cells or could be shed from epithelial cells into the tumor stroma. It is possible that the release of syndecan-1 to the stromal compartment contributes to the pancreatic malignant phenotype by binding to growth factors that support cell proliferation, as happens in breast cancer [116,117].

The KRAS mutation is the most frequent mutation in PDAC and is believed to be an initiating step for pancreatic carcinogenesis. It seems that 95% of late-stage pancreatic cancers present with a mutated and highly overexpressed KRAS. Interestingly, syndecan-1 has been described to cooperate with KRas to induce the malignant phenotype. In particular, a recent study demonstrates that syndecan-1 expression serves as a KRas effector, inducing macropinocytosis in PDAC [23]. Macropinocytosis is a type of endocytosis, which involves the non-specific intake of extracellular material, like soluble molecules, nutrients, and antigens. The study demonstrated that in low-glutamine medium, upregulated KRas cells decreased proliferative capacity. In the presence of albumin as a substitute of glutamine, cell proliferation was rescued as albumin was incorporated into the cell by macropinocytosis. SDC1 knock-out cells reduced albumin intake capacity and consequently, reduced cell proliferation in low-glutamine conditions [23]. Thus, this new study reveals that syndecan-1 plays a crucial role in macropinocytosis in KRAS-driven pancreatic cancer. In another study, the authors investigated whether there was an association between SDC1 and KRAS expression and patient survival. They found that patients carrying KRAS somatic mutations had a higher SDC1 mRNA expression than those without mutations, and that this gene signature elevated mortality [118]. Both studies suggest that targeting KRAS and SDC1 in combination could improve patient outcomes [23,118].

4.1.2. Syndecan-2

Syndecan-2 also plays a significant role in pancreatic cancer, working as an invasive-associated gene that, as well as syndecan-1, cooperates with KRas to induce the invasive phenotype [27]. Oliveira et al. [27] reported high syndecan-2 expression levels in various pancreatic cancer cell lines, like T3M4, Panc-1, and SU8686. When SDC2 was silenced by iRNA, migrating and invading cells were reduced significantly, although cell growth was not. In a similar way, some of the KRas/MAPK signaling pathway components, like phosphorylated Src and phosphorylated ERK, were also reduced when reducing SDC2 expression levels. Therefore, they demonstrated that syndecan-2 is an important mediator in PDAC and that it cooperates with KRas to increase malignancy and perineural invasion. The upregulation of syndecan-2 indirectly interferes with the KRas/MAPK signaling pathway enhancing the mutated KRAS gene upregulation and in consequence, the Ras protein GTP-phosphorylation. In particular, p120-GAP and RACK1 proteins compete for the binding to the syndecan-2 cytoplasmatic domain [119]. RACK1 plays an important role regulating the phosphorylation of Src and preventing it when Src is associated to syndecan-2. However, when p120-GAP, instead of RACK1, binds to syndecan-2, Src is phosphorylated using this binding as a switch signal [119]. When phosphorylated, Src interacts with several receptors, G-proteins, signal transducers, and transcription molecules, resulting in biological functions involving cell proliferation, growth, and differentiation [120]. The early steps for p120-GAP binding to syndecan-2 are related to the KRAS mutation. It has been demonstrated that in pancreatic cancer cell lines with wild-type KRAS, RACK1 and syndecan-2 are interacting. Alternatively, p120-GAP binds to syndecan-2 in those cell lines with mutated KRAS [27] (Figure 4). This study shows the importance of syndecan-2 in pancreatic ductal adenocarcinoma and its cooperation with oncogenic KRAS gene to aggravate this malignant phenotype. Even though syndecan-1 and -2 participate in the regulation of different processes in PDAC, both enhance KRas signaling. This fact opens a new window in PDAC treatment, as syndecan targeting could downregulate the KRas/MAPK pathway.

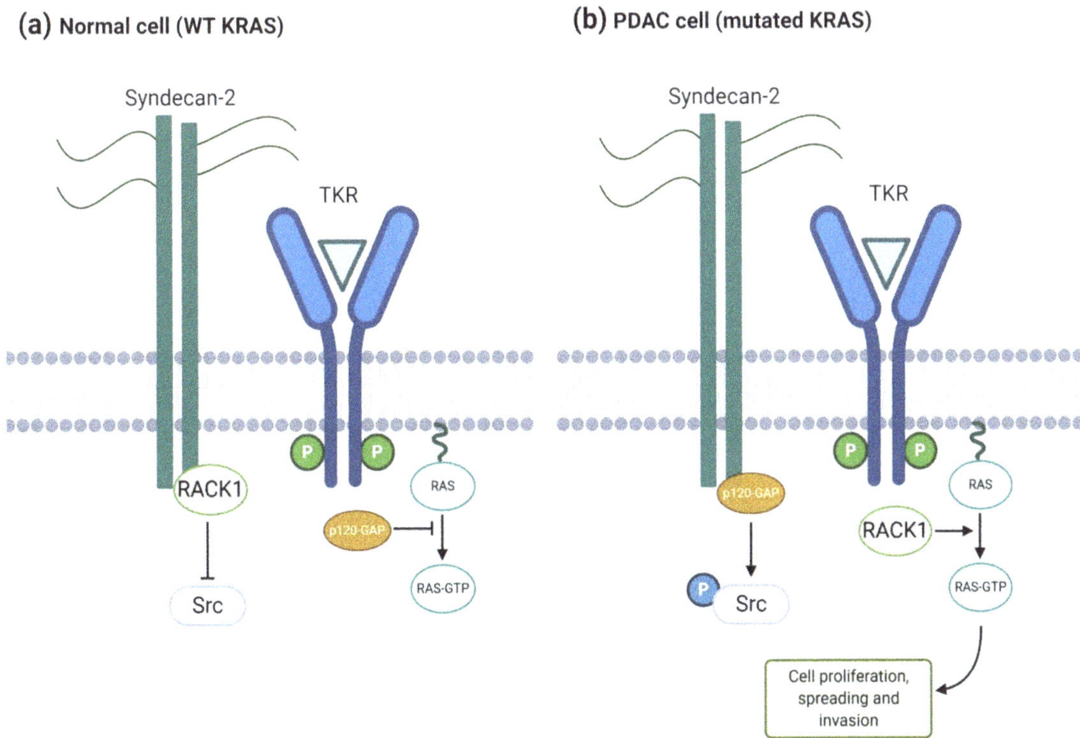

Figure 4. Syndecan-2 and KRas cooperate to induce an invasive phenotype in pancreatic cancer cells. (**a**) In normal cells, RACK1 binding to syndecan-2 prevents Src activation and free p120-GAP inhibits Ras signaling triggered by TKRs (**b**) In pancreatic ductal adenocarcinoma (PDAC) cells, p120-GAP binding to syndecan-2 activates Src and free RACK1 enhances TKR-mediated Ras activation, promoting cell proliferation, spreading, and invasion.

4.1.3. Syndecan-3 and -4

Syndecan-3 is also increased in PDAC and its expression has been positively correlated with tumor size in an orthotopic mice model [121]. It is associated with Midkine (MK), a type of neurotrophic factor that triggers different responses, like neurite outgrowth, neuronal survival, carcinogenesis, and tumor progression. It has been shown that the interaction between MK and syndecan-3 generates perineural invasion and poor prognosis [122].

Less is known about the presence and role of syndecan-4 in the pancreas. It has been detected in pancreatic islet β-cells of mice, rats, and humans, and also in the pancreatic β-cell line MIN6 [123,124]. Importantly, syndecan-4 expression was found to be negative in other pancreatic islet cells as well as in exocrine cells, suggesting its specific role in the regulation of insulin secretion in β-cells. Moreover, SDC4 mRNA expression in β-cells is transiently upregulated by IL-1β via the Src-STAT3 pathway [123]. The presence and/or dysregulation of syndecan-4 has not been specifically described in PDAC and it does not seem to be relevant. However, even if not described in pancreatic duct epithelial cells, it is present in activated cultured pancreatic stellate cells (PSCs) [38]. PSCs are responsible for producing the desmoplastic stroma in which cancer cells are embedded, and therefore, for promoting PDAC progression [125]. Syndecan-4 has been involved in focal adhesions (FAs) formation through the binding and activation of PKCα [126]. The absence of syndecan-4 not only generates smaller FAs, but also shows an impaired actin-cytoskeleton and defective smooth muscle actin incorporation [127]. Chronopoulos et al. [38] demonstrated that applying an apical force to induce a syndecan-4 response in PSCs generates an increase in talin-1 and kindlin-2 basal accumulation. These are both focal adhesion proteins that bind the integrin cytoplasmic domain, recruit cytoskeletal, and signaling proteins involved in mechanotransduction, and enhance the integrin activation for ECM binding [128]. However, this recruitment requires PI3K action. Localized force on syndecan-4 triggers PI3K activation, which enriches talin-1 and kindlin-2 concentration inducing a cell-global response, even at distal sites from the force application point. A direct link between the presence of syndecan-4 in PSCs and pancreatic cancer progression has not been described yet. However, considering the function of syndecan-4 in regulating focal adhesion formation [39] and cytoskeleton organization [40,41], and since pancreatic cancer tissue can be several folds stiffer than its healthy counterpart [129–131], it would be interesting to study the role of syndecan-4 in modulating the mechanical response to the increased tissue stiffness.

5. Syndecan-2 in Angiogenesis

Angiogenesis is the process through which new blood vessels are formed from preexisting ones. It mainly occurs during embryonic development, wound healing processes, and organism growth, in which it is highly regulated. However, angiogenesis is present in many pathologies such as rheumatoid arthritis, diabetes, tumor growth, and metastasis, acting as a non-regulated process [132]. Angiogenesis depends on the final balance between pro- and antiangiogenic molecules. In normal tissues, antiangiogenic factors predominate over the proangiogenic factors. In malignant cells, an event called the "angiogenic switch" breaks the normal equilibrium between pro- and antiangiogenic factors, boosting the angiogenic process [133]. Blood vessels have a fundamental role in tumor metastasis to other organs [134], and therefore, the inhibition of angiogenesis is a promising anti-cancer strategy.

Syndecan-2 is a key angiogenic element [25]. Its downregulation in human endothelial cells prevents angiogenesis, reducing cell adhesion and spreading [26]. In fact, syndecan-2 expression represents 70% over the total HSPGs content in human microvascular endothelial cells (HMVECs) [26]. Syndecan-2 expression is usually regulated by growth factors [135]. It has been reported that treatment of microvascular endothelial cells with FGF, VEGF, and PMA increases syndecan-2 expression, and that FGF was the growth factor that enhanced this expression the most, probably due to its co-receptor behavior in some signaling pathways [136]. Moreover, SDC2 gene silencing resulted in reduced cell proliferation and spreading, while enhancing cell migration. Angiogenesis is a multi-step process which includes endothelial cell signaling for the activation of cytoskeletal rearrangements and gene transcription. Moreover, syndecan-2 seems to be involved in this whole process by activating cell proliferation and adhesion, which are necessary for the generation of new blood vessels. Further studies demonstrated that syndecan-2 downregulation generates a decrease and reorganization of focal adhesions. This happened while actin was increasing its assembly, impairing a migratory behavior [137].

One of the principal roles of syndecan-2 in endothelial cells is the interaction with growth factors (VEGFA and FGF) and promotion of signaling pathways via their receptors, VEGFR and FGFR [26]. VEGFR2 is the main signal receptor for VEGFA$_{165}$, a well-known circulating heparin-binding VEGFA isoform. Syndecans act as co-receptors for growth factors binding, increasing the membrane local concentration and enhancing the binding to VEGFR2 [62,138]. Surprisingly, this Syndecan-2-VEGFA$_{165}$-VEGFR2 interaction complex had never been studied until Corti et al. [139] generated a global and endothelial-specific

SDC2 knock-out mouse, where this trimolecular complex association was studied [139]. While SDC2 knock-down in zebrafish caused severe problems during embryonic development [25], SDC2$^{-/-}$ mice were born alive. Deep analysis of mice development showed a reduction in retinal vessel outgrowth and decreased vascular branching. Moreover, these mice showed problems with wound healing (in which angiogenesis is a key process) (Table 2). As syndecan-4 structurally and evolutionarily is almost identical to syndecan-2 [140], the authors performed the same experiment for SDC4$^{-/-}$ mice lines. Whereas SDC2$^{-/-}$ mice showed retinal vascular problems, SDC4$^{-/-}$ did not. To reveal the importance of VEGFA$_{165}$ as the key growth factor interacting with syndecan-2, they performed a cornea assay model in which angiogenic response to growth factor addition was examined. They also studied the function of VEGFR2 and syndecan-2 in endothelial cells (ECs) and human umbilical vein endothelial cells (HUVECs). The results suggested that VEGFA$_{165}$ needs syndecan-2 cooperation to enhance VEGFR2 activation. VEGFA$_{165}$ binds to syndecan-2 ectodomain, which facilitates the posterior binding to VEGFR2, generating the syndecan-2-VEGFA$_{165}$-VEGFR2 trimolecular complex (Figure 5a). Further experiments were focused on elucidating through which syndecan domain this molecular interaction happened. They found that the complex formation was due to the binding of VEGFA$_{165}$ to the HS chains of syndecan-2, in particular the 6-O sulfations of heparan sulfate, which was previously reported [141,142]. Collectively, all these considerations demonstrate that syndecan-2 is a crucial angiogenic element, and therefore, syndecan-2 shedding could act as an angiogenic inhibitor. To demonstrate this, Whiteford et al. [143] generated an expression system in HEK293T cells in which the syndecan-2 ectodomain (S2ED) was constitutively released from the cells, and the anti-angiogenic properties of the shed syndecan-2 in a xenograft tumor model were investigated. Mice injected with empty vector cells (control) showed larger tumors compared to those mice injected with HEK293T cells expressing the released syndecan-2 ectodomain. They also demonstrated the inducing activity of TNF-α to shed syndecan-2 and that the anti-angiogenic activity of syndecan-2 was exclusively on its shed form. Moreover, 3D co-cultures of fibroblasts and HUVECs in the presence of S2ED showed a significant reduction in tubule length and branching compared to controls, demonstrating that shed syndecan-2 inhibited endothelial cells invasion through the direct inhibition of cell migration. In particular, S2ED binds CD148 receptor in endothelial cells and reduces the angiogenic response by inactivating β1-integrin (Figure 5b). It is remarkable how syndecan-2 can induce contrary cell signals. When working as a cell membrane receptor, it promotes angiogenesis acting as a co-receptor promoting the association of some ligands to bind its receptors [139]. However, when shed from the cell surface, the syndecan-2 ectodomain can trigger an anti-angiogenic pathway through the inactivation of β1-integrins driven by the association with CD148 [143].

Table 2. Mice models to study syndecan function.

Mice KO	Phenotype	Conclusion
SDC1$^{-/-}$	Reduced Wnt1-induced hyperplasia and inhibition of the Wnt pathway	Syndecan-1 promotes tumorigenesis via the Wnt pathway in breast cancer [18]
SDC2$^{-/-}$	Retinal vascular problems Deficient wound healing	Syndecan-2 is a key angiogenic element [139]
SDC4$^{-/-}$	Delayed skin wound healing and angiogenesis after injury	Syndecan-4 plays important roles in wound healing and angiogenesis [89]

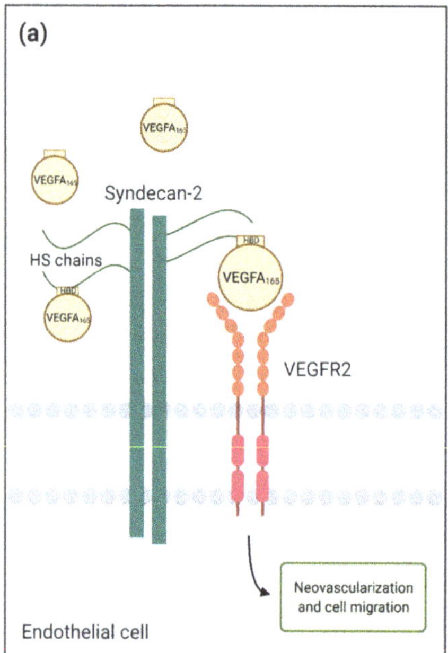
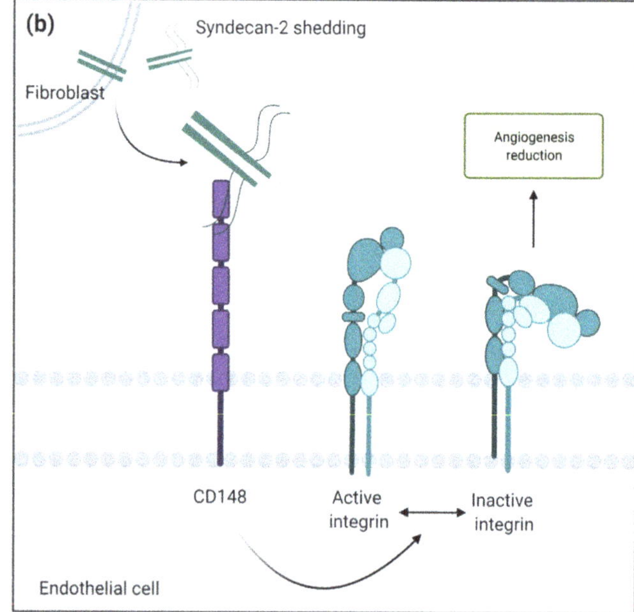

Figure 5. Roles of syndecan-2 in promoting and inhibiting angiogenesis. (**a**) Syndecan-2-VEGFA$_{165}$-VEGFR2 tricomplex promotes angiogenesis. Heparan sulfate chains present in syndecan-2 associate to VEGFA$_{165}$ through a heparin binding domain (HBD) increasing its local concentration and enhancing the binding to its receptor VEGFR2 (**b**) Syndecan-2 shedding inhibits angiogenesis. Shed syndecan-2 binds to CD148 which promotes the reduction of β1-integrin activity in endothelial cells. The inactivation of β1-integrin reduces cell migration and consequently inhibits angiogenesis.

6. Future Perspectives

Pancreatic Ductal Adenocarcinoma is a fatal malignancy with a high mortality rate and an extremely low five-year survival rate of 9% [144]. The initial diagnosis is usually done when the cancer is really advanced, leading to poor prognosis [1]. For an improvement in early PDAC diagnosis, there is a need to find novel biomarkers and to be able to treat the pancreatic cancer in the early stages. In the late 1980s, glycan carbohydrate antigen (CA 19-9) was found to be present at high concentrations in the serum of pancreatic cancer patients [145]. Nowadays, it is the most clinically used biomarker for PDAC detection [146]. Additional biomarkers such as plasma thrombospondin-2 (THBS2) are being studied. THBS2 is a promising biomarker that enables distinguishing between stages of PDAC carcinogenesis. Moreover, combination of CA 19-9 and THBS2 biomarkers improves the discrimination between PDAC patients from those with chronic pancreatitis [147].

Four genes are identified as the predominant effectors of PDAC: KRAS, CDKN2A, TP53, and SMAD4 [4]. KRAS mutation is one of the first events in PanIN progression into PDAC. Constitutive activation of KRas protein promotes tumor development through the activation of signaling events such as the MAPK and PI3K/Akt pathways [4]. Even though mutant KRAS is not a suitable target for PDAC therapy [148], the inhibition of some of its activators or mediators could serve as treatment. In this way, syndecan-1 and syndecan-2, even at different levels, both participate in KRas signaling. It has been recently described that syndecan-1 is a critical mediator of macropinocytosis in KRAS-driven pancreatic cancer [23]. Pharmacological inhibition of macropinocytosis has not been achieved to date. However, this new finding opens a window and calls for targeting syndecan-1 as a therapeutic treatment to inhibit macropinocytosis in pancreatic cancer cells, and thus preventing them from nourishing.

Because syndecan-1 is overexpressed on the surface of different cancer cells, antibody-drug conjugates are considered as an option to target cells expressing this syndecan. In particular, indatuximab ravtansine (BT062) is a monoclonal antibody linked to the cytotoxic agent DM4 (ravtansine) that targets syndecan-1-expressing cells. The internalization of the conjugate by the target cell and the subsequent DM4 release results in its activation and cytotoxic activity. Moreover, its high specificity results in low systemic toxicity. BT062 has been used in different clinical trials to target multiple myeloma (ClinicalTrials.gov (accessed on 1 March 2021), Identifier: NCT00723359 and NCT01001442) [97], triple negative metastatic breast cancer and metastatic bladder cancer (ClinicalTrialsRegister.eu (accessed on 1 March 2021), Identifier: 2013-003252-20).

Synstatin (SSTN$_{92-119}$) is a short peptide that competitively blocks the triple interaction between syndecan-1, $\alpha v\beta 3$ or $\alpha v\beta 5$ integrins, and IGF-1R, preventing the activation of these last two receptors [66–68]. Synstatin has been demonstrated to be a potent inhibitor of mammary tumor growth and angiogenesis in a xenograft mice model with no evident toxic effects [66]. Thus, it is a promising therapeutic agent for diseases that involve $\alpha v\beta 3$ or $\alpha v\beta 5$ integrins and IGF1-R. It is possible that this triple interaction between syndecan-1, $\alpha v\beta 3$ or $\alpha v\beta 5$ integrins, and IGF1-R may also be present in pancreatic cancer, since these receptors are also expressed in this kind of tumor [149,150], and if so, synstatin could also be a possible agent for the treatment of PDAC.

The SDC1 gene has been recently reported to be a target of the microRNA-494 [151]. miR-494 decreased mRNA and protein expression levels of syndecan-1 in the pancreatic cancer cell line SW1990 and inhibited EMT, cell proliferation, and migration in vitro. Moreover, it delayed tumor growth in a xenograft mouse model. This study may provide the basis for the application of miR-494 in pancreatic oncology. However, miR-494 also targets other genes such as FOXM1, SIRT1, and c-Myc [152,153], and therefore, further studies would be needed before it can be considered a therapeutic approach against pancreatic cancer.

Instead of being used as a direct target for inhibition, syndecan-1 could also be useful to target the pancreatic tumor itself. For example, syndecan-1 probes have been used to localize and image orthotopic pancreatic tumors in mice using multispectral optoacoustic tomography (MSOT) [154]. This technique promises to be useful in monitoring tumor development, location, and response to therapy, and syndecan-1 has been demonstrated to be a good target for imaging of pancreatic tumors with minimal probe accumulation in off-target organs [154]. The fact that syndecan-1 accumulates in pancreatic tumors could also be exploited for nanoparticle direction to the pancreas for drug delivery.

Syndecan-2 also has a high potential as a therapeutic target for PDAC treatment. Its expression in this type of cancer has been demonstrated [155] and its roles are related to many tumorigenic aspects. Importantly, syndecan-2 enhances pancreatic cancer cell invasion via the Ras/MAPK pathway [27]. Moreover, its association to the actin cytoskeleton confers the ability to respond to localized mechanical tensions, inducing a global response that regulates cell mechanics through different pathways that promote cell adhesion, migration, and invasion. Syndecan-2 also plays key roles in angiogenesis, promoting the binding of VEGFA$_{165}$ to its receptor VEGFR2 [139], and therefore it may be necessary for tumor vascularization. Syndecan-2 downregulation could contribute to the inhibition of tumor growth, angiogenesis, cell invasion and metastasis. Unfortunately, to date, a way to target syndecan-2 has not been described, so more research in this direction would be valuable.

It is noteworthy that most of the in vitro experiments that aim to study syndecan interactions and signaling are performed in traditional 2D cell cultures, which do not recreate the biological ECM architecture. Given the complex interactions within a tissue, important biological features may be missed if they are only studied in unnatural and constraining 2D cell cultures. It has been extensively reported that cell proliferation, morphology, behavior, and signaling are dramatically modified when culturing cells in 2D versus 3D conditions [156]. Therefore, there is a need for 3D cell models that could bridge the gap between 2D cell culture and animal models. Gagliano et al. [157] studied the effect of culturing HPAF-II, HPAC, and PL45 pancreatic ductal adenocarcinoma cell lines in 3D spheroids. They found that SDC1 expression increased in HPAF-II 3D spheroids while it decreased in HPAC spheroids and remained constant in PL45 spheroids compared to 2D cultures. The expression pattern of HPA was also cell line-dependent, while MMP7 was upregulated in 3D spheroids compared to 2D culture in all the cell lines analyzed [157]. This study highlights the importance of 3D cultures to better mimic the malignant phenotype of pancreatic cancer cells found in vivo. 3D cultures not only provide another dimension but also can contribute to mimicking the tumor microenvironment [158]. For example, in 2D cultures biomolecules diffuse away very quickly while in 3D nanofiber-based scaffolds, growth factors bind to the matrix fibers [159] and are likely to establish a local molecular gradient, which is critical for studying syndecan function as co-receptors in a closer real scenario. Therefore, the development of new 3D cancer models that better mimic not only the in vivo phenotype of pancreatic cancer cells but also its microenvironment could help to identify new signaling pathways involved in pancreatic tumorigenesis.

Author Contributions: Conceptualization, N.B., J.B.-M. and C.E.S.; writing—original draft preparation, N.B. and J.B.-M.; writing—review and editing, N.B., A.A. and C.E.S.; supervision, C.E.S. All authors have read and agreed to the published version of the manuscript.

Funding: This research was funded by the Department of Bioengineering (IQS-School of Engineering, URL) and by the Ministry of Science and Innovation (MICINN) of the Spanish Government, grant number RTI2018-096455-B-I00.

Data Availability Statement: The data presented in this study are openly available.

Acknowledgments: Figures were created with BioRender.com (accessed on 24 February 2021).

Conflicts of Interest: The authors declare no conflict of interest.

References

1. Rawla, P.; Sunkara, T.; Gaduputi, V. Epidemiology of Pancreatic Cancer: Global Trends, Etiology and Risk Factors. *World J. Oncol.* **2019**, *10*, 10–27. [CrossRef] [PubMed]
2. Yuan, C.; Morales-Oyarvide, V.; Babic, A.; Clish, C.B.; Kraft, P.; Bao, Y.; Qian, Z.R.; Rubinson, D.A.; Ng, K.M.; Giovannucci, E.L.; et al. Cigarette Smoking and Pancreatic Cancer Survival. *J. Clin. Oncol.* **2017**, *35*, 1822–1828. [CrossRef] [PubMed]
3. Goral, V. Pancreatic Cancer: Pathogenesis and Diagnosis. *Asian Pac. J. Cancer Prev.* **2015**, *16*, 5619–5624. [CrossRef]
4. Grant, T.J.; Hua, K.; Singh, A. Molecular Pathogenesis of Pancreatic Cancer. *Prog. Mol. Biol. Transl. Sci.* **2016**, *144*, 241–275. [PubMed]
5. Lemmon, M.A.; Schlessinger, J. Cell Signaling by Receptor-Tyrosine Kinases. *Cell* **2010**, *141*, 1117–1134. [CrossRef] [PubMed]
6. Alberts, B.; Bray, D.; Lewis, J.; Raff, M.; Roberts, K.; Watson, J.D. *Molecular Biology of the Cell*, 3rd ed.; W.W Norton & Co.: New York, NY, USA, 1994.
7. Hynes, R.O. The Extracellular Matrix: Not Just Pretty Fibrils. *Science* **2009**, *326*, 1216–1219. [CrossRef]
8. Klein, G.; Vellenga, E.; Fraaije, M.W.; Kamps, W.A.; de Bont, E.S.J.M. The Possible Role of Matrix Metalloproteinase (MMP)-2 and MMP-9 in Cancer, e.g., Acute Leukemia. *Crit. Rev. Oncol. Hematol.* **2004**, *50*, 87–100. [CrossRef]
9. Laklai, H.; Miroshnikova, Y.A.; Pickup, M.W.; Collisson, E.A.; Kim, E.; Barrett, A.S.; Hill, R.C.; Lakins, N.J.; Schlaepfer, D.; Mouw, J.K.; et al. Genotype Tunes Pancreatic Ductal Adenocarcinoma Tissue Tension to Induce Matricellular-Fibrosis and Tumor Progression. *Nat. Med.* **2016**, *22*, 497–505. [CrossRef] [PubMed]
10. Weniger, M.; Honselmann, K.C.; Liss, A.S. The Extracellular Matrix and Pancreatic Cancer: A Complex Relationship. *Cancers* **2018**, *10*, 316. [CrossRef] [PubMed]
11. Nagarajan, A.; Malvi, P.; Wajapeyee, N. Heparan Sulfate and Heparan Sulfate Proteoglycans in Cancer Initiation and Progression. *Front. Endocrinol.* **2018**, *9*, 483. [CrossRef]
12. Christianson, H.C.; Belting, M. Heparan Sulfate Proteoglycan as a Cell-Surface Endocytosis Receptor. *Matrix Biol.* **2014**, *35*, 51–55. [CrossRef] [PubMed]
13. Bishop, J.R.; Schuksz, M.; Esko, J.D. Heparan Sulphate Proteoglycans Fine-Tune Mammalian Physiology. *Nature* **2007**, *446*, 1030–1037. [CrossRef]
14. Knelson, E.H.; Nee, J.C.; Blobe, G.C. Heparan Sulfate Signalling in Cancer. *Trends Biochem. Sci.* **2014**, *39*, 277–288. [CrossRef] [PubMed]
15. Xian, X.; Gopal, S.; Couchman, J.R. Syndecans as Receptors and Organizers of the Extracellular Matrix. *Cell Tissue Res.* **2010**, *339*, 31–46. [CrossRef] [PubMed]
16. Saunders, S.; Jalkanen, M.; O'Farrell, S.; Bernfield, M. Molecular Cloning of Syndecan, an Integral Membrane Proteoglycan. *J. Cell Biol.* **1989**, *108*, 1547–1556. [CrossRef] [PubMed]

17. Beauvais, D.L.M.; Rapraeger, A.C. Syndecan-1-Mediated Cell Spreading Requires Signaling by αVβ3 Integrins in Human Breast Carcinoma Cells. *Exp. Cell Res.* **2003**, *286*, 219–232. [CrossRef]

18. Alexander, C.M.; Reichsman, F.; Hinkes, M.T.; Lincecum, J.; Becker, K.A.; Cumberledge, S. Syndecan-1 Is Required for Wnt-1-Induced Mammary Tumorigenesis in Mice. *Nat. Genet.* **2000**, *25*, 329–332. [CrossRef]

19. Malek-Hosseini, Z.; Jelodar, S.; Talei, A.; Ghaderi, A.; Doroudchi, M. Elevated Syndecan-1 Levels in the Sera of Patients with Breast Cancer Correlate with Tumor Size. *Breast Cancer* **2017**, *24*, 742–747. [CrossRef]

20. Szarvas, T.; Reis, H.; vom Dorp, F.; Tschirdewahn, S.; Niedworok, C.; Nyirady, P. Soluble Syndecan-1 (SDC1) Serum Level as an Independent Pre-Operative Predictor of Cancer-Specific Survival in Prostate Cancer. *Prostate* **2016**, *76*, 977–985. [CrossRef]

21. Wei, H.T.; Guo, E.N.; Dong, B.G.; Chen, L.S. Prognostic and Clinical Significance of Syndecan-1 in Colorectal Cancer: A Meta-Analysis. *BMC Gastroenterol.* **2015**, *15*, 152. [CrossRef] [PubMed]

22. Conejo, J.R.; Kleeff, J.; Koliopanos, A.; Matsuda, K.; Zhu, Z.W.; Goecke, H. Syndecan-1 Expression Is up-Regulated in Pancreatic but Not in Other Gastrointestinal Cancers. *Int. J. Cancer* **2000**, *88*, 12–20. [CrossRef]

23. Liu, J.; Yan, L.; Kapoor, A.; Hou, P.; Chen, Z.; Feng, N. Syndecan1 Is a Critical Mediator of Macropinocytosis in Pancreatic Cancer. *Nature* **2019**, *568*, 410–414.

24. Marynen, P.; Zhang, J.; Cassiman, J.J.; Van den Berghe, H.; David, G. Partial Primary Structure of the 48- and 90-Kilodalton Core Proteins of Cell Surface-Associated Heparan Sulfate Proteoglycans of Lung Fibroblasts. Prediction of an Integral Membrane Domain and Evidence for Multiple Distinct Core Proteins at the Cell Surfa. *J. Biol. Chem.* **1989**, *264*, 7017–7024. [CrossRef]

25. Chen, E.; Hermanson, S.; Ekker, S.C. Syndecan-2 Is Essential for Angiogenic Sprouting During Zebrafish Development. *Blood* **2004**, *103*, 1710–1719. [CrossRef]

26. Noguer, O.; Villena, J.; Lorita, J.; Vilaró, S.; Reina, M. Syndecan-2 Downregulation Impairs Angiogenesis in Human Microvascular Endothelial Cells. *Exp. Cell Res.* **2009**, *315*, 795–808. [CrossRef] [PubMed]

27. De Oliveira, T.; Abiatari, I.; Raulefs, S.; Sauliunaite, D.; Erkan, M.; Kong, B.; Fries, H.; Michalski, C.W.; Kleeff, J. Syndecan-2 Promotes Perineural Invasion and Cooperates With K-Ras to Induce an Invasive Pancreatic Cancer Cell Phenotype. *Mol. Cancer* **2012**, *11*, 19. [CrossRef] [PubMed]

28. Vicente, C.M.; Ricci, R.; Nader, H.B.; Toma, L. Syndecan-2 Is Upregulated in Colorectal Cancer Cells Through Interactions with Extracellular Matrix Produced by Stromal Fibroblasts. *BMC Cell Biol.* **2013**, *14*, 25. [CrossRef]

29. Munesue, S.; Kusano, Y.; Oguri, K.; Itano, N.; Yoshitomi, Y.; Nakanishi, H.; Yamashina, I.; Okayama, M. The Role of Syndecan-2 in Regulation of Actin-Cytoskeletal Organization of Lewis Lung Carcinoma-Derived Metastatic Clones. *Biochem. J.* **2002**, *363*, 201–209. [CrossRef] [PubMed]

30. Carey, D.J.; Evans, D.M.; Stahl, R.C.; Asundi, V.K.; Conner, K.J.; Garbes, P. Molecular Cloning and Characterization of N-Syndecan, a Novel Transmembrane Heparan Sulfate Proteoglycan. *J. Cell Biol.* **1992**, *117*, 191–201. [CrossRef]

31. Gould, S.E.; Upholt, W.B.; Kosher, R.A. Syndecan 3: A Member of the Syndecan Family of Membrane-Intercalated Proteoglycans That Is Expressed in High Amounts at the Onset of Chicken Limb Cartilage Differentiation. *Proc. Natl. Acad. Sci. USA* **1992**, *89*, 3271–3275. [CrossRef]

32. Reizes, O.; Lincecum, J.; Wang, Z.; Goldberger, O.; Huang, L.; Kaksonen, M. Transgenic Expression of Syndecan-1 Uncovers a Physiological Control of Feeding Behavior by Syndecan-3. *Cell* **2001**, *106*, 105–116. [CrossRef]

33. Patterson, A.M.; Cartwright, A.; David, G.; Fitzgeraldm, O.; Bresnihan, B.; Ashton, B.A.; Middleton, J. Differential Expression of Syndecans and Glypicans in Chronically Inflamed Synovium. *Ann. Rheum. Dis.* **2008**, *67*, 592–601. [CrossRef] [PubMed]

34. Arokiasamy, S.; Balderstone, M.J.M.; De Rossi, G.; Whiteford, J.R. Syndecan-3 in Inflammation and Angiogenesis. *Front. Immunol.* **2020**, *10*, 1–7. [CrossRef] [PubMed]

35. De Witte, L.; Bobardt, M.; Chatterji, U.; Degeest, G.; David, G.; Geijtenbeek, T.B.H. Syndecan-3 Is a Dendritic Cell-Specific Attachment Receptor For HIV-1. *Proc. Natl. Acad. Sci. USA* **2007**, *104*, 19464–19469. [CrossRef] [PubMed]

36. David, G.; van der Schueren, B.; Marynen, P.; Cassiman, J.J.; van den Berghe, H. Molecular Cloning of Amphiglycan, a Novel Integral Membrane Heparan Sulfate Proteoglycan Expressed by Epithelial and FI-Broblastic Cells. *J. Cell Biol.* **1992**, *118*, 961–969. [CrossRef]

37. Kojima, T.; Shworak, N.W.; Rosenberg, R.D. Molecular Cloning and Expression of Two Distinct cDNA-Encoding Heparan Sulfate Proteoglycan Core Proteins from a Rat en-Dothelial Cell Line. *J. Biol. Chem.* **1992**, *267*, 4870–4877. [CrossRef]

38. Chronopoulos, A.; Thorpe, S.D.; Cortes, E.; Lachowski, D.; Rice, A.J.; Mykuliak, V.V. Syndecan-4 Tunes Cell Mechanics by Activating the Kindlin-Integrin-RhoA Pathway. *Nat. Mater.* **2020**, *19*, 669–678. [CrossRef] [PubMed]

39. Couchman, J.R. Syndecans: Proteoglycan Regulators of Cell-Surface Microdomains? *Nat. Rev. Mol. Cell Biol.* **2003**, *4*, 926–937. [CrossRef]

40. Keum, E.; Kim, Y.; Kim, J.; Kwon, S.; Lim, Y.; Han, I. Syndecan-4 Regulates Localization, Activity and Stability of Protein Kinase C-α. *Biochem. J.* **2004**, *378*, 1007–1014. [CrossRef]

41. Greene, D.K.; Tumova, S.; Couchman, J.R.; Woods, A. Syndecan-4 Associates With α-Actinin. *J. Biol. Chem.* **2003**, *278*, 7617–7623. [CrossRef]

42. Afratis, N.; Gialeli, C.; Nikitovic, D.; Tsegenidis, T.; Karousou, E.; Theocharis, A.D.; Pavao, M.S.; Tzanakakis, G.N.; Karamanos, N.K. Glycosaminoglycans: Key Players in Cancer Cell Biology and Treatment. *FEBS J.* **2012**, *279*, 1177–1197. [CrossRef]

43. Yoneda, A.; Couchman, J.R. Regulation of Cytoskeletal Organization by Syndecan Transmembrane Proteoglycans. *Matrix Biol.* **2003**, *22*, 25–33. [CrossRef]

44. Whiteford, J.R.; Behrends, V.; Kirby, H.; Kusche-Gullberg, M.; Muramatsu, T.; Couchman, J.R. Syndecans Promote Integrin-Mediated Adhesion of Mesenchymal Cells in Two Distinct Pathways. *Exp. Cell. Res.* **2007**, *313*, 3902–3913. [CrossRef]

45. Roper, J.A.; Williamson, R.C.; Bass, M.D. Syndecan and Integrin Interactomes: Large Complexes in Small Spaces. *Curr Opin. Struct Biol.* **2012**, *22*, 583–590. [CrossRef]

46. Beauvais, D.L.M.; Burbach, B.J.; Rapraeger, A.C. The Syndecan-1 Ectodomain Regulates αVβ3 Integrin Activily in Human Mammary Carcinoma Cells. *J. Cell Biol.* **2004**, *167*, 171–181. [CrossRef] [PubMed]

47. McQuade, K.J.; Beauvais, D.L.M.; Burbach, B.J.; Rapraeger, A.C. Syndecan-1 Regulates αvβ5 Integrin Activity in B82L Fibroblasts. *J. Cell Sci.* **2006**, *119*, 2445–2456. [CrossRef]

48. Whiteford, J.R.; Xian, X.; Chaussade, C.; Vanhaesebroeck, B.; Nourshargh, S.; Couchman, J.R. Syndecan-2 Is a Novel Ligand for the Protein Tyrosine Phosphatase Receptor CD148. *Mol. Biol. Cell* **2011**, *22*, 3609–3624. [CrossRef] [PubMed]

49. Choi, S.; Lee, E.; Kwon, S.; Park, H.; Yi, J.Y.; Kim, S.; Han, I.E.; Yun, Y.; Oh, E.S. Transmembrane Domain-Induced Oligomerization Is Crucial for the Functions of Syndecan-2 and Syndecan-4. *J. Biol. Chem.* **2005**, *280*, 42573–42579. [CrossRef] [PubMed]

50. Multhaupt, H.A.; Yoneda, A.; Whiteford, J.R.; Oh, E.S.; Lee, W.; Couchman, J.R. Syndecan Signaling: When, Where and Why? *J. Physiol. Pharmacol.* **2009**, *60*, 31–38. [PubMed]

51. Kinnunen, A.; Kinnunen, T.; Kaksonen, M.; Nolo, R.; Panula, P.; Rauvala, H. N-syndecan and HB-GAM (Heparin-Binding Growth-Associated Molecule) associate with early axonal tracts in the rat brain. *Eur. J. Neurosci.* **1998**, *10*, 635–648. [CrossRef]

52. Granés, F.; Berndt, C.; Roy, C.; Mangeat, P.; Reina, M.; Vilaró, S. Identification of a Novel Ezrin-Binding Site in Syndecan-2 Cytoplasmic Domain. *FEBS Lett.* **2003**, *547*, 212–216. [CrossRef]

53. Kinnunen, T.; Kaksonen, M.; Saarinen, J.; Kalkkinen, N.; Peng, H.B.; Rauvala, H. Cortactin-Src Kinase Signaling Pathway Is Involved in N-Syndecan- Dependent Neurite Outgrowth. *J. Biol. Chem.* **1998**, *273*, 10702–10708. [CrossRef]

54. Zhang, H.G.; Grizzle, W.E. Exosomes: A Novel Pathway of Local and Distant Intercellular Communication That Facilitates the Growth and Metastasis of Neoplastic Lesions. *Am. J. Pathol.* **2014**, *184*, 28–41. [CrossRef] [PubMed]

55. Horowitz, A.; Murakami, M.; Gao, Y.; Simons, M. Phosphatidylinositol-4,5-Bisphosphate Mediates the Interaction of Syndecan-4 With Protein Kinase C. *Biochemistry* **1999**, *38*, 15871–15877. [CrossRef]

56. Whiteford, J.R.; Ko, S.; Lee, W.; Couchman, J.R. Structural and Cell Adhesion Properties of Zebrafish Syndecan-4 Are Shared with Higher Vertebrates. *J. Biol. Chem.* **2008**, *283*, 29322–29330. [CrossRef]

57. Lim, S.T.; Longley, R.L.; Couchman, J.R.; Woods, A. Direct Binding of Syndecan-4 Cytoplasmic Domain to the Catalytic Domain of Protein Kinase Cα (PKCα) Increases Focal Adhesion Localization of PKCα. *J. Biol Chem.* **2003**, *278*, 13795–13802. [CrossRef] [PubMed]

58. Dovas, A.; Choi, Y.; Yoneda, A.; Multhaupt, H.A.B.; Kwon, S.H.; Kang, D. Serine 34 Phosphorylation of Rho Guanine Dissociation Inhibitor (RhoGDIα) Links Signaling from Conventional Protein Kinase C to RhoGTPase in Cell Adhesion. *J. Biol. Chem.* **2010**, *285*, 23296–23308. [CrossRef]

59. Koo, B.K.; Jung, Y.S.; Shin, J.; Han, I.; Mortier, E.; Zimmermann, P. Structural Basis of Syndecan-4 Phosphorylation as a Molecular Switch to Regulate Signaling. *J. Mol. Biol.* **2006**, *355*, 651–663. [CrossRef] [PubMed]

60. Choi, Y.; Kim, S.; Lee, J.; Ko, S.G.; Lee, W.; Han, I.O.; Woods, A.; Oh, E.S. The Oligomeric Status of Syndecan-4 Regulates Syndecan-4 Interaction With α-Actinin. *Eur. J. Cell Biol.* **2008**, *87*, 807–815. [CrossRef]

61. Couchman, J.R. Transmembrane Signaling Proteoglycans. *Annu. Rev. Cell Dev. Biol.* **2010**, *26*, 89–114. [CrossRef]

62. Gitay-Goren, H.; Soker, S.; Vlodavsky, I.; Neufeld, G. The Binding of Vascular Endothelial Growth Factor to Its Receptors Is Dependent on Cell Surface-Associated Heparin-Like Molecules. *J. Biol. Chem.* **1992**, *267*, 6093–6098. [CrossRef]

63. Allen, B.L.; Filla, M.S.; Rapraeger, A.C. Role of Heparan Sulfate as a Tissue-Specific Regulator of FGF-4 and FGF Receptor Recognition. *J. Cell. Biol.* **2001**, *155*, 845–857. [CrossRef]

64. Sarrazin, S.; Lamanna, W.C.; Esko, J.D. Heparan Sulfate Proteoglycans. *Cold Spring Harb. Perspect. Biol.* **2011**, *3*, 1–33. [CrossRef]

65. Wang, H.; Jin, H.; Rapraeger, A.C. Syndecan-1 and Syndecan-4 Capture Epidermal Growth Factor Receptor Family Members and the α3β1 Integrin via Binding Sites in Their Ecto-Domains: Novel Synstatins Prevent Kinase Capture and Inhibitα6β4-Integrindependent Epithelial Cell Motility. *J. Biol. Chem.* **2015**, *290*, 26103–26113. [CrossRef]

66. Beauvais, D.M.; Ell, B.J.; McWhorter, A.R.; Rapraeger, A.C. Syndecan-1 Regulates αvβ3 and αvβ5 Integrin Activation During Angiogenesis and Is Blocked by Synstatin, a Novel Peptide Inhibitor. *J. Exp. Med.* **2009**, *206*, 691–705. [CrossRef] [PubMed]

67. Beauvais, D.L.M.; Rapraeger, A.C. Syndecan-1 Couples the Insulin-Like Growth Factor-1 Receptor to Inside-Out Integrin Activation. *J. Cell Sci.* **2010**, *123*, 3796–3807. [CrossRef]

68. Rapraeger, A.C. Synstatin: A Selective Inhibitor of the Syndecan-1-Coupled IGF1R-αvβ3 Integrin Complex in Tumorigenesis and Angiogenesis. *FEBS J.* **2013**, *280*, 2207–2215. [CrossRef]

69. Avalos, A.M.; Valdivia, A.D.; Muñoz, N.; Herrera-Molina, R.; Tapia, J.C.; Lavandero, S. Neuronal Thy-1 Induces Astrocyte Adhesion by Engaging Syndecan-4 in a Cooperative Interaction With αVβ3 Integrin That Activates PKCα and RhoA. *J. Cell Sci.* **2009**, *122*, 3462–3471. [CrossRef]

70. Mostafavi-Pour, Z.; Askari, J.A.; Parkinson, S.J.; Parker, P.J.; Ng, T.T.; Humphries, M.J. Integrin-Specific Signaling Pathways Controlling Focal Adhesion Formation and Cell Migration. *J. Cell Biol.* **2003**, *161*, 155–167. [CrossRef] [PubMed]

71. Fiore, V.F.; Ju, L.; Chen, Y.; Zhu, C.; Barker, T.H. Dynamic Catch of a Thy-1-α5 β1 + Syndecan-4 Trimolecular Complex. *Nat. Commun.* **2014**, *5*. [CrossRef] [PubMed]

72. Mao, D.; Zhang, Y.; Lu, H.; Zhang, H. Molecular Basis Underlying Inhibition of Metastasis of Gastric Cancer by Anti-VEGFa Treatment. *Tumor Biol.* **2014**, *35*, 8217–8223. [CrossRef]
73. Rao, V.H.; Kansal, V.; Stoupa, S.; Agrawal, D.K. MMP-1 and MMP-9 Regulate Epidermal Growth Factor-Dependent Collagen Loss in Human Carotid Plaque Smooth Muscle Cells. *Physiol. Rep.* **2014**, *2*, e00224. [CrossRef]
74. Cui, N.; Hu, M.; Khalil, R.A. Biochemical and Biological Attributes of Matrix Metalloproteinases. *Prog. Mol. Biol. Transl. Sci.* **2017**, *147*, 1–73.
75. Nagase, H.; Visse, R.; Murphy, G. Structure and Function of Matrix Metalloproteinases and TIMPs. *Cardiovasc. Res.* **2006**, *69*, 562–573. [CrossRef]
76. Visse, R.; Nagase, H. Matrix Metalloproteinases and Tissue Inhibitors of Metalloproteinases: Structure, Function, and Biochemistry. *Circ. Res.* **2003**, *92*, 827–839. [CrossRef]
77. Gopal, S. Syndecans in Inflammation at a Glance. *Front. Immunol.* **2020**, *11*, 1–8. [CrossRef]
78. Subramanian, S.V.; Fitzgerald, M.L.; Bernfield, M. Regulated Shedding of Syndecan-1 and-4 Ectodomains by Thrombin and Growth Factor Receptor Activation. *J. Biol. Chem.* **1997**, *272*, 14713–14720. [CrossRef]
79. Brule, S.; Charnaux, N.; Sutton, A.; Ledoux, D.; Chaigneau, T.; Saffar, L.; Gattegno, L. The Shedding of Syndecan-4 and Syndecan-1 From Hela Cells and Human Primary Macrophages Is Accelerated by SDF-1/CXCL12 and Mediated by the Matrix Metalloproteinase-9. *Glycobiology* **2006**, *16*, 488–501. [CrossRef] [PubMed]
80. Chen, Y.; Hayashida, A.; Bennett, A.E.; Hollingshead, S.K.; Pyong, W.P. Streptococcus Pneumoniae Sheds Syndecan-1 Ectodomains Through ZmpC, a Metalloproteinase Virulence Factor. *J. Biol. Chem.* **2007**, *282*, 159–167. [CrossRef]
81. Jalkanen, M.; Rapraeger, A.; Saunders, S.; Bernfield, M. Cell Surface Proteoglycan of Mouse Mammary Epithelial Cells Is Shed by Cleavage of Its Matrix-Binding Ectodomain From Its Membrane-Associated Domain. *J. Cell. Biol.* **1987**, *105*, 3087–3096. [CrossRef]
82. Wang, J.B.; Guan, J.; Shen, J.; Zhou, L.; Zhang, Y.J.; Si, Y.F.; Yang, L.; Jian, X.-h.; Sheng, Y. Insulin Increases Shedding of Syndecan-1 in the Serum of Patients with Type 2 Diabetes Mellitus. *Diabetes Res. Clin. Pr.* **2009**, *86*, 83–88. [CrossRef]
83. Yang, Y.; MacLeod, V.; Miao, H.Q.; Theus, A.; Zhan, F.; Shaughnessy, J.D.; Sawyer, J.; Li, J.P.; Zcharia, E.; Vlodavsky, I.; et al. Heparanase Enhances Syndecan-1 Shedding: A Novel Mechanism for Stimulation of Tumor Growth and Metastasis. *J. Biol. Chem.* **2007**, *282*, 13326–13333. [CrossRef] [PubMed]
84. Manon-Jensen, T.; Itoh, Y.; Couchman, J.R. Proteoglycans in Health and Disease: The Multiple Roles of Syndecan Shedding. *FEBS J.* **2010**, *277*, 3876–3889. [CrossRef]
85. Götte, M.; Joussen, A.M.; Klein, C.; Andre, P.; Wagner, D.D.; Hinkes, M.T. Role of Syndecan-1 in Leukocyte-Endothelial Interactions in the Ocular Vasculature. *Investig. Ophthalmol. Vis. Sci.* **2002**, *43*, 1135–1141.
86. Götte, M. Syndecans in Inflammation. *FASEB J.* **2003**, *17*, 575–591. [CrossRef] [PubMed]
87. Hong, H.; Song, H.K.; Hwang, E.S.; Lee, A.R.; Han, D.S.; Kim, S.E. Up-Regulation of Syndecan-2 in Proximal Colon Correlates with Acute Inflammation. *FASEB J.* **2019**, *33*, 11381–11395. [CrossRef] [PubMed]
88. Vuong, T.T.; Reine, T.M.; Sudworth, A.; Jenssen, T.G.; Kolset, S.O. Syndecan-4 Is a Major Syndecan in Primary Human Endothelial Cells In Vitro, Modulated by Inflammatory Stimuli and Involved in Wound Healing. *J. Histochem. Cytochem.* **2015**, *63*, 280–292. [CrossRef] [PubMed]
89. Echtermeyer, F.; Streit, M.; Wilcox-Adelman, S.; Saoncella, S.; Denhez, F.; Detmar, M. Delayed Wound Repair and Impaired Angiogenesis in Mice Lacking Syndecan-4. *J Clin Investig.* **2001**, *107*, 9–14. [CrossRef]
90. Chen, P.; Abacherli, L.E.; Nadler, S.T.; Wang, Y.; Li, Q.; Parks, W.C. MMP7 Shedding of Syndecan-1 Facilitates Re-Epithelialization by Affecting α2β1 Integrin Activation. *PLoS ONE* **2009**, *4*, e6565. [CrossRef] [PubMed]
91. Purushothaman, A.; Uyama, T.; Kobayashi, F.; Yamada, S.; Sugahara, K.; Rapraeger, A.C. Heparanase-Enhanced Shedding of Syndecan-1 by Myeloma Cells Promotes Endothelial Invasion and Angiogenesis. *Blood* **2010**, *115*, 2449–2457. [CrossRef] [PubMed]
92. Bertrand, J.; Bollmann, M. Soluble Syndecans: Biomarkers for Diseases and Therapeutic Options. *Br. J. Pharmacol.* **2019**, *176*, 67–81. [CrossRef]
93. Park, P.W.; Foster, T.J.; Nishi, E.; Duncan, S.J.; Klagsbrun, M.; Chen, Y. Activation of Syndecan-1 Ectodomain Shedding by Staphylococcus aureus α-Toxin and β-Toxin. *J. Biol. Chem.* **2004**, *279*, 251–258. [CrossRef]
94. Haynes, A.; Ruda, F.; Oliver, J.; Hamood, A.N.; Griswold, J.A.; Park, P.W. Syndecan 1 Shedding Contributes to Pseudomonas Aeruginosa Sepsis. *Infect Immun.* **2005**, *73*, 7914–7921. [CrossRef]
95. Sanderson, R.D.; Yang, Y. Syndecan-1: A Dynamic Regulator of the Myeloma Microenvironment. *Clin. Exp. Metastasis* **2008**, *25*, 149–159. [CrossRef] [PubMed]
96. Maeda, T.; Alexander, C.M.; Friedl, A. Induction of Syndecan-1 Expression in Stromal Fibroblasts Promotes Proliferation of Human Breast Cancer Cells. *Cancer Res.* **2004**, *64*, 612–621. [CrossRef]
97. Jagannath, S.; Heffner, L.T.; Ailawadhi, S.; Munshi, N.C.; Zimmerman, T.M.; Rosenblatt, J. Indatuximab Ravtansine (BT062) Monotherapy in Patients with Relapsed and/or Refractory Multiple Myeloma. *Clin. Lymphoma Myeloma Leuk.* **2019**, *19*, 372–380. [CrossRef]
98. Lim, H.C.; Couchman, J.R. Syndecan-2 Regulation of Morphology in Breast Carcinoma Cells Is Dependent on RhoGTPases. *Biochim. Biophys. Acta Gen. Subj.* **2014**, *1840*, 2482–2490. [CrossRef]
99. Farnedi, A.; Rossi, S.; Bertani, N.; Gulli, M.; Silini, E.M.; Mucignat, M.T.; Poli, T.; Sesenna, E.; Lanfranco, D.; Montebugnoli, L.; et al. Proteoglycan-Based Diversification of Disease Outcome in Head and Neck Cancer Patients Identifies NG2/CSPG4 and Syndecan-2 as Unique Relapse and Overall Survival Predicting Factors. *BMC Cancer* **2015**, *15*, 352. [CrossRef]

100. Diamantopoulou, Z.; Kitsou, P.; Menashi, S.; Courty, J.; Katsoris, P. Loss of Receptor Protein Tyrosine Phosphatase β/ζ (RPTPβ/ζ) Promotes Prostate Cancer Metastasis. *J. Biol. Chem.* **2012**, *287*, 40339–40349. [CrossRef] [PubMed]
101. Roskams, T.; De Vos, R.; David, G.; Van Damme, B.; Desmet, V. Heparan Sulphate Proteoglycan Expression in Human Primary Liver Tumours. *J. Pathol.* **1998**, *185*, 290–297. [CrossRef]
102. Tsonis, A.I.; Afratis, N.; Gialeli, C.; Ellina, M.I.; Piperigkou, Z.; Skandalis, S.S. Evaluation of the Coordinated Actions of Estrogen Receptors with Epidermal Growth Factor Receptor and Insulin-Like Growth Factor Receptor in the Expression of Cell Surface Heparan Sulfate Proteoglycans and Cell Motility in Breast Cancer Cells. *FEBS J.* **2013**, *280*, 2248–2259. [CrossRef] [PubMed]
103. Mundhenke, C.; Meyer, K.; Drew, S.; Friedl, A. Heparan Sulfate Proteoglycans as Regulators of Fibroblast Growth Factor-2 Receptor Binding in Breast Carcinomas. *Am. J. Pathol.* **2002**, *160*, 185–194. [CrossRef]
104. Storz, P.; Crawford, H.C. Carcinogenesis of Pancreatic Ductal Adenocarcinoma. *Gastroenterology* **2020**, *158*, 2072–2081. [CrossRef]
105. Juuti, A.; Nordling, S.; Lundin, J.; Louhimo, J.; Haglund, C. Syndecan-1 Expression—A Novel Prognostic Marker in Pancreatic Cancer. *Oncology* **2005**, *68*, 97–106. [CrossRef]
106. Koliopanos, A.; Friess, H.; Kleeff, J.; Shi, X.; Liao, Q.; Pecker, I. Heparanase Expression in Primary and Metastatic Pancreatic Cancer. *Cancer Res.* **2001**, *61*, 4655–4659. [PubMed]
107. Rohloff, J.; Zinke, J.; Schoppmeyer, K.; Tannapfel, A.; Witzigmann, H.; Mössner, J. Heparanase Expression Is a Prognostic Indicator for Postoperative Survival in Pancreatic Adenocarcinoma. *Br. J. Cancer* **2002**, *86*, 1270–1275. [CrossRef]
108. Hoffmann, A.C.; Mori, R.; Vallbohmer, D.; Brabender, J.; Drebber, U.; Baldus, S.E. High Expression of Heparanase Is Significantly Associated with Dedifferentiation and Lymph Node Metastasis in Patients With Pancreatic Ductal Adenocarcinomas and Correlated to PDGFA and via HIF1a to HB-EGF and bFGF. *J. Gastrointest. Surg.* **2008**, *12*, 1674–1681. [CrossRef] [PubMed]
109. Meirovitz, A.; Hermano, E.; Lerner, I.; Zcharia, E.; Pisano, C.; Peretz, T.; Elkin, M. Role of Heparanase in Radiation-Enhanced Invasiveness of Pancreatic Carcinoma. *Cancer Res.* **2011**, *71*, 2772–2780. [CrossRef]
110. Chen, X.; Zhao, H.; Chen, C.; Li, J.; He, J.; Fu, X. The HPA/SDC1 Axis Promotes Invasion and Metastasis of Pancreatic Cancer Cells by Activating EMT via FGF2 Upregulation. *Oncol. Lett.* **2019**, *19*, 211–220. [CrossRef]
111. Ding, K.; Lopez-Burks, M.; Sánchez-Duran, J.A.; Korc, M.; Lander, A.D. Growth Factor-Induced Shedding of Syndecan-1 Confers Glypican-1 Dependence on Mitogenic Responses of Cancer Cells. *J. Cell Biol.* **2005**, *171*, 729–738. [CrossRef]
112. Ramani, V.C.; Sanderson, R.D. Chemotherapy Stimulates Syndecan-1 Shedding: A Potentially Negative Effect of Treatment That May Promote Tumor Relapse. *Matrix Biol.* **2014**, *35*, 215–222. [CrossRef] [PubMed]
113. Joensuu, H.; Anttonen, A.; Eriksson, M.; Mäkitaro, R.; Alfthan, H.; Kinnula, V.; Sirpa, L. Soluble Syndecan-1 and Serum Basic Fibroblast Growth Factor Are New Prognostic Factors in Lung Cancer. *Cancer Res.* **2002**, *62*, 5210–5217. [PubMed]
114. Yamamoto, H.; Itoh, F.; Iku, S.; Adachi, Y.; Fukushima, H.; Sasaki, S.; Mukaiya, M.; Hirata, K.; Imai, K. Expression of Matrix Metalloproteinases and Tissue Inhibitors of Metalloproteinases in Human Pancreatic Adenocarcinomas: Clinicopathologic and Prognostic Significance of Matrilysin Expression. *J. Clin. Oncol.* **2001**, *19*, 1118–1127. [CrossRef]
115. Crawford, H.C.; Scoggins, C.R.; Washington, M.K.; Matrisian, L.M.; Leach, S.D. Matrix Metalloproteinase-7 Is Expressed by Pancreatic Cancer Precursors and Regulates Acinar-to-Ductal Metaplasia in Exocrine Pancreas. *J. Clin. Investig.* **2002**, *109*, 1437–1444. [CrossRef] [PubMed]
116. Su, G.; Blaine, S.A.; Qiao, D.; Friedl, A. Shedding of Syndecan-1 by Stromal Fibroblasts Stimulates Human Breast Cancer Cell Proliferation via FGF2 Activation. *J. Biol. Chem.* **2007**, *282*, 14906–14915. [CrossRef]
117. Nikolova, V.; Koo, C.-Y.; Ibrahim, S.A.; Wang, Z.; Spillmann, D.; Dreier, R. Differential Roles for Membrane-Bound and Soluble Syndecan-1 (CD138) in Breast Cancer Progression. *Carcinogenesis* **2009**, *30*, 397–407. [CrossRef]
118. Wu, Y.; Huang, H.; Fervers, B.; Lu, L. Syndecan-1 and KRAS Gene Expression Signature Associates with Patient Survival in Pancreatic Cancer. *Pancreas* **2020**, *49*, 1187–1194. [CrossRef]
119. Huang, J.W.; Chen, C.L.; Chuang, N.N. P120-GAP Associated with Syndecan-2 to Function as an Active Switch Signal for Src Upon Transformation with Oncogenic Ras. *Biochem. Biophys. Res. Commun.* **2005**, *329*, 855–862. [CrossRef]
120. Wheeler, D.L.; Iida, M.; Dunn, E.F. The Role of Src in Solid Tumors. *Oncologist* **2009**, *14*, 667. [CrossRef]
121. Yao, J.; Zhang, L.L.; Huang, X.M.; Li, W.Y.; Gao, S.G. Pleiotrophin and N-Syndecan Promote Perineural Invasion and Tumor Progression in an Orthotopic Mouse Model of Pancreatic Cancer. *World J. Gastroenterol.* **2017**, *23*, 3907–3914. [CrossRef]
122. Yao, J.; Li, W.Y.; Li, S.G.; Feng, X.S.; Gao, S.G. Midkine Promotes Perineural Invasion in Human Pancreatic Cancer. *World J. Gastroenterol.* **2014**, *20*, 3018–3024. [CrossRef] [PubMed]
123. Brioudes, E.; Alibashe-Ahmed, M.; Lavallard, V.; Berney, T.; Bosco, D. Syndecan-4 Is Regulated by IL-1β in β-Cells and Human Islets. *Mol. Cell Endocrinol.* **2020**, *510*, 110815. [CrossRef] [PubMed]
124. Cheng, J.Y.C.; Whitelock, J.; Poole-Warren, L. Syndecan-4 Is Associated with Beta-Cells in the Pancreas and the min6 Beta-Cell Line. *Histochem. Cell Biol.* **2012**, *138*, 933–944. [CrossRef] [PubMed]
125. Apte, M.V.; Park, S.; Phillips, P.A.; Santucci, N.; Goldstein, D.; Kumar, R.K.; Ramm, G.A.; Buchler, M.; Friess, H.; McCarroll, J.A.; et al. Desmoplastic Reaction in Pancreatic Cancer: Role of Pancreatic Stellate Cells. *Pancreas* **2004**, *29*, 179–187. [CrossRef]
126. Fogh, B.S.; Multhaupt, H.A.B.; Couchman, J.R. Protein Kinase C, Focal Adhesions and the Regulation of Cell Migration. *J. Histochem. Cytochem.* **2014**, *62*, 172–184. [CrossRef] [PubMed]
127. Gopal, S.; Bober, A.; Whiteford, J.R.; Multhaupt, H.A.B.; Yoneda, A.; Couchman, J.R. Heparan Sulfate Chain Valency Controls Syndecan-4 Function in Cell Adhesion. *J Biol. Chem.* **2010**, *285*, 14247–14258. [CrossRef] [PubMed]

128. Calderwood, D.A.; Campbell, I.S.; Critchley, D.R. Talins and Kindlins: Partners in Integrin-Mediated Adhesion. *Nat. Rev. Mol.* **2013**, *14*, 503–517. [CrossRef] [PubMed]

129. Itoh, Y.; Takehara, Y.; Kawase, T.; Terashima, K.; Ohkawa, Y.; Hirose, Y.; Koda, A.; Hyodo, N.; Ushio, T.; Hirai, Y.; et al. Feasibility of Magnetic Resonance Elastography for the Pancreas at 3T. *J. Magn. Reson. Imaging* **2016**, *43*, 384–390. [CrossRef]

130. Rubiano, A.; Delitto, D.; Han, S.; Gerber, M.; Galitz, C.; Trevino, J.; Thomas, R.M.; Hughes, S.J.; Simmons, C.S. Viscoelastic Properties of Human Pancreatic Tumors and in Vitro Constructs to Mimic Mechanical Properties. *Acta Biomater.* **2017**, *67*, 331–340. [CrossRef]

131. Rice, A.J.; Cortes, E.; Lachowski, D.; Cheung, B.C.H.; Karim, S.A.; Morton, J.P.; Hernandez, A.D.R. Matrix Stiffness Induces Epithelial—Mesenchymal Transition and Promotes Chemoresistance in Pancreatic Cancer Cells. *Oncogenesis* **2017**, *6*, e342–e352. [CrossRef] [PubMed]

132. Folkman, J.; Shing, Y. Angiogenesis. *J. Biol. Chem.* **1992**, *267*, 10931–10934. [CrossRef]

133. Hanahan, D.; Folkman, J. Patterns and Emerging Mechanisms of the Angiogenic Switch during Tumorigenesis. *Cell* **1996**, *86*, 353–364. [CrossRef]

134. Yancopoulos, G.D.; Davis, S.; Gale, N.W.; Rudge, J.S.; Wiegand, S.J.; Holash, J. Vascular-Specific Growth Factors and Blood Vessel Formation. *Nature* **2000**, *407*, 242–248. [CrossRef] [PubMed]

135. Essner, J.J.; Chen, E.; Ekker, S.C. Syndecan-2. *Int. J. Biochem. Cell Biol.* **2006**, *38*, 152–156.

136. Clasper, S.; Vekemans, S.; Fiore, M.; Plebanski, M.; Wordsworth, P.; David, G. Inducible Expression of the Cell Surface Heparan Sulfate Proteoglycan Syndecan-2 (Fibroglycan) on Human Activated Macro-Phages Can Regulate Fibroblast Growth Factor Action. *J. Biol. Chem.* **1999**, *274*, 24113–24123. [CrossRef] [PubMed]

137. Noguer, O.; Reina, M. Is Syndecan-2 a Key Angiogenic Element? *Sci. World J.* **2009**, *9*, 729–732. [CrossRef]

138. Grünewald, F.S.; Prota, A.E.; Giese, A.; Ballmer-Hofer, K. Structure-Function Analysis of VEGF Receptor Activation and the Role of Coreceptors in Angiogenic Signaling. *Biochim. Biophys. Acta Proteins Proteom.* **2010**, *1804*, 567–580. [CrossRef] [PubMed]

139. Corti, F.; Wang, Y.; Rhodes, J.M.; Atri, D.; Archer-Hartmann, S.; Zhang, J. N-Terminal Syndecan-2 Domain Selectively Enhances 6-O Heparan Sulfate Chains Sulfation and Promotes VEGFA165-Dependent Neovascularization. *Nat. Commun.* **2019**, *10*, 1–14.

140. Oh, E.S.; Couchman, J.R. Syndecans-2 and-4; Close Cousins, but Not Identical Twins. *Mol. Cells* **2004**, *17*, 181–187.

141. Ono, K.; Hattori, H.; Takeshita, S.; Kurita, A.; Ishihara, M. Structural Features in Heparin That Interact with VEGF165 and Modulate Its Biological Activity. *Glycobiology* **1999**, *9*, 705–711. [CrossRef]

142. Robinson, C.J.; Mulloy, B.; Gallagher, J.T.; Stringer, S.E. VEGF165-Binding Sites Within Heparan Sulfate Encompass Two Highly Sulfated Domains and Can Be Liberated by K5 Lyase. *J. Biol. Chem.* **2006**, *281*, 1731–1740. [CrossRef] [PubMed]

143. De Rossi, G.; Evans, A.R.; Kay, E.; Woodfin, A.; McKay, T.R.; Nourshargh, S. Shed Syndecan-2 Inhibits Angiogenesis. *J. Cell Sci.* **2014**, *127*, 4788–4799. [CrossRef] [PubMed]

144. Siegel, R.L.; Miller, K.D.; Jemal, A. Cancer Statistics, 2019. *CA Cancer J. Clin.* **2019**, *69*, 7–34. [CrossRef]

145. Malesci, A.; Tommasini, M.A.; Bonato, C.; Bocchia, P.; Bersani, M.; Zerbi, A. Determination of CA 19-9 Antigen in Serum and Pancreatic Juice for Differential Diagnosis of Pancreatic Adenocarcinoma from Chronic Pancreatitis. *Gastroenterology* **1987**, *92*, 60–67. [CrossRef]

146. Torphy, R.J.; Fujiwara, Y.; Schulick, R.D. Pancreatic Cancer Treatment: Better, but a Long Way to Go. *Surg. Today* **2020**, *50*, 1117–1125. [CrossRef] [PubMed]

147. Kim, J.; Bamlet, W.R.; Oberg, A.L.; Chaffee, K.G.; Donahue, G.; Cao, J.; Chari, S.; Garcia, B.A.; Petersen, G.M.; Zaret, K.S. Detection of Early Pancreatic Ductal Adenocarcinoma With thrombospondin-2 and CA19-9 Blood Markers. *Sci. Transl. Med.* **2017**, *9*, eaah5583. [CrossRef] [PubMed]

148. Regel, I.; Mayerle, J.; Mahajan, U.M. Current Strategies and Future Perspectives for Precision Medicine in Pancreatic Cancer. *Cancers* **2020**, *12*, 1024. [CrossRef]

149. Iovanna, J.; Mallmann, M.C.; Gonçalves, A.; Turrini, O.; Dagorn, J.C. Current Knowledge on Pancreatic Cancer. *Front. Oncol.* **2012**, *2*. [CrossRef]

150. Mutgan, A.C.; Besikcioglu, H.E.; Wang, S.; Friess, H.; Ceyhan, G.O.; Demir, I.E. Insulin/IGF-Driven Cancer Cell-Stroma Crosstalk as a Novel Therapeutic Target in Pancreatic Cancer. *Mol. Cancer* **2018**, *17*, 1–11. [CrossRef] [PubMed]

151. Yang, Y.; Tao, X.; Li, C.B.; Wang, C.M. MicroRNA-494 Acts as a Tumor Suppressor in Pancreatic Cancer, Inhibiting Epithelial-Mesenchymal Transition, Migration and Invasion by Binding to SDC1. *Int. J. Oncol.* **2018**, *53*, 1204–1214. [CrossRef]

152. Li, L.; Li, Z.; Kong, X.; Xie, D.; Jia, Z.; Jiang, W. Down-Regulation of MicroRNA-494 via Loss of SMAD4 Increases FOXM1 and β-Catenin Signaling in Pancreatic Ductal Adenocarcinoma Cells. *Gastroenterology* **2014**, *147*, 485–497. [CrossRef] [PubMed]

153. Liu, Y.; Li, X.; Zhu, S.; Zhang, J.G.; Yang, M.; Qin, Q. Ectopic Expression of MIR-494 Inhibited the Proliferation, Invasion and Chemoresistance of Pancreatic Cancer by Regulating SIRT1 and c-Myc. *Gene Ther.* **2015**, *22*, 729–738. [CrossRef]

154. Kimbrough, C.W.; Hudson, S.; Khanal, A.; Egger, M.E.; McNally, L.R. Orthotopic Pancreatic Tumors Detected by Optoacoustic Tomography Using Syndecan-1. *J. Surg. Res.* **2015**, *193*, 246–254. [CrossRef]

155. Hrabar, D.; Aralica, G.; Gomerčić, M.; Ljubičić, N.; Krušlin, B.; Tomas, D. Epithelial and Stromal Expression of Syndecan-2 in Pancreatic Carcinoma. *Anticancer Res.* **2010**, *30*, 2749–2753.

156. Griffith, L.G.; Swartz, M.A. Capturing Complex 3d Tissue Physiology In Vitro. *Nat. Rev. Mol. Cell Biol.* **2006**, *7*, 211–224. [CrossRef] [PubMed]

157. Gagliano, N.; Sforza, C.; Sommariva, M.; Menon, A.; Conte, V.; Sartori, P. 3D-Spheroids: What Can They Tell Us About Pancreatic Ductal Adenocarcinoma Cell Phenotype? *Exp. Cell. Res.* **2017**, *357*, 299–309. [CrossRef] [PubMed]
158. Betriu, N.; Semino, C.E. Development of a 3D Co-Culture System as a Cancer Model Using a Self-Assembling Peptide Scaffold. *Gels* **2018**, *4*, 65. [CrossRef] [PubMed]
159. Fernández-Muiños, T.; Recha-Sancho, L.; López-Chicón, P.; Castells-Sala, C.; Mata, A.; Semino, C.E. Bimolecular Based Heparin and Self-Assembling Hydrogel for Tissue Engineering Applications. *Acta Biomater.* **2015**, *16*, 35–48. [CrossRef] [PubMed]

 *biomolecules*

MDPI

Review

Lumican in Carcinogenesis—Revisited

Eirini-Maria Giatagana [1], Aikaterini Berdiaki [1], Aristidis Tsatsakis [2], George N. Tzanakakis [1] and Dragana Nikitovic [1,*

[1] Laboratory of Histology-Embryology, Department of Morphology, School of Medicine, University of Crete, 71003 Heraklion, Greece; eirini_gt@hotmail.com (E.-M.G.); berdiaki@uoc.gr (A.B.); tzanakak@uoc.gr (G.N.T.)

[2] Laboratory of Toxicology, Department of Morphology, School of Medicine, University of Crete, 71003 Heraklion, Greece; tsatsaka@uoc.gr

* Correspondence: nikitovic@uoc.gr; Tel.: +30-281-039-4557

Abstract: Carcinogenesis is a multifactorial process with the input and interactions of environmental, genetic, and metabolic factors. During cancer development, a significant remodeling of the extracellular matrix (ECM) is evident. Proteoglycans (PGs), such as lumican, are glycosylated proteins that participate in the formation of the ECM and are established biological mediators. Notably, lumican is involved in cellular processes associated with tumorigeneses, such as EMT (epithelial-to-mesenchymal transition), cellular proliferation, migration, invasion, and adhesion. Furthermore, lumican is expressed in various cancer tissues and is reported to have a positive or negative correlation with tumor progression. This review focuses on significant advances achieved regardingthe role of lumican in the tumor biology. Here, the effects of lumican on cancer cell growth, invasion, motility, and metastasis are discussed, as well as the repercussions on autophagy and apoptosis. Finally, in light of the available data, novel roles for lumican as a cancer prognosis marker, chemoresistance regulator, and cancer therapy target are proposed.

Keywords: lumican; cancer; extracellular matrix; proteoglycans; metastasis; cancer cell growth; motility; biomarker

Citation: Giatagana, E.-M.; Berdiaki, A.; Tsatsakis, A.; Tzanakakis, G.N.; Nikitovic, D. Lumican in Carcinogenesis—Revisited. *Biomolecules* **2021**, *11*, 1319. https://doi.org/10.3390/biom11091319

Academic Editor: Davide Vigetti

Received: 10 August 2021
Accepted: 4 September 2021
Published: 6 September 2021

Publisher's Note: MDPI stays neutral with regard to jurisdictional claims in published maps and institutional affiliations.

1. Introduction-Cancer and ECM

Carcinogenesis is a multifactorial process with the input and interactions of environmental, genetic, and metabolic factors [1,2]. During this process, normal cells are transformed and exhibit enhanced survival, aggressive growth, motility, and invasion, as well as the capability to remodel their microenvironment. Indeed, the altered tumor microenvironment (TME) and the interactions therein facilitate cancer expansion [3,4]. The complex milieu of the TME, in addition to tumor cells, contains blood vessels, tissue non-malignant cells, stromal cells, infiltrating immune cells, and the modified extracellular matrix (ECM) characteristic for each phase of cancer progression [4,5]. Thus, the complex ECM structure consisting of fibrillar proteins, proteoglycans (PGs), and glycosaminoglycans (GAGs) is extensively reorganized [6] and, together with the cellular compartment, forms the new neoplastic organ [7].

Furthermore, the remodeled ECM creates a permissive environment supporting all tumor cell functions [4,8–10]. Importantly, ECM cues coordinate the different effectors of the TME and modulate the plethora of signaling pathways involved in the propagation of the "hallmarks of cancer" [2,11]. Moreover, the tumor ECM created by specific stromal cell subsets boosts the tumor immune escape mechanisms, triggering and sustaining an immunosuppressive networkof immunoregulatory cues [12]. In addition, tumors commonly exhibit desmoplasia, an increased deposition and cross binding of the ECM proteins where the cancer-associated fibroblasts (CAFs) as primary ECM producers are the main cell type [13].

Likewise, fibrosis, characterized by the abnormal accumulation of collagen perpetrated and supported by mechanisms including wound healing, ECM degradation, and epithelial-to-mesenchymal transition (EMT), significantly impacts cancer progression and putative therapeutical strategies [14]. Indeed, the resulting ECM "stiffening" is a vital regulator of tumor cell functions.

Therefore, both the cellular and ECM components of the formed neoplastic "organ" are extensively modified during tumorigenesis and regulate cancer progression as previously discussed [1,8,15].

2. SLRPs Structure and Function—Focus on Lumican

PGs are glycosylated proteins that participate in the formation of the ECM. These hybrid molecules consist of a protein core into which one or more GAG chains are covalently bound. Four different types of GAGs can be attached to PGs'protein core, heparan sulfate (HS), chondroitin sulfate/dermatan sulfate (CS/DS), and keratan sulfate (KS) chains.

Forty-five PGs have been identified up until now, and they exhibit extensive variability in their protein core composition and glycosylation pattern. Three criteria are considered when classifying PGs: their topology (cellular or subcellular), overall gene/protein homology, and the presence of specific protein modules within their respective protein cores [16].

Specific structural features determine PG functions, such as the core protein structure, GAG chains' composition, and sulfation pattern [17]. Indeed, their protein cores characterize the existence of unique protein modules thatmembers of a given class often share, such as the PDZ-like, laminin-like, and EGF-like domains [16]. Furthermore, it has been shown that specific sulphation motif sequences within the CS/HS chainscarry biological information to the cells [17,18]. Notably, the binding partners for GAG chains remain partly uncharacterized, and different PG classes seem to function by utilizing overlapping signaling with various outcomes [17].

Although PGs play an essential role in regulating cellular processes like tissue homeostasis and development [19], their expression pattern and functions are changed during tumorigenesis and are correlated with cancer development and progression [1,8]. Thus, solid tumors' behavior and differentiation status are closely associated with altered PG expression profiles, with epithelial tumors exhibiting a more discrete PG phenotype than mesenchymal tumors [20].

The small leucine-rich proteoglycans (SLRPs) are a distinct family of 18 proteins with unique characteristics. They consist of a small protein core (36–42 kDa) with a variable number of central leucine-rich repeats (LRRs) variously substituted with GAG chains [2,16]. LRRs exhibit different amino acid sequences in discrete SLRPs, their size varying between 20 and 29 residues, while the N and C-terminal regions of the protein core bear numerous cysteine residues [21,22].

SLRP classification is based on the conservation of the amino acid residues of the protein core, the organization of disulfide bonds at the molecule's N- and C-terminal regions, and their gene/protein homology. They are categorized into five different classes [23]. PGs belonging to classes I, II, and III are canonical, while classes IV and V are non-canonical [24]. Class I SLRPs, like biglycan and decorin, are mainly substituted with CS/DS chains; class II SLRPs, like lumican (LUM), are covalently bound with KS chains. In contrast, class III members can bear KS chains (osteoglycin), CS/DS chains (epiphycan), or do not carry GAG chains (opticin) like classes IV and V SLRPs [16,25].

Many studies have shown that SLRPs interact with diverse cell membrane receptors, cytokines, chemokines, and ECM molecules [16,26]. Notably, most SLRP family members undergo different post-translational glycosylation [27] and are competent to regulate signal transduction mechanisms, and affect various cellular functions, like proliferation, migration, and differentiation [2,28]. In addition, many studies have also reported that SLRPs' interaction with growth factors or tyrosine kinase receptors affects cellular behavior and tumor progression [29–32].

3. Lumican Structure, Function, and Expression-Correlation with Carcinogenesis

Lumican, a class II SLRP, has a 38 kDa protein core exhibiting four distinct regions: a 16 amino acid peptide, a negatively charged N-terminal region containing tyrosine sulfate and disulfide bonds, a 6–10 LRR motifs characterized by common molecular architecture that supports protein interactions, and a C-terminal region consisting of two conserved cysteine residues [22]. Amino acid sequencing data revealed the presence of four possible substitution positions with KS chains or oligosaccharides within the LRR region [33]. However, it seems that not all of these positions can be used for glycosylation of the protein by KS chains [34]. Moreover, it has been suggested that there is an increase in non-glycosylated forms of lumican with age due to the decrease of KS synthesis [35].

The lumican gene is located on chromosome 12q21.3-q22 [33], and its expression is significantly altered between tissues during different developmental stages. For example, its expression is early detectable in the chicken cornea during fetal development [36]. Still, it is not expressed until birth in human cartilage [37], indicating species-dependent roles of lumicanduring embryogenesis.

Lumican participates in the structural organization of tissues. Thus, lumican-deficient mice collagen fibrils exhibit an increased diameter forming a disorganized matrix [38]. Early studies in the mouse model showed that lumican is widely distributed in most interstitial connective tissues [39]. Indeed, this SLRP is an important PG of the bone matrix, and its' expression is positively correlated to the bone differentiation stage [40]. Furthermore, lumican is highly expressed in the skin and cornea [41,42], where lumicandeficiency is translated into tissue disfiguration with resulting skin laxity and a decrease in corneal clarity [41,42]. Lumican expression in parenchymal cells such as urothelial and colon epithelium, albeit at lower levels, has been determined [43,44]. The role of lumican, however, partly overlaps that of fibromodulin [40]. Indeed, these two class II PGs are extensively expressed in collagenous connective tissues where they significantly affect tissue integrity [45].

Notably, lumican is involved in cellular processes associated with tumorigeneses, such as EMT (epithelial-to-mesenchymal transition), cellular proliferation, migration, invasion, and adhesion [32,46,47]. Furthermore, lumican is expressed in various cancer tissues and is reported to have a positive or negative correlation with tumor progression [26]. More specifically, immunohistochemistry demonstrated a much higher expression of lumican in cancerous gastric tissues than normal tissues.In this cancer type, the lumican expression was correlated with histological classification, cancer dissemination to secondary sites, and lymphatic metastasis [48]. Furthermore, the TCGA database analysis showed a higher expression of lumican in the gastric cancer tissues than the neighboring non-tumor tissues [49]. This was correlated, as probed by the Kaplan−Meier analysis, with a poor prognosis. Moreover, a multivariate analysis demonstrated a strong positive association between a high LUM expression and poor overall survival. Notably, lumican enhanced 14 signaling pathwayspotentially correlated with this cancer progression [49]. On the other hand, the expression of lumican and versican by cancer-associated fibroblastswas associated with a poor relapse-free and overall survival of esophageal squamous cell carcinoma [50].

In colon cancer, the lumican expression was correlated with lymph node metastasis and a lower survival rate [51]. Specifically, lumican was detected in the cytoplasm of cancer cells in 62.7% of 158 patients undergoing curative surgery for advanced colorectal cancer with lymph node metastasis. Notably, lumican expression was positively associated with the spread of lymph node metastasis and had lower survival rates [51]. This study is in accordance with the UALCAN database analysis, which determined a high lumican mRNA expression in colorectal adenocarcinoma tissues [52]. The application of the univariate and multivariate COX analysis and Kaplan−Meier method to this dataset identified the lumican expression as a poor prognosis marker.

Moreover, LinkedOmics demonstrated that the LUM expression was strongly associated with miR200 family expression and tumor immune escape. Indeed, it was determined that lumican facilitated colon cancer progression through a miRNA200-dependent epithelial-to-mesenchymal progression. Zang et al. suggested that lumican is a potential target in colon cancer [52]. Furthermore, when the tissue microarrays and tissue sections were analyzed, lumican was found to be expressed by both transformed cells and the stroma of colon adenomas and carcinomas. Notably, it was more frequently detected in carcinoma than adenoma cells and in carcinomas and high-risk adenomas combined compared withlow-risk adenomas [44]. On the other hand, the lumican expression by the colon cancer cells was positively correlated with a longer disease-specific and disease-free survival in stage II colon cancer patients, and a more prolonged disease-specific survival in microsatellite-stable stage II colon cancer patients, suggesting a disease stage dependence [53].

In pancreatic cancer, the expression of lumican was demonstrated by a immunohistological analysis [54], where pancreatic stellate cells were identified as a major source of this PG [55]. Notably, a small fraction of the PDAC tumor mass is attributed to cancer cells, the majority consisting of desmoplastic TME with abundant activated fibroblasts, leukocytes, and pancreatic stellate cells [56].

Melanoma cells do not express lumican, but the increased expression of lumican to the peritumoral stroma is negatively correlated withthis tumor growth [57]. The correlation of the lumican expression and various tumor progression is summarized in Table 1.

Table 1. Lumican expression in tumor tissues and correlation to carcinogenesis.

Cancer Type	Detected Expression (Protein/mRNA)	Level of Expression	Clinical Correlation	Ref.
Gastric cancer	Protein	Overexpressed in cancerous gastric tissues compared to normal tissues	Cancer dissemination to secondary sites and lymphatic metastasis	[48]
Gastric cancer	mRNA	Higher expression of lumican in the gastric cancer tissues than neighboring non-tumor tissues	Poor overall survival	[49]
Colon cancer	Protein	Overexpressed by cancer cells	Lymph node metastasis and a lower survival rate	[51]
Colon cancer	mRNA	Overexpressed	Poor prognosis	[52]
Adenoma to colon cancer transition	Protein	Increased expression during the transition process	Cancer stage	[44]
Colon cancer	Protein	Overexpressed	Positively correlated to a longer disease-specific and disease-free survival in stage II colon cancer patients and a more prolonged disease-specific survival in microsatellite-stable stage II colon cancer patients	[53]
Pancreatic ductal adenocarcinoma (PDAC)	Protein	Overexpressed	Associated with prolonged survival after surgery	[54]
Melanoma	Protein	Not expressed by tumor cells, expressed at peritumoral stroma	Negatively associated with melanoma growth	[57]

Lumican Regulates Cancer Cell Growth, Invasion, and Metastasis

Many studies have shown that lumican modulates tumor cells' proliferation, invasion, and metastasis with different mechanisms, either enhancing or preventing cancer progression. A characteristic example is the regulation of the growth factor activity in mesenchymal tumors and the effects on these cancer cell functions [32,46,58]. Lumican is the most abundant SLRP produced by HTB94 human chondrosarcoma cells and a positive regulator of these cells' growth. Indeed, lumican deficiency significantly inhibits basal and IGF-I induced HTB94 cell growth. The oncogenic action of IGF-I is mediated by its receptor, IGF-IR, whose phosphorylation levels are strongly attenuated in lumican-deficient HTB94 cells. Furthermore, lumican affects ERK1/2 activation, which seems crucial to IGF-I-dependent HTB94 cell growth [32].

Likewise, lumican expression and secretion by osteosarcoma Saos-2 and MG63 cells are correlated with their differentiation [46]. Indeed, the well-differentiated Saos-2 cells had a negative growth response to lumican, while their migration and the chemotactic response to fibronectin were enhanced. Moreover, the mechanism was mediated by Smad-2 downstream signaling. On the other hand, these cellular functions of poorly differentiated MG63 cells are not affected by low endogenous lumican levels [46]. Further studies revealed that lumican-deficient Saos-2 cells exhibited increased adhesion onto fibronectin, which was abolished upon neutralization of the endogenous transforming growth factor β2 (TGF-β2) activity. On the other hand, treatment with exogenous TGF-β2 was shown to stimulate Saos-2 cell fibronectin-dependent adhesion [58]. Nikitovic et al. thus suggested that lumican is an upstream regulator of the TGF-β2/Smad 2 signaling pathway in an osteosarcoma cell model.

Lumican pro-tumorigenic effects are also observed in gastric, bladder, colon, clear cell renal, and liver cancers [59–62]. A high lumican expression in gastric cancer tissues indicates a poor patient prognosis [48]. Indeed, Wang et al. showed that the increased expression of lumican by human gastric cancer-associated fibroblasts is positively associated with lymph node metastasis, TNM stage, depth of invasion, and a poor survival rate of gastric cancer. Indeed, lumican promotes gastric cancer cell growth by activating the integrin β1/FAK signaling axis [63].

In human colon adenocarcinoma cells, lumican overexpression was found to be accompanied by changes in the actin polymerization state, immediately associated with cancer cells migration and higher metastatic potential [61,64]. In addition, hepatic cancer HepG2 and MHCC97H cells express more lumican in comparison with normal Lo02 hepatocytes. Transfection of hepatic cancer cells with shRNAs specific for lumican resulted in decreased invasion and migration mediated by reducing the ERK-1 and JNK activation status [65].

In a neuroblastoma model, lumican was a downstream mediator of FOXO3 transcription factor action and enhanced these cells' migration. FOXO3 is correlated with a poor outcome in high-stage neuroblastoma due to its' chemoprotective and angiogenesis-stimulating properties [62]. Notably, upon inhibiting FOXO3 by the small molecular weight compound repaglinide, the binding of FOXO3 to the LUM promoter was attenuated, abrogating the FOXO3-dependent lumican expression and decreasing neuroblastoma cell 2D- and 3D-migration [62]. Regarding clear renal cell carcinoma (cRCC), the microarray analysis demonstrated a higher expression of matrix regulators lumican and CEACAM6 in metastatic tissues than patient-matched primary tissues [60]. Indeed, these authors conclude that the ECM genes a re crucial triggers resulting in visceral, bone, and soft tissue metastases in cRCC.

On the other hand, the lumican expression is suggested to attenuate discrete tumor progression, including pancreatic cancerand melanoma, as recently discussed [59]. Thus, lumican was shown to inhibit cancer cell proliferation inthe early stages of pancreatic ductal adenocarcinoma (PDAC) [58]. Indeed, it was demonstrated that exogenous lumican induces features of a quiescent state, including growth arrest, apoptosis, and chemoresistance [66]. Interestingly, this was partly executed through an EGFR-dependent mechanism, as lumican induced the dimerization of the EGFR receptors and the subsequent uptake and degradation [66]. Indeed, the interactions of lumican with growth factors/growth factor receptors and the effects on tumor cell functions are schematically depicted in Figure 1. A study with patient tumor tissues, ex-vivo cultures of patient-derived xenografts (PDX), pancreatic ductal adenocarcinoma (PDAC) stellate, and tumor cells was conducted to investigate whether hypoxia within the tumor microenvironment alters stromal lumican expression and secretion [67]. Li et al. demonstrated that hypoxia significantly reduced lumican secretion from pancreatic stellate cells and induced autophagy in these cells, as well as in ex vivo cultures of PDX, but not cancer cells cultured under 2D conditions [67].

Regarding melanoma, lumican also seems to be negatively correlated with its progression. In vivo experiments in lumican-null mice revealed that lumican is an endogenous inhibitor of melanoma growth and modulates the response to TAX2, an anticancer cyclic peptide. Notably, the null mice tumors were twice as large as the wild-type animal tumors [68]. Furthermore, the lumican protein core was shown to inhibit melanoma cells' migration. Indeed, lumican induced changes in actin filaments and $\beta 1$ integrin ligation, and enhanced vinculin accumulation in the cell cytoplasm, destabilizing focal adhesion complexes. In addition, the phosphorylation levels of FAK were significantly decreased. Combining these alterations in the cytoskeleton and the adhesion molecules' activation status may contribute to the lumican anticancer effect in A375 melanoma [69].

Moreover, lumican was shown to affect the signaling of Snail, the main EMT trigger, cancer-facilitating molecules. Thus, when the Snail1 overexpressing B16F1 melanoma cells and the Mock-B16F1 cells were inoculated in $Lum^{+/+}$ and $Lum^{-/-}$ mice, a significantly higher number of metastatic nodes were detected in the lungs of $Lum^{-/-}$ mice inoculated with Snail-overexpressing B16F1 cells. These data suggest that endogenous lumican of the wild-type mice markedly attenuates melanoma metastasis to the lungs. Notably, the expression and activities of molecules, including ECM mediators, correlated to the invasive phenotype were altered in in vitro models [70]. Another study, in an immunocompetent model of melanoma, implanted in $Lum^{-/-}$ vs. wild type syngeneic mice, concluded that endogenous lumican modulates the organization of the tumor matrix regarding the intratumoral distribution of matrix proteins, growth factors, and stromal cells in a manner correlated with disease progression [68].

Furthermore, lumican attenuated the growth of melanoma cells and downregulated the response to the anticancer validated peptide TAX2. Indeed, Jeanne et al. identified lumican as an essential regulator of the tumor matrix structure and function [68]. Recently, in a mouse model of primary melanoma, the lumican-derived L9Mc peptide abrogated the growth and increased the apoptosis of B16F1 cells, as determined by infrared spectral imaging and histopathology [71].

The transcription factor FOXO3 is associated with a poor outcome in high-stage neuroblastoma (NB), facilitating chemoprotection and tumor angiogenesis. In addition, FOXO3 stimulates metastasis formation in other tumor entities, one of the biggest challenges in treating aggressive NB. The SLRP member lumican has been determined as a FOXO3-regulated gene that stimulates cellular migration in NB [62].

For some cancer types, such as lung cancer, contrasting roles of lumican were reported. Non-small lung cancer cell lines growth is negatively impaired by lumican, as lumican-deficient H460 and A549 cells exhibit a prolonged doubling time and retarded growth. Specifically, lumican deficiency affected central spindle and midbody formation, resulting in chromosome missegregation, multinucleated cells, increased chromosome instability, and retarded cell growth [72]. In contrast, a separate study reported that the depletion of lumican increased lung cancer cell invasion. Upon lumican downregulation, its colocalization with p120 catenin (p120ctn), an intracellular scaffolding protein of the catenin family, is decreased, leading to morphological changes and actin cytoskeleton remodeling, which accelerated cell invasion [73].

Tumor aggressiveness is connected to EMT, as the differentiation state of cancer cells defines their invasive properties [74]. A recent study in breast cancer in vitro showed that lumican treatment in combination with the knockdown of ERα and the suppression of ERβ can regulate these cells' differentiation state, morphology, expression of matrix effectors, and cell behavior [75]. Indeed, the effects of lumican seem to be hormone-receptor dependent as the aggressive metastatic ERβ-positive MDA-MB-231, the ERβ-suppressed (shERβMDA-MB-231) cells, and the ERα-positive MCF-7/c breast cancer cells of a low metastatic ability exhibit varying responses to lumican. Thus, exogenous lumican increases the expression of α2 and β1 integrins in MDA-MB-231 and in shERβMDA-MB-231 compared withMCF-7/c cells. Furthermore, specific integrin-dependent downstream signaling pathways, including FAK, ERK 1/2 MAPK 42/44, and Akt, were attenuated by lumican [76]. Moreover, Karamanou et al. suggested that treating breast cancer cells seeded to 3D collagen cultures with lumican enhancedcell−cell contacts and cell grouping, initiating a less invasive phenotype [47]. A separate study showed that this SLRP might inhibit or even reverse the metastatic features that breast cancer cells acquire undergoing EMT by increasing the gene expression of the EMT inhibitor miR-200b [70]. On the other hand, Leygue et al. showed that lumican expression differs during breast tumorigenesis, andlumican mRNA, identified in the tumor stroma, is correlated with a higher tumor grade and lower expression of estrogen receptors and younger age of the patients [77].

Lumican effects on different cellular functions of cancer cells are summarized in Table 2.

Table 2. Lumican's role in various cancer types and the mechanism of action.

	Cancer Type	Model	Alterations in Signaling Pathways	Effect on Cell Function	Ref.
	Chondrosarcoma	HTB94 human cell line (in vitro)	IGF-I/IGF-IR/ERK1/2	Cell growth	[32]
	Osteosarcoma	Saos-2 human cell line (in vitro)	TGF-β2/Smad2	Migration and adhesion to fibronectin substrate	[46,58]
Tumorigenic action	Gastric cancer	MKN45 human cell line, primary cell cultures, tissue biopsies (in vitro), and ice model (in vivo)	Integrin-β1/FAK	Cell growth, migration, and invasion	[63]
	Liver cancer	HepG2 and MHCC97H human cell lines (in vitro)	ERK1/JNK	Migration and invasion	[65]
	Neuroblastoma	SH-EP, SK-N-SH, and ZMR32 human cell lines (in vitro)	FoxO	Migration	[62]

Table 2. *Cont.*

	Cancer Type	Model	Alterations in Signaling Pathways	Effect on Cell Function	Ref.
Anti-tumorigenic action	Lung cancer	A549, H460, H1975, H157, and H838 human cell lines (in vitro)	p120 catenin	Cadherin-mediated invasion	[73]
	Pancreatic ductal adenocarcinoma (PDAC)	PANC-1 human cell line, PancO2 murine cell line, primary PDAC cells from PDX models (in vitro), and mice model (in vivo)	EGFR and TGF-β/p38/Smads	Cell growth	[66]
	Pancreatic ductal adenocarcinoma (PDAC)	PANC-1 human cell line, primary cell cultures (in vitro), and tissue biopsies from PDX model (ex vivo)	HIF-1a and AMPK	Cell growth	[67]
	Melanoma	A375 human cell line (in vitro)	Integrin-β1/FAK/vinculin	Migration	[69]
	Melanoma	B16F1 human cell line (in vitro) and mice model (in vivo)	Snail1	Metastasis and invasion	[66]
	Breast cancer	MCF-7/c and MDA-MB-231 human cell lines (in vitro)	CD44/Hyaluronan synthase and Integrin-α1 and -β1/FAK/ERK1/2/MAPK 42/44/Akt	EMT metastasis	[47]

The multifaceted signaling roles of lumican in carcinogencsis are schematically depicted in Figure 1.

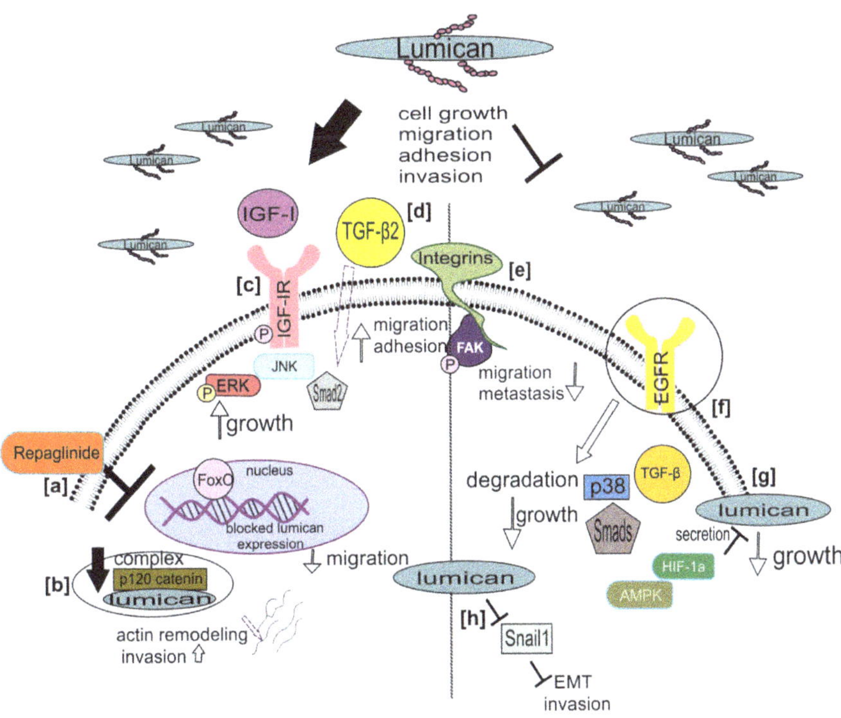

Figure 1. Schematic representation of lumican's signaling in carcinogenesis. (**a**) Inhibiting the binding of FOXO3 to the lumican promoter by the small molecular weight leads to decreased FOXO3-dependent lumican expression and neuroblastoma cell migration. (**b**) Upon lumican downregulation, its colocalization with p120 catenin (p120ctn) decreases, leading to actin cytoskeleton remodeling and accelerated lung cancer cell invasion. (**c**) Lumican-deficient hepatic cancer cells show decreased invasion and migration mediated by reducing IGF-IR, ERK-1, and JNK activation status. (**d**) Lumican is an upstream regulator of the TGF-β2/Smad 2 signaling pathway in an osteosarcoma cell model, regulating cell adhesion. (**e**) Lumican interacts with the integrin β1/FAK signaling axis, affecting tumor progression positively or negatively. (**f**) Lumican induces the dimerization of the EGFR receptors and their subsequent uptake and degradation, leading to attenuated PDAC cell growth. (**g**) Hypoxia significantly reduces lumican secretion from pancreatic stellate cells and results in attenuated PDAC cell growth. (**h**) Lumican affects the signaling of Snail, an EMT trigger molecule that facilitates cancer metastasis, attenuating melanoma metastasis to the lungs.

4. Lumican Modulates Cancer Cell Motility

The mechanisms presented in this section involve the interplay of lumican with specific cell membrane receptors, which leads to the activation of downstream signaling pathways. A crucial downstream mediator is focal adhesion kinase (FAK), which participates in focal adhesion turnover, actin cytoskeleton reorganization, and MMP expression, and regulates cell motility and, therefore, metastasis. One of the key examples is the role of lumican on melanoma cell adhesion and motility. Initially, the lumican protein core was shown to inhibit melanoma cells' migration. Indeed, lumican induced changes in actin filaments and β1 integrin ligation, and enhanced vinculin accumulation in the cell cytoplasm, destabilizing focal adhesion complexes. In addition, the phosphorylation levels of FAK were significantly decreased. Combining these alterations in the cytoskeleton and adhesion molecules' activation status is suggested to contribute to the lumican anticancer effect in A375melanoma [69].

Moreover, the potential anti-metastatic role of lumican in melanoma by inhibiting the membrane-type matrix metalloproteinase (MMP)-14 activity and melanoma cell migration in vitro has been studied in vitro and in vivo [78–82]. MMP-14 is necessary for cell migration, because it modulates the activity and expression of downstream MMPs; activates integrins and CD44 [73]; and regulates intracellular signaling involving MAPK, FAK, Src, and Rac [69,76,83–86]. Importantly, the glycosylated full-length lumican was likewise shown to block the MMP-14 activity, behaving as a competitive inhibitor [79]. Indeed, lumican inhibits the degradation of ECM by inhibiting MMP-14, then influencing integrin clustering, modulating focal adhesion site stability and FAK phosphorylation at Tyr-397, leading to the inhibition of melanoma cell migration [81]. Moreover, the lumcorin peptide corresponding to a sequence of 17 amino acids carried by the core protein of lumican inhibits melanoma cell chemotaxis in a manner similar to lumican protein [87]. Interestingly, lumcorin triggered the expression of an intermediate form of MMP-14 (~59 kDa) and attenuated its activity [88].

During EMT, where cancer, including melanoma cells, acquire enhanced motility, vital participation of Snail signaling has been shown [89]. Notably, lumican attenuated the Snail-induced MMP-14 activity and migration in B16F1, but not in HT-29 cells. In Snail overexpressing Snail-B16F1 cells, lumican significantly inhibits and melanoma primary tumor development. Thus, a lumican-based strategy targeting the Snail-induced MMP-14 activity might be helpful for melanoma treatment [89]. Lumican actions involving processes like reduced formations of cytoskeletal projections such as lamellipodia and invadopodia were also associated with decreased ZO-1, keratin 8/18, integrin β1, and MT1-MMP expression/activity [90].

Indeed, lumican can affect the biological roles of various downstream mediators, including integrins, cyclin D1, cortactin, vinculin, hyaluronan synthase 2, heparanase, and the phosphorylation of AKT, p130 Cas, and GSK3α/β [70,79,80,88,91–93].

Other cancer cell types that have been studied regarding the lumican-dependent motility effects include lung, breast, colon, liver, bladder, and pancreatic cancer, as well as neuroblastomas [3,61,62,65,69,73,79,80,88,94,95].

Thus, it has been determined that type I collagen promotes the most robust adhesion and migration of eight pancreatic cancer cell lines, explicitly mediated by the alpha2beta1integrin [94]. In continuation, Zeltz et al. determined that lumican is a specific inhibitor of alpha2beta1 integrin, attenuating the ability of A375 melanoma cells to migrate. This effect was verified in a study on Chinese hamster ovary (CHO) cells expressing the α2 integrin subunit (CHO-A2), whose ability to migrate was attenuated by lumican in contrast with the wild-type CHO cells (CHO-WT) lacking this subunit. Moreover, in the presence of recombinant lumican, the pFAK/FAK ration was strongly downregulated in CHO-A2 cells [92]. Likewise, in breast cancer, lumican significantly downregulates the migratory abilities of tumor cells in a mannerdependent on their hormone receptor status [76].

On the other hand, it was shown that lumican enhanced the adhesion and migration on the collagen of both pancreatic cancer cells and pancreatic stellate cells in a manner dependent on TGF-B [55]. Yang et al. indicated an interplay between lumican and microtubules that acts as a molecular switch to coordinate the balance between cell adhesion and migration. Indeed, it is suggested that lumican propagates these effects through p120-catenin signaling and cytoskeletal remodeling, as well as the activities of Rac and Rho [73].

Downregulation of the lumican expression attenuated lung osteotropic cancer cell's adhesion to various ECM components, ultimately decreasing these cells' migration. On the other hand, the introduction of exogenous lumican restored the motility of lumican knockdown cells and enhanced the invasion of lung cancer cells in the bone niche [3].

In liver cancer cells, silencing lumican by shRNA reduced cell invasion and migration via inhibiting the activation of the ERK1/JNK pathway, suggesting that lumican is a positive regulator of these cells' migratory abilities [65]. In the neuroblastoma, the silencing of the lumican gene in FOXO3 expressing IMR32 and SK-N-SH neuroblastoma cells or adding a FOXO3 inhibitor that restricted lumican transcription resulted in these cells' reduced migration capacity [62]. These results suggest that FOXO3 is a lumican biological partner that is important to neuroblastoma development. Likewise, a recent study on bladder cancer showed that lumicans' expression was more prominent in bladder cancer tissue and cell lines than in healthy adjunct tissues. Moreover, in in vitro models, the downregulation of lumican decreased bladder cancer cells' migration by attenuating the downstream MAPK signaling [43].

The motility of colon cancer cells is also upregulated by lumican. Thus, Radwanska et al. showed that human LS180 colon cancer cells that overexpress lumican tend to create podosome-like structures. This was noted due to the redistribution of vinculin and its simultaneous colocalization with actin and gelsolin in the cells' submembrane region [61]. Therefore, these authors conclude that the secreted lumican enhances LS180 cells' motility. Likewise, lumican upregulates gastric cancer cell migration through an integrin β1-FAK downstream signaling pathway, as depicted in Figure 1 [63].

In summary, the effect of lumican on cancer cell motility seems to be cancer-type dependent, as both positive [3,43,55,61–63,68,73], negative [69,76,78–81,89,92,94], or no effect has been determined [32]. Therefore, an in-depth study of the utilized mechanisms is imminent in order to identify the genotypic phenotype of lumican-responsive cancer to develop target therapeutic strategies.

5. Lumican at the Crossroad between Apoptosis and Autophagy

The suppression of apoptosis together with deregulated cell growth provides two essential criteria for cancer progression [95]. Early studies have demonstrated the ability of lumican to regulate the apoptosis of corneal and embryonic fibroblasts [96]. Lumicans' mechanism of action incorporates Fas-FasL signaling and the modulation of cell growth and apoptosis mediators, including p21 and p53 [96]. In continuation, the effects of lumican on cancer cell apoptosis were determined. Thus, B16F1 cells transfected to overexpress lumican present an initiation and/or increase ofapoptosis [97]. Moreover, in a mouse model of B16F1 melanoma primary tumor growth, lumican treatment with the L9Mc peptide increased cancer cell apoptosis [71].

In a separate cancer model, lumican secreted by stromal cells was shown to attenuate the expression and activity of hypoxia-inducible factor-1α (HIF1α)via Akt signaling, leading to the enhanced apoptosis of pancreatic cancer cells [54]. Furthermore, lumican was determined to trigger, characterized with apoptosis, a quiescent pancreatic cancer state [66]. Moreover, lumican was shown to facilitate endothelial cell apoptosis through Fas-dependent signaling. Thus, lumican-overexpressing murine fibrosarcoma (MCA102) and pancreatic adenocarcinoma (Pan02) cells providesmaller tumors in vivo compared withwild-type cancer cells [98]. This was correlated withattenuated neoplasm tissue vascular density. Therefore, lumican repressed tumor growth in this model, increasing endothelial cell apoptosis [98]. Likewise, the intensity of VEGF immunostaining and the abundance of blood vessels in melanoma lung metastasis nodules were decreased in lumican-expressing tumors. Therefore, in addition to inducing the apoptosis of melanoma cells, lumican inhibited tumor-associated angiogenesis [69]. Furthermore, it has been suggested that lumican inhibits angiogenesis through MMP14 and integrin α2β1 signaling.

Autophagy is an ancient catabolic process in which cells sequester damaged organelles and protein aggregates to process their degradation [99]. Indeed, under normal conditions, autophagy is a beneficial process dynamically regulated by starvation and other stresses [99]. However, autophagy can facilitate the viability and chemoresistance of cancer cells, the maintenance of cancer stem cells, and, in a context-dependent manner, have an inhibiting effect on tumor growth [100]. Thus, enhancing autophagy may attenuate inflammatory responses that support carcinogenesis and abrogate tumor escape from the host immune system defense mechanisms [101]. On the other hand, the upregulation of autophagy may have a pro-survival effect on cancer cells [102].

The cancer-microenvironment niche and its components, such as proteoglycans, decidedly regulate autophagy [2]. To date, a primarily an anti-autophagic role has been attributed to the lumican [103]. Indeed, chemotherapeutic agents increase the secretion of lumican in PDAC, which, by inhibiting autophagy, enhances chemotherapy-induced growth inhibition. Indeed, this effect of lumican was verified in both in vitro and in vivo PDAC models, including patient-derived xenografts [104]. However, a feedback mechanism seems to be emerging, as it was recently demonstrated that hypoxia induces autophagy in pancreatic cancer stellate cells, accomplished through an AKMP/TOR/p70S6K/4EBP signaling pathway-mediated protein degradation and synthesis inhibition [67]. Furthermore, the mechanism strongly downregulates lumican secretion through post-transcriptional regulation of pancreatic stellate cells, not cancer cells [67]. Notably, pancreatic stellate cells exhibit significant regulatory roles in tumor immunology, paracrine signaling, and metabolism in pancreatic ductal carcinoma [105].

6. Implications of Lumican in Cancer-Associated Inflammation

The process of tumorigenesis is intimately correlated with chronic inflammation, with a significant 20% of cancer incidences directly related to chronic infections [106]. All tumor types separately of etiology specifically interact with the immune system at all stages of carcinogenesis [107]. The ECM components play a significant role in these interactions. Indeed, the tumor microenvironment, extensive remodeling modulates the immune response [108]. Other SLRPs such as biglycan have been implicated in cancer-associated inflammation [109,110].

The available knowledge on the role of lumican in the processes of tumor-associated inflammation is restricted [26]. Some research, however, indicates a connection between lumicans' biological effects and inflammation. Thus, for example, when using a mouse colitis model, Lohr et al. showed that lumican exacerbates the immune and inflammatory responses [111]. Specifically, Lum$^{-/-}$ mice had a decreased secretion of cytokines such as tumor necrosis factor-alpha (TNF-α); CXCL1 secretion; correlated to the retarded translocation of and NF-κB translocation to the nucleus; and attenuated neutrophil infiltration [111]. In contrast, the Lum$^{-/-}$ mice presented substantial weight loss and extended tissue damage compare withthe wild-type mice. These authors suggest that lumican supports the homeostasis of the intestine by facilitating the inflammatory response found to be beneficial to the initial stages of colitis [111]. Similar results were obtained in the LPS-induced sepsis murine Lum$^{-/-}$ model. Indeed, lumican-deficient mice had a strongly downregulated inflammatory response to sterile inflammation. This was evident as an attenuated secretion of pro-inflammatory mediators, including TNF-α, IL-6,and IL-1β cytokines [112]. The correlation of lumican action to TLR signaling has been, likewise, implicated in a murine pathogen-induced inflammatory response. Thus, in mice, lumican seems to facilitate the bindingof bacteria to CD14 and the subsequent presentation of the comp0lex to TLR4 [113]. Indeed, CD-14 is a glycosyl phosphatidyl inositol-linked membrane protein that enhances TLR-2 and TLR-4 downstream signaling [114]. In this manner, lumican upregulatedbacteria phagocytosis [113]. Therefore, lumican has been characterized as a promotor of TLR4- and CD14-dependent pathogen sensing [115]. Likewise, lumican was demonstrated to modulate peripheral monocyte extravasation via Fas–FasL signaling [116]. Therefore, it seems feasible that the mechanisms mentioned above of lumican action can participate in the tumor inflammatory milieu.

7. Lumican as Prognosis Marker, Chemoresistance Regulator, and Cancer Target

The expression of lumican has been correlated with prognosis and disease stage in various cancer types. For example, in gastric and colon cancer, lumican expression was associated with cancer dissemination to secondary sites, lymphatic metastasis, and a poor overall survival of patients [48,51]. Moreover, in colon cancer, the expression of lumican was positively correlated with the disease stage [52]. In RCC, lumican exhibited a higher expression of lumican in metastatic tissue than patient-matched primary tissues, and is suggested as a metastases marker [60].

In melanoma and breast cancer, lumican exerts anticancer properties, and lumican-based therapeutic strategies have been examined [69,70].

Notably, lumican is suggested to modulate the response to chemotherapeutic agents. Thus, chemotherapeutic agents increased the secretion of lumican in PDAC cells, which was correlated with the extent of the therapy response. In various PDAC models, including cell lines, patient-derived xenografts, and lumican knockout mice, lumican was found to enhance the anticancer chemotherapy effect [104]. Specifically, chemotherapeutic agents in PDAC cells facilitate autophagosome formation and enhance LC3 expression through the ROS-mediated AMP-activated kinase (AMPK) signaling pathway. Conversely, lumican attenuates AMPK activity, abrogating the protective mechanism of chemotherapy-induced autophagy in in vitro and in vivo PDAC models [104]. This was correlatedwithDNA damage, apoptosis, downregulated cell viability, lactate production, glucose consumption, and release of vascular endothelial growth factor [104].

In rectal cancer patients treated with radiotherapy, apoptosis inducers (lumican, thrombospondin 2, and galectin-1) exhibited a higher expression in responders than in non-responding patients [117]. Therefore, gene expression profiling may benefit from radiotherapy response predictionand provide insights into developing novel therapeutic targets for rectal cancer.

Notably, leukemia stem cells (LSCs) have been correlated withtherapeutic failure and the relapse of acute lymphoblastic leukemia. The interaction between LSCs and bone marrow mesenchymal stem cells (BM-MSCs) results in a decreased expression of lumican by BM-MSCs cells. Importantly, a downregulated lumican expression results in attenuated apoptosis and enhanced chemoresistance to VP-16 in human pre-B cell leukemia Nalm-6 cells [118]. Therefore, reduced lumican expression by cells BM-MSCs may facilitate cancer cells' escape from chemotherapy and immune surveillance and support leukemia relapse [119].

Based on the available data and in a cancer-type specific manner, lumican roles as anticancer agents have been proposed. Thus, as lumican may attenuate or even annulate specific EMT-correlated metastatic features in breast cancer cells, a lumican-based anticancer therapy targeting EMT could be beneficial [47]. Furthermore, as lumican modulates the response to the matrix-targeting therapy peptide TAX inhibiting tumor growth, it presents a feasible therapeutic option for neuroblastoma [68].Indeed, the attenuation of the FOXO3/LUM axis by the small molecular weight compound repaglinide is suggested as a novel strategy for neuroblastoma and other FOXO3-dependent neoplasms [62]. The latter is depicted in Figure 1.

8. Conclusions

In conclusion, lumican is an important mediator of tumorigenesis and cancer progression involving the cellular functions of proliferation, motility, apoptosis, autophagy, and angiogenesis regulation, as represented in Figure 2. Lumican has been characterized as both an anticancer molecule and a tumor promoter. However, recent research efforts have shed more light on the characterization of its roles, which seem to depend on the tumor origin and type and disease stage.Therefore, lumican has been proposed as both a therapy target and anticancer agent. Moreover, by modulating specific biological functions, lumican can affect the response to chemotherapy and predict the response to radiotherapy. Considering that the tumor microenvironment is a complex network of different cell types, ECM components, and signaling molecules, with the ability to modulate cell growth and metastasis, defining its critical modulators is essential. The novel findings on the multivalent roles of lumicans have the potential to be translated into effective therapeutic strategies, and thus it is necessary to continue research efforts in this direction.

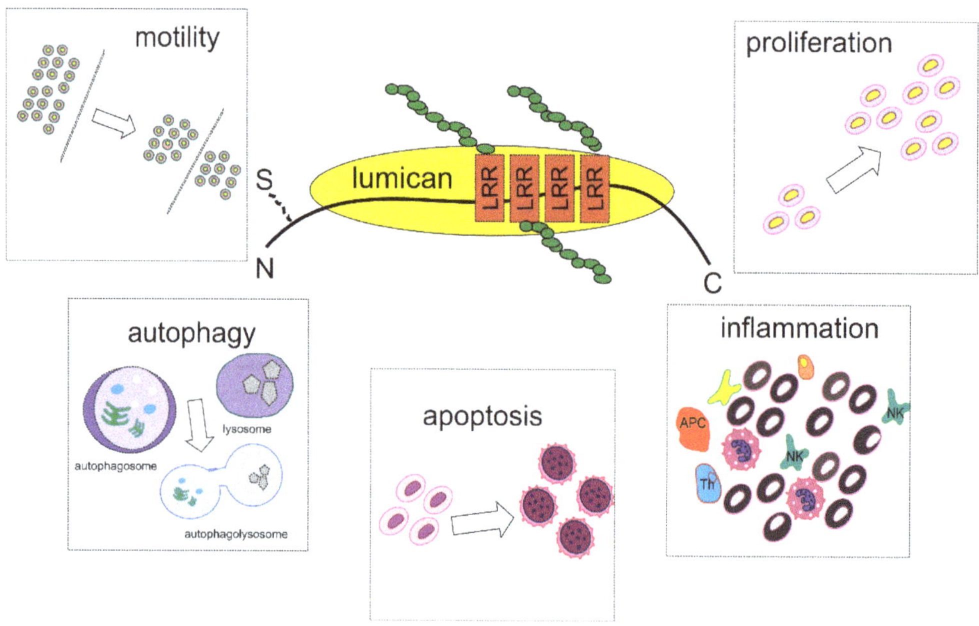

Figure 2. Lumican affects cancer cell behavior.Lumican alters cancer cell proliferation, migration, adhesion, invasion, metastasis, and apoptosis, and affects autophagy and inflammation signaling pathways with different mechanisms.

Author Contributions: Conceptualization, D.N. and E.-M.G.; writing—original draft preparation, E.-M.G., D.N., A.B., and A.T. writing—review and editing, D.N. and G.N.T. All authors have read and agreed to the published version of the manuscript.

Funding: D.N. was partially funded by the Research Committee of the University of Crete (ELKE), grant number (KA:10028; 10648).

Acknowledgments: This article is part of the Innogly Cost action (18103) initiative.

Conflicts of Interest: The authors declare no conflict of interest.

References

1. Tzanakakis, G.; Neagu, M.; Tsatsakis, A.; Nikitovic, D. Proteoglycans and Immunobiology of Cancer—Therapeutic Implications. *Front. Immunol.* **2019**, *10*, 875. [CrossRef] [PubMed]
2. Tzanakakis, G.; Giatagana, E.-M.; Kuskov, A.; Berdiaki, A.; Tsatsakis, A.; Neagu, M.; Nikitovic, D. Proteoglycans in the Pathogenesis of Hormone-Dependent Cancers: Mediators and Effectors. *Cancers* **2020**, *12*, 2401. [CrossRef] [PubMed]
3. Hsiao, K.C.; Chu, P.Y.; Chang, G.C.; Liu, K.J. Elevated Expression of Lumican in Lung Cancer Cells Promotes Bone Metastasis through an Autocrine Regulatory Mechanism. *Cancers* **2020**, *12*, 233. [CrossRef] [PubMed]
4. Winkler, J.; Abisoye-Ogunniyan, A.; Metcalf, K.J.; Werb, Z. Concepts of extracellular matrix remodelling in tumour progression and metastasis. *Nat. Commun.* **2020**, *11*, 1–19. [CrossRef]
5. Anderson, N.M.; Simon, M.C. The tumor microenvironment. *Curr. Biol.* **2020**, *30*, R921–R925. [CrossRef] [PubMed]
6. Hynes, R.O. The Extracellular Matrix: Not Just Pretty Fibrils. *Science* **2009**, *326*, 1216. [CrossRef] [PubMed]
7. Nikitovic, D.; Etzardi, M.; Eberdiaki, A.; Tsatsakis, A.; Tzanakakis, G.N. Cancer Microenvironment and Inflammation: Role of Hyaluronan. *Front. Immunol.* **2015**, *6*, 169. [CrossRef]
8. Nikitovic, D.; Berdiaki, A.; Spyridaki, I.; Krasanakis, T.; Tsatsakis, A.; Tzanakakis, G.N. Proteoglycans—Biomarkers and Targets in Cancer Therapy. *Front. Endocrinol.* **2018**, *9*, 69. [CrossRef]
9. Tampa, M.; Georgescu, S.; Mitran, M.; Mitran, C.; Matei, C.; Caruntu, A.; Scheau, C.; Nicolae, I.; Matei, A.; Caruntu, C.; et al. Current Perspectives on the Role of Matrix Metalloproteinases in the Pathogenesis of Basal Cell Carcinoma. *Biomolecules* **2021**, *11*, 903. [CrossRef]
10. Ahmad, R.; Eubank, T.; Lukomski, S.; Boone, B. Immune Cell Modulation of the Extracellular Matrix Contributes to the Pathogenesis of Pancreatic Cancer. *Biomolecules* **2021**, *11*, 901. [CrossRef]

11. Hanahan, D.; Weinberg, R.A. The hallmarks of cancer. *Cell* **2000**, *100*, 57–70. [CrossRef]

12. Wu, S.Z.; Roden, D.L.; Wang, C.; Holliday, H.; Harvey, K.; Cazet, A.S.; Murphy, K.J.; Pereira, B.; Al-Eryani, G.; Bartonicek, N.; et al. Stromal cell diversity associated with immune evasion in human triple-negative breast cancer. *EMBO J.* **2020**, *39*, e104063. [CrossRef]

13. Zeltz, C.; Primac, I.; Erusappan, P.; Alam, J.; Noel, A.; Gullberg, D. Cancer-associated fibroblasts in desmoplastic tumors: Emerging role of integrins. *Semin. Cancer Biol.* **2019**, *62*, 166–181. [CrossRef]

14. Thomas, D.; Radhakrishnan, P. Tumor-stromal crosstalk in pancreatic cancer and tissue fibrosis. *Mol. Cancer* **2019**, *18*, 1–15. [CrossRef] [PubMed]

15. Onyeisi, J.; Lopes, C.; Götte, M. Syndecan-4 as a Pathogenesis Factor and Therapeutic Target in Cancer. *Biomolecules* **2021**, *11*, 503. [CrossRef] [PubMed]

16. Iozzo, R.V.; Schaefer, L. Proteoglycan form and function: A comprehensive nomenclature of proteoglycans. *Matrix Biol.* **2015**, *42*, 11–55. [CrossRef] [PubMed]

17. Yu, P.; Pearson, C.S.; Geller, H.M. Flexible Roles for Proteoglycan Sulfation and Receptor Signaling. *Trends Neurosci.* **2017**, *41*, 47–61. [CrossRef]

18. Hayes, A.; Sugahara, K.; Farrugia, B.; Whitelock, J.M.; Caterson, B.; Melrose, J. Biodiversity of CS-proteoglycan sulphation motifs: Chemical messenger recognition modules with roles in information transfer, control of cellular behaviour and tissue morphogenesis. *Biochem. J.* **2018**, *475*, 587–620. [CrossRef]

19. Couchman, J.R.; Pataki, C.A. An Introduction to Proteoglycans and Their Localization. *J. Histochem. Cytochem.* **2012**, *60*, 885–897. [CrossRef]

20. Mytilinaiou, M.; Nikitovic, D.; Berdiaki, A.; Kostouras, A.; Papoutsidakis, A.; Tsatsakis, A.; Tzanakakis, G.N. Emerging roles of syndecan 2 in epithelial and mesenchymal cancer progression. *IUBMB Life* **2017**, *69*, 824–833. [CrossRef]

21. McEwan, P.A.; Scott, P.G.; Bishop, P.; Bella, J. Structural correlations in the family of small leucine-rich repeat proteins and proteoglycans. *J. Struct. Biol.* **2006**, *155*, 294–305. [CrossRef]

22. Kobe, B.; Kajava, A.V. The leucine-rich repeat as a protein recognition motif. *Curr. Opin. Struct. Biol.* **2001**, *11*, 725–732. [CrossRef]

23. Schaefer, L.; Iozzo, R.V. Biological Functions of the Small Leucine-rich Proteoglycans: From Genetics to Signal Transduction. *J. Biol. Chem.* **2008**, *283*, 21305–21309. [CrossRef] [PubMed]

24. Zhan, S.; Li, J.; Ge, W. Multifaceted Roles of Asporin in Cancer: Current Understanding. *Front. Oncol.* **2019**, *9*, 948. [CrossRef] [PubMed]

25. Sanders, E.J.; Walter, M.A.; Parker, E.; Ara'mburo, C.; Harvey, S. Opticin binds retinal growth hormone in the embryonic vitreous. *Investig. Opthalmol. Vis. Sci.* **2003**, *44*, 5404–5409. [CrossRef] [PubMed]

26. Nikitovic, D.; Papoutsidakis, A.; Karamanos, N.; Tzanakakis, G. Lumican affects tumor cell functions, tumor–ECM interactions, angiogenesis and inflammatory response. *Matrix Biol.* **2014**, *35*, 206–214. [CrossRef] [PubMed]

27. Kram, V.; Kilts, T.M.; Bhattacharyya, N.; Li, L.; Young, M.F. Small leucine rich proteoglycans, a novel link to osteoclastogenesis. *Sci. Rep.* **2017**, *7*, 1–17. [CrossRef]

28. Tzanakakis, G.; Giatagana, E.-M.; Berdiaki, A.; Spyridaki, I.; Hida, K.; Neagu, M.; Tsatsakis, A.; Nikitovic, D. The Role of IGF/IGF-IR-Signaling and Extracellular Matrix Effectors in Bone Sarcoma Pathogenesis. *Cancers* **2021**, *13*, 2478. [CrossRef] [PubMed]

29. Zafiropoulos, A.; Nikitovic, D.; Katonis, P.; Tsatsakis, A.; Karamanos, N.K.; Tzanakakis, G.N. Decorin-Induced Growth Inhibition Is Overcome through Protracted Expression and Activation of Epidermal Growth Factor Receptors in Osteosarcoma Cells. *Mol. Cancer Res.* **2008**, *6*, 785–794. [CrossRef]

30. Voudouri, K.; Nikitovic, D.; Berdiaki, A.; Kletsas, D.; Karamanos, N.; Tzanakakis, G.N. IGF-I/EGF and E2 signaling crosstalk through IGF-IR conduit point affects breast cancer cell adhesion. *Matrix Biol.* **2016**, *56*, 95–113. [CrossRef] [PubMed]

31. Aggelidakis, J.; Berdiaki, A.; Nikitovic, D.; Papoutsidakis, A.; Papachristou, D.J.; Tsatsakis, A.M.; Tzanakakis, G.N. Biglycan Regulates MG63 Osteosarcoma Cell Growth Through a LPR6/beta-Catenin/IGFR-IR Signaling Axis. *Front. Oncol.* **2018**, *8*, 470. [CrossRef]

32. Papoutsidakis, A.; Giatagana, E.M.; Berdiaki, A.; Spyridaki, I.; Spandidos, D.A.; Tsatsakis, A.; Tzanakakis, G.N.; Nikitovic, D. Lumican mediates HTB94 chondrosarcoma cell growth via an IGFIR/Erk1/2 axis. *Int. J. Oncol.* **2020**, *57*, 791–803. [CrossRef]

33. Chakravarti, S.; Stallings, R.L.; Sundarraj, N.; Cornuet, P.K.; Hassell, J.R. Primary Structure of Human Lumican (Keratan Sulfate Proteoglycan) and Localization of the Gene (LUM) to Chromosome 12q21.3–q22. *Genomics* **1995**, *27*, 481–488. [CrossRef]

34. Dunlevy, J.R.; Neame, P.J.; Vergnes, J.-P.; Hassell, J.R. Identification of the N-Linked Oligosaccharide Sites in Chick Corneal Lumican and Keratocan That Receive Keratan Sulfate. *J. Biol. Chem.* **1998**, *273*, 9615–9621. [CrossRef]

35. Roughley, P.J.; White, R.J.; Cs-Szabó, G.; Mort, J.S. Changes with age in the structure of fibromodulin in human articular cartilage. *Osteoarthr. Cartil.* **1996**, *4*, 153–161. [CrossRef]

36. Cornuet, P.K.; Blochberger, T.C.; Hassell, J.R. Molecular polymorphism of lumican during corneal development. *Investig. Ophthalmol. Vis. Sci.* **1994**, *35*, 870–877.

37. Grover, J.; Chen, X.N.; Korenberg, J.R.; Roughley, P.J. The human lumican gene. Organization, chromosomal location, and expression in articular cartilage. *J. Biol. Chem.* **1995**, *270*, 21942–21949. [CrossRef]

38. Chakravarti, S.; Magnuson, T.; Lass, J.H.; Jepsen, K.J.; LaMantia, C.; Carroll, H. Lumican Regulates Collagen Fibril Assembly: Skin Fragility and Corneal Opacity in the Absence of Lumican. *J. Cell Biol.* **1998**, *141*, 1277–1286. [CrossRef] [PubMed]

39. Ying, S.; Shiraishi, A.; Kao, C.W.-C.; Converse, R.L.; Funderburgh, J.L.; Swiergiel, J.; Roth, M.R.; Conrad, G.W.; Kao, W.W.-Y. Characterization and Expression of the Mouse Lumican Gene. *J. Biol. Chem.* **1997**, *272*, 30306–30313. [CrossRef]

40. Raouf, A.; Ganss, B.; McMahon, C.; Vary, C.; Roughley, P.J.; Seth, A. Lumican is a major proteoglycan component of the bone matrix. *Matrix Biol.* **2002**, *21*, 361–367. [CrossRef]

41. Quantock, A.J.; Meek, K.; Chakravarti, S. An x-ray diffraction investigation of corneal structure in lumican-deficient mice. *Investig. Ophthalmol. Vis. Sci.* **2001**, *42*, 1750–1756.

42. Chakravarti, S.; Petroll, W.M.; Hassell, J.R.; Jester, J.; Lass, J.H.; Paul, J.; Birk, D.E. Corneal opacity in lumican-null mice: Defects in collagen fibril structure and packing in the posterior stroma. *Investig. Ophthalmol. Vis. Sci.* **2000**, *41*, 3365–3373.

43. Mao, W.; Luo, M.; Huang, X.; Wang, Q.; Fan, J.; Gao, L.; Zhang, Y.; Geng, J. Knockdown of Lumican Inhibits Proliferation and Migration of Bladder Cancer. *Transl. Oncol.* **2019**, *12*, 1072–1078. [CrossRef]

44. De Wit, M.; Carvalho, B.; Diemen, P.M.D.-V.; Van Alphen, C.; Beliën, J.A.M.; Meijer, G.A.; Fijneman, R.J.A. Lumican and versican protein expression are associated with colorectal adenoma-to-carcinoma progression. *PLoS ONE* **2017**, *12*, e0174768. [CrossRef] [PubMed]

45. Jepsen, K.J.; Wu, F.; Peragallo, J.H.; Paul, J.; Roberts, L.; Ezura, Y.; Oldberg, A.; Birk, D.E.; Chakravarti, S. A Syndrome of Joint Laxity and Impaired Tendon Integrity in Lumican- and Fibromodulin-deficient Mice. *J. Biol. Chem.* **2002**, *277*, 35532–35540. [CrossRef] [PubMed]

46. Nikitovic, D.; Berdiaki, A.; Zafiropoulos, A.; Katonis, P.; Tsatsakis, A.; Karamanos, N.; Tzanakakis, G. Lumican expression is positively correlated with the differentiation and negatively with the growth of human osteosarcoma cells. *FEBS J.* **2007**, *275*, 350–361. [CrossRef] [PubMed]

47. Karamanou, K.; Franchi, M.; Vynios, D.; Brézillon, S. Epithelial-to-mesenchymal transition and invadopodia markers in breast cancer: Lumican a key regulator. *Semin. Cancer Biol.* **2019**, *62*, 125–133. [CrossRef] [PubMed]

48. Chen, L.; Zhang, Y.; Zuo, Y.; Ma, F.; Song, H. Lumican expression in gastric cancer and its association with biological behavior and prognosis. *Oncol. Lett.* **2017**, *14*, 5235–5240. [CrossRef]

49. Chen, X.; Li, X.; Hu, X.; Jiang, F.; Shen, Y.; Xu, R.; Wu, L.; Wei, P.; Shen, X. LUM Expression and Its Prognostic Significance in Gastric Cancer. *Front. Oncol.* **2020**, *10*, 605. [CrossRef]

50. Yamauchi, N.; Kanke, Y.; Saito, K.; Okayama, H.; Yamada, S.; Nakajima, S.; Endo, E.; Kase, K.; Yamada, L.; Nakano, H.; et al. Stromal expression of cancer-associated fibroblast-related molecules, versican and lumican, is strongly associated with worse relapse-free and overall survival times in patients with esophageal squamous cell carcinoma. *Oncol. Lett.* **2021**, *21*, 445. [CrossRef]

51. Seya, T.; Tanaka, N.; Shinji, S.; Yokoi, K.; Koizumi, M.; Teranishi, N.; Yamashita, K.; Tajiri, T.; Ishiwata, T.; Naito, Z. Lumican expression in advanced colorectal cancer with nodal metastasis correlates with poor prognosis. *Oncol. Rep.* **2006**, *16*, 1225–1230. [CrossRef] [PubMed]

52. Zang, Y.; Dong, Q.; Lu, Y.; Dong, K.; Wang, R.; Liang, Z. Lumican inhibits immune escape and carcinogenic pathways in colorectal adenocarcinoma. *Aging* **2021**, *13*, 4388–4408. [CrossRef]

53. De Wit, M.; Belt, E.J.T.; Diemen, P.M.D.-V.; Carvalho, B.; Coupé, V.M.H.; Stockmann, H.B.A.C.; Bril, H.; Beliën, J.A.M.; Fijneman, R.J.A.; Meijer, G.A. Lumican and Versican Are Associated with Good Outcome in Stage II and III Colon Cancer. *Ann. Surg. Oncol.* **2013**, *20* (Suppl. 3), S348–S359. [CrossRef] [PubMed]

54. Dcyali, C.; Truty, M.A.; Kang, Y.; Chopin-Laly, X.; Zhang, R.; Roife, D.J.; Chatterjee, D.; Lin, E.; Thomas, R.M.; Wang, H.; et al. Extracellular Lumican Inhibits Pancreatic Cancer Cell Growth and Is Associated with Prolonged Survival after Surgery. *Clin. Cancer Res.* **2014**, *20*, 6529–6540. [CrossRef]

55. Kang, Y.; Roife, D.; Lee, Y.; Lv, H.; Suzuki, R.; Ling, J.; Rios Perez, M.V.; Li, X.; Dai, B.; Pratt, M.; et al. Transforming Growth Factor-beta Limits Secretion of Lumican by Activated Stellate Cells within Primary Pancreatic Adenocarcinoma Tumors. *Clin. Cancer Res.* **2016**, *22*, 4934–4946. [CrossRef] [PubMed]

56. Lu, P.; Weaver, V.M.; Werb, Z. The extracellular matrix: A dynamic niche in cancer progression. *J. Cell Biol.* **2012**, *196*, 395–406. [CrossRef] [PubMed]

57. Brézillon, S.; Ventéo, L.; Ramont, L.; D'Onofrio, M.-F.; Perreau, C.; Pluot, M.; Maquart, F.-X.; Wegrowski, Y. Expression of lumican, a small leucine-rich proteoglycan with antitumour activity, in human malignant melanoma. *Clin. Exp. Dermatol.* **2007**, *32*, 405–416. [CrossRef]

58. Nikitovic, D.; Chalkiadaki, G.; Berdiaki, A.; Aggelidakis, J.; Katonis, P.; Karamanos, N.K.; Tzanakakis, G.N. Lumican regulates osteosarcoma cell adhesion by modulating TGFbeta2 activity. *Int. J. Biochem. Cell Biol.* **2011**, *43*, 928–935. [CrossRef]

59. Appunni, S.; Rubens, M.; Ramamoorthy, V.; Anand, V.; Khandelwal, M.; Saxena, A.; McGranaghan, P.; Odia, Y.; Kotecha, R.; Sharma, A. Lumican, pro-tumorigenic or anti-tumorigenic: A conundrum. *Clin. Chim. Acta* **2020**, *514*, 1–7. [CrossRef]

60. Ho, T.H.; Serie, D.J.; Parasramka, M.; Cheville, J.C.; Bot, B.M.; Tan, W.; Wang, L.; Joseph, R.W.; Hilton, T.; Leibovich, B.C.; et al. Differential gene expression profiling of matched primary renal cell carcinoma and metastases reveals upregulation of extracellular matrix genes. *Ann. Oncol.* **2016**, *28*, 604–610. [CrossRef]

61. Radwanska, A.; Litwin, M.; Nowak, D.; Baczynska, D.; Wegrowski, Y.; Maquart, F.-X.; Malicka-Blaszkiewicz, M. Overexpression of lumican affects the migration of human colon cancer cells through up-regulation of gelsolin and filamentous actin reorganization. *Exp. Cell Res.* **2012**, *318*, 2312–2323. [CrossRef]

62. Salcher, S.; Spoden, G.; Huber, J.M.; Golderer, G.; Lindner, H.; Ausserlechner, M.J.; Kiechl-Kohlendorfer, U.; Geiger, K.; Obexer, P. Repaglinide Silences the FOXO3/Lumican Axis and Represses the Associated Metastatic Potential of Neuronal Cancer Cells. *Cells* **2019**, *9*, 1. [CrossRef] [PubMed]

63. Wang, X.; Zhou, Q.; Yu, Z.; Wu, X.; Chen, X.; Li, J.; Li, C.; Yan, M.; Zhu, Z.; Liu, B.; et al. Cancer-associated fibroblast-derived Lumican promotes gastric cancer progression via the integrin beta1-FAK signaling pathway. *Int. J. Cancer* **2017**, *141*, 998–1010. [CrossRef] [PubMed]

64. Nowak, D.; Krawczenko, A.; Duś, D.; Malicka-Błaszkiewicz, M. Actin in human colon adenocarcinoma cells with different metastatic potential. *Acta Biochim. Pol.* **2002**, *49*, 823–828. [CrossRef] [PubMed]

65. Mu, Q.-M.; He, W.; Hou, G.-M.; Liang, Y.; Wang, G.; Li, C.-L.; Liao, B.; Liu, X.; Ye, Z.; Lu, J.-L.; et al. Interference of Lumican Regulates the Invasion and Migration of Liver Cancer Cells. *Sichuan Da Xue Xue Bao Yi Xue Ban* **2018**, *49*, 358–363. [PubMed]

66. Li, X.; Kang, Y.; Roife, D.; Lee, Y.; Pratt, M.; Perez, M.R.; Dai, B.; Koay, E.J.; Fleming, J.B. Prolonged exposure to extracellular lumican restrains pancreatic adenocarcinoma growth. *Oncogene* **2017**, *36*, 5432–5438. [CrossRef]

67. Li, X.; Lee, Y.; Kang, Y.; Dai, B.; Perez, M.R.; Pratt, M.; Koay, E.J.; Kim, M.; Brekken, R.A.; Fleming, J.B. Hypoxia-induced autophagy of stellate cells inhibits expression and secretion of lumican into microenvironment of pancreatic ductal adenocarcinoma. *Cell Death Differ.* **2018**, *26*, 382–393. [CrossRef]

68. Jeanne, A.; Untereiner, V.; Perreau, C.; Proult, I.; Gobinet, C.; Boulagnon-Rombi, C.; Terryn, C.; Martiny, L.; Brézillon, S.; Dedieu, S. Lumican delays melanoma growth in mice and drives tumor molecular assembly as well as response to matrix-targeted TAX2 therapeutic peptide. *Sci. Rep.* **2017**, *7*, 1–16. [CrossRef]

69. Brézillon, S.; Radwanska, A.; Zeltz, C.; Malkowski, A.; Ploton, D.; Bobichon, H.; Perreau, C.; Malicka-Blaszkiewicz, M.; Maquart, F.-X.; Wegrowski, Y. Lumican core protein inhibits melanoma cell migration via alterations of focal adhesion complexes. *Cancer Lett.* **2009**, *283*, 92–100. [CrossRef]

70. Karamanou, K.; Franchi, M.; Proult, I.; Rivet, R.; Vynios, D.; Brézillon, S. Lumican Inhibits In Vivo Melanoma Metastasis by Altering Matrix-Effectors and Invadopodia Markers. *Cells* **2021**, *10*, 841. [CrossRef]

71. Brézillon, S.; Untereiner, V.; Mohamed, H.; Ahallal, E.; Proult, I.; Nizet, P.; Boulagnon-Rombi, C.; Sockalingum, G.D. Label-Free Infrared Spectral Histology of Skin Tissue Part II: Impact of a Lumican-Derived Peptide on Melanoma Growth. *Front. Cell Dev. Biol.* **2020**, *8*. [CrossRef]

72. Yang, C.-T.; Hsu, P.-C.; Chow, S.-E. Downregulation of lumican enhanced mitotic defects and aneuploidy in lung cancer cells. *Cell Cycle* **2019**, *19*, 97–108. [CrossRef]

73. Yang, C.-T.; Li, J.-M.; Chu, W.-K.; Chow, S.-E. Downregulation of lumican accelerates lung cancer cell invasion through p120 catenin. *Cell Death Dis.* **2018**, *9*, 414. [CrossRef] [PubMed]

74. Saitoh, M. Involvement of partial EMT in cancer progression. *J. Biochem.* **2018**, *164*, 257–264. [CrossRef]

75. Karamanou, K.; Franchi, M.; Piperigkou, Z.; Perreau, C.; Maquart, F.-X.; Vynios, D.H.; Brézillon, S. Lumican effectively regulates the estrogen receptors-associated functional properties of breast cancer cells, expression of matrix effectors and epithelial-to-mesenchymal transition. *Sci. Rep.* **2017**, *7*, srep45138. [CrossRef] [PubMed]

76. Karamanou, K.; Franchi, M.; Onisto, M.; Passi, A.; Vynios, D.H.; Brézillon, S. Evaluation of lumican effects on morphology of invading breast cancer cells, expression of integrins and downstream signaling. *FEBS J.* **2020**, *287*, 4862–4880. [CrossRef]

77. Leygue, E.; Snell, L.; Dotzlaw, H.; Hole, K.; Hiller-Hitchcock, T.; Roughley, P.J.; Watson, P.H.; Murphy, L.C. Expression of lumican in human breast carcinoma. *Cancer Res.* **1998**, *58*, 1348–1352. [PubMed]

78. Wang, Y.; Shi, J.; Chai, K.; Ying, X.; Zhou, B.P. The Role of Snail in EMT and Tumorigenesis. *Curr. Cancer Drug Targets* **2013**, *13*, 963–972. [CrossRef] [PubMed]

79. Pietraszek-Gremplewicz, K.; Chatron-Colliet, A.; Brézillon, S.; Perreau, C.; Jakubiak-Augustyn, A.; Krotkiewski, H.; Maquart, F.-X.; Wegrowski, Y. Lumican: A new inhibitor of matrix metalloproteinase-14 activity. *FEBS Lett.* **2014**, *588*, 4319–4324. [CrossRef] [PubMed]

80. Pietraszek-Gremplewicz, K.; Karamanou, K.; Niang, A.; Dauchez, M.; Belloy, N.; Maquart, F.X.; Baud, S.; Brezillon, S. Small leucine-rich proteoglycans and matrix metalloproteinase-14: Key partners? *Matrix Biol.* **2019**, *75*, 271–285. [CrossRef]

81. Karamanou, K.; Perrot, G.; Maquart, F.-X.; Brézillon, S. Lumican as a multivalent effector in wound healing. *Adv. Drug Deliv. Rev.* **2018**, *129*, 344–351. [CrossRef]

82. Malinowski, M.; Pietraszek, K.; Perreau, C.; Boguslawski, M.; Decot, V.; Stoltz, J.-F.; Vallar, L.; Niewiarowska, J.; Cierniewski, C.; Maquart, F.-X.; et al. Effect of Lumican on the Migration of Human Mesenchymal Stem Cells and Endothelial Progenitor Cells: Involvement of Matrix Metalloproteinase-14. *PLoS ONE* **2012**, *7*, e50709. [CrossRef]

83. Seomun, Y.; Kim, J.-T.; Joo, C.-K. MMP-14 mediated MMP-9 expression is involved in TGF-beta1-induced keratinocyte migration. *J. Cell. Biochem.* **2008**, *104*, 934–941. [CrossRef]

84. Takino, T.; Suzuki, T.; Seiki, M. Isolation of Highly Migratory and Invasive Cells in Three-Dimensional Gels. *Curr. Protoc. Cell Biol.* **2020**, *86*, e103. [CrossRef]

85. Long, W.; Yi, P.; Amazit, L.; LaMarca, H.L.; Ashcroft, F.; Kumar, R.; Mancini, M.A.; Tsai, S.Y.; Tsai, M.J.; O'Malley, B.W. SRC-3Delta4 mediates the interaction of EGFR with FAK to promote cell migration. *Mol. Cell* **2010**, *37*, 321–332. [CrossRef] [PubMed]

86. Brézillon, S.; Pietraszek-Gremplewicz, K.; Maquart, F.-X.; Wegrowski, Y. Lumican effects in the control of tumour progression and their links with metalloproteinases and integrins. *FEBS J.* **2013**, *280*, 2369–2381. [CrossRef]

87. Zeltz, C.; Brézillon, S.; Perreau, C.; Ramont, L.; Maquart, F.-X.; Wegrowski, Y. Lumcorin: A leucine-rich repeat 9-derived peptide from human lumican inhibiting melanoma cell migration. *FEBS Lett.* **2009**, *583*, 3027–3032. [CrossRef]

88. Pietraszek, K.; Brézillon, S.; Perreau, C.; Malicka-Błaszkiewicz, M.; Maquart, F.-X.; Wegrowski, Y. Lumican—Derived Peptides Inhibit Melanoma Cell Growth and Migration. *PLoS ONE* **2013**, *8*, e76232. [CrossRef] [PubMed]

89. Kuphal, S.; Palm, H.G.; Poser, I.; Bosserhoff, A. Snail-regulated genes in malignant melanoma. *Melanoma Res.* **2005**, *15*, 305–313. [CrossRef]

90. Coulson-Thomas, V.J.; Coulson-Thomas, Y.M.; Gesteira, T.F.; de Paula, C.A.A.; Carneiro, C.R.; Ortiz, V.; Toma, L.; Kao, W.W.-Y.; Nader, H.B. Lumican expression, localization and antitumor activity in prostate cancer. *Exp. Cell Res.* **2013**, *319*, 967–981. [CrossRef] [PubMed]

91. Stasiak, M.; Boncela, J.; Perreau, C.; Karamanou, K.; Chatron-Colliet, A.; Proult, I.; Przygodzka, P.; Chakravarti, S.; Maquart, F.-X.; Kowalska, M.A.; et al. Lumican Inhibits SNAIL-Induced Melanoma Cell Migration Specifically by Blocking MMP-14 Activity. *PLoS ONE* **2016**, *11*, e0150226. [CrossRef]

92. Zeltz, C.; Brezillon, S.; Kapyla, J.; Eble, J.A.; Bobichon, H.; Terryn, C.; Perreau, C.; Franz, C.M.; Heino, J.; Maquart, F.X.; et al. Lumican inhibits cell migration through alpha2beta1 integrin. *Exp. Cell Res.* **2010**, *316*, 2922–2931. [CrossRef]

93. Zhou, K.; Ge, M. Effect of lumican gene over-expression on proliferation of lung adenocarcinoma cell A549 and its mechanism. *Xi Bao Yu Fen Zi Mian Yi Xue Za Zhi* **2013**, *29*, 492–495. [PubMed]

94. Grzesiak, J.J.; Bouvet, M. The alpha2beta1 integrin mediates the malignant phenotype on type I collagen in pancreatic cancer cell lines. *Br. J. Cancer* **2006**, *94*, 1311–1319. [CrossRef]

95. Evan, G.I.; Vousden, K.H. Proliferation, cell cycle and apoptosis in cancer. *Nature* **2001**, *411*, 342–348. [CrossRef] [PubMed]

96. Vij, N.; Roberts, L.; Joyce, S.; Chakravarti, S. Lumican suppresses cell proliferation and aids Fas–Fas ligand mediated apoptosis: Implications in the cornea. *Exp. Eye Res.* **2004**, *78*, 957–971. [CrossRef]

97. Vuillermoz, B.; Khoruzhenko, A.; D'Onofrio, M.F.; Ramont, L.; Venteo, L.; Perreau, C.; Antonicelli, F.; Maquart, F.X.; Wegrowski, Y. The small leucine-rich proteoglycan lumican inhibits melanoma progression. *Exp. Cell Res.* **2004**, *296*, 294–306. [CrossRef] [PubMed]

98. Williams, K.E.; Fulford, L.A.; Albig, A.R. Lumican Reduces Tumor Growth Via Induction of Fas-Mediated Endothelial Cell Apoptosis. *Cancer Microenviron.* **2010**, *4*, 115–126. [CrossRef]

99. Mizushima, N.; Yoshimori, T.; Ohsumi, Y. The Role of Atg Proteins in Autophagosome Formation. *Annu. Rev. Cell Dev. Biol.* **2011**, *27*, 107–132. [CrossRef]

100. Levy, J.M.M.; Towers, C.G.; Thorburn, A. Targeting autophagy in cancer. *Nat. Rev. Cancer* **2017**, *17*, 528–542. [CrossRef]

101. Zhong, Z.; Sanchez-Lopez, E.; Karin, M. Autophagy, Inflammation, and Immunity: A Troika Governing Cancer and Its Treatment. *Cell* **2016**, *166*, 288–298. [CrossRef]

102. Gu, Y.; Li, P.; Peng, F.; Zhang, M.; Zhang, Y.; Liang, H.; Zhao, W.; Qi, L.; Wang, H.; Wang, C.; et al. Autophagy-related prognostic signature for breast cancer. *Mol. Carcinog.* **2015**, *55*, 292–299. [CrossRef] [PubMed]

103. Neill, T.; Kapoor, A.; Xie, C.; Buraschi, S.; Iozzo, R.V. A functional outside-in signaling network of proteoglycans and matrix molecules regulating autophagy. *Matrix Biol.* **2021**, *100*, 118–149. [CrossRef]

104. Li, X.; Roife, D.; Kang, Y.; Dai, B.; Pratt, M.; Fleming, J.B. Extracellular lumican augments cytotoxicity of chemotherapy in pancreatic ductal adenocarcinoma cells via autophagy inhibition. *Oncogene* **2016**, *35*, 4881–4890. [CrossRef] [PubMed]

105. Fu, Y.; Liu, S.; Zeng, S.; Shen, H. The critical roles of activated stellate cells-mediated paracrine signaling, metabolism and onco-immunology in pancreatic ductal adenocarcinoma. *Mol. Cancer* **2018**, *17*, 1–14. [CrossRef]

106. Aggarwal, B.B.; Vijayalekshmi, R.V.; Sung, B. Targeting Inflammatory Pathways for Prevention and Therapy of Cancer: Short-Term Friend, Long-Term Foe. *Clin. Cancer Res.* **2009**, *15*, 425–430. [CrossRef] [PubMed]

107. Grivennikov, S.I.; Greten, F.R.; Karin, M. Immunity, inflammation, and cancer. *Cell* **2010**, *140*, 883–899. [CrossRef]

108. Avgustinova, A.; Iravani, M.; Robertson, D.; Fearns, A.; Gao, Q.; Klingbeil, P.; Hanby, A.M.; Speirs, V.; Sahai, E.; Calvo, F.; et al. Tumour cell-derived Wnt7a recruits and activates fibroblasts to promote tumour aggressiveness. *Nat. Commun.* **2016**, *7*, 10305. [CrossRef]

109. Schaefer, L.; Tredup, C.; Gubbiotti, M.A.; Iozzo, R.V. Proteoglycan neofunctions: Regulation of inflammation and autophagy in cancer biology. *FEBS J.* **2016**, *284*, 10–26. [CrossRef]

110. Guo, D.; Zhang, W.; Yang, H.; Bi, J.; Xie, Y.; Cheng, B.; Wang, Y.; Chen, S. Celastrol Induces Necroptosis and Ameliorates Inflammation via Targeting Biglycan in Human Gastric Carcinoma. *Int. J. Mol. Sci.* **2019**, *20*, 5716. [CrossRef]

111. Lohr, K.; Sardana, H.; Lee, S.; Wu, F.; Huso, D.L.; Hamad, A.R.; Chakravarti, S. Extracellular matrix protein lumican regulates inflammation in a mouse model of colitis. *Inflamm. Bowel Dis.* **2012**, *18*, 143–151. [CrossRef]

112. Lu, X.-M.; Ma, L.; Jin, Y.-N.; Yu, Y.-Q. Lumican overexpression exacerbates lipopolysaccharide-induced renal injury in mice. *Mol. Med. Rep.* **2015**, *12*, 4089–4094. [CrossRef]

113. Skjesol, A.; Yurchenko, M.; Bösl, K.; Gravastrand, C.; Nilsen, K.E.; Grøvdal, L.M.; Agliano, F.; Patane, F.; Lentini, G.; Kim, H.; et al. The TLR4 adaptor TRAM controls the phagocytosis of Gram-negative bacteria by interacting with the Rab11-family interacting protein 2. *PLOS Pathog.* **2019**, *15*, e1007684. [CrossRef]

114. Jiang, Z.; Georgel, P.; Du, X.; Shamel, L.; Sovath, S.; Mudd, S.; Huber, M.; Kalis, C.; Keck, S.; Galanos, C.; et al. CD14 is required for MyD88-independent LPS signaling. *Nat. Immunol.* **2005**, *6*, 565–570. [CrossRef]

115. Maiti, G.; Frikeche, J.; Lam, C.Y.-M.; Biswas, A.; Shinde, V.; Samanovic, M.; Kagan, J.C.; Mulligan, M.J.; Chakravarti, S. Matrix lumican endocytosed by immune cells controls receptor ligand trafficking to promote TLR4 and restrict TLR9 in sepsis. *Proc. Natl. Acad. Sci. USA* **2021**, *118*, e2100999118. [CrossRef]
116. Wu, F.; Vij, N.; Roberts, L.; López-Briones, S.; Joyce, S.; Chakravarti, S. A Novel Role of the Lumican Core Protein in Bacterial Lipopolysaccharide-induced Innate Immune Response. *J. Biol. Chem.* **2007**, *282*, 26409–26417. [CrossRef] [PubMed]
117. Watanabe, T.; Komuro, Y.; Kiyomatsu, T.; Kanazawa, T.; Kazama, Y.; Tanaka, J.; Tanaka, T.; Yamamoto, Y.; Shirane, M.; Muto, T.; et al. Prediction of Sensitivity of Rectal Cancer Cells in Response to Preoperative Radiotherapy by DNA Microarray Analysis of Gene Expression Profiles. *Cancer Res.* **2006**, *66*, 3370–3374. [CrossRef] [PubMed]
118. Yu, Z.; Liu, L.; Shu, Q.; Li, D.; Wang, R. Leukemia stem cells promote chemoresistance by inducing downregulation of lumican in mesenchymal stem cells. *Oncol. Lett.* **2019**, *18*, 4317–4327. [CrossRef] [PubMed]
119. Dick, J.E.; Bhatia, M.; Gan, O.; Kapp, U.; Wang, J.C.Y. Assay of human stem cells by repopulation of NOD/SCID mice. *STEM CELLS* **1997**, *15* (Suppl. 1), 199–207. [CrossRef] [PubMed]

Review

Immune Cell Modulation of the Extracellular Matrix Contributes to the Pathogenesis of Pancreatic Cancer

Ramiz S. Ahmad [1], Timothy D. Eubank [2,3], Slawomir Lukomski [2,3] and Brian A. Boone [1,2,3,*]

[1]   Department of Surgery, West Virginia University, Morgantown, WV 26506, USA; ramiz.ahmad@hsc.wvu.edu
[2]   Department of Microbiology, Immunology and Cell Biology, West Virginia University, Morgantown,
      WV 26506, USA; tdeubank@hsc.wvu.edu (T.D.E.); slukomski@hsc.wvu.edu (S.L.)
[3]   West Virginia University Cancer Institute, West Virginia University, Morgantown, WV 26506, USA
*   Correspondence: brian.boone@hsc.wvu.edu

Abstract: Pancreatic ductal adenocarcinoma (PDAC) is a highly lethal malignancy with a five-year survival rate of only 9%. PDAC is characterized by a dense, fibrotic stroma composed of extracellular matrix (ECM) proteins. This desmoplastic stroma is a hallmark of PDAC, representing a significant physical barrier that is immunosuppressive and obstructs penetration of cytotoxic chemotherapy agents into the tumor microenvironment (TME). Additionally, dense ECM promotes hypoxia, making tumor cells refractive to radiation therapy and alters their metabolism, thereby supporting proliferation and survival. In this review, we outline the significant contribution of fibrosis to the pathogenesis of pancreatic cancer, with a focus on the cross talk between immune cells and pancreatic stellate cells that contribute to ECM deposition. We emphasize the cellular mechanisms by which neutrophils and macrophages, specifically, modulate the ECM in favor of PDAC-progression. Furthermore, we investigate how activated stellate cells and ECM influence immune cells and promote immunosuppression in PDAC. Finally, we summarize therapeutic strategies that target the stroma and hinder immune cell promotion of fibrogenesis, which have unfortunately led to mixed results. An enhanced understanding of the complex interactions between the pancreatic tumor ECM and immune cells may uncover novel treatment strategies that are desperately needed for this devastating disease.

Keywords: pancreatic ductal adenocarcinoma; extracellular matrix; fibrosis; immune cell modulation; neutrophils; neutrophil extracellular trap; macrophages

Citation: Ahmad, R.S.; Eubank, T.D.; Lukomski, S.; Boone, B.A. Immune Cell Modulation of the Extracellular Matrix Contributes to the Pathogenesis of Pancreatic Cancer. *Biomolecules* **2021**, *11*, 901. https://doi.org/10.3390/biom11060901

Academic Editor: Vladimir N. Uversky

Received: 21 May 2021
Accepted: 13 June 2021
Published: 17 June 2021

Publisher's Note: MDPI stays neutral with regard to jurisdictional claims in published maps and institutional affiliations.

1. Introduction

Pancreatic ductal adenocarcinoma (PDAC) is one of the most lethal malignancies of the gastrointestinal system, with a five-year survival rate of only 9% [1]. It is currently the seventh-leading cause of deaths among cancers worldwide [2], and the incidence continues to rise [1,3]. At the time of pancreatic cancer diagnosis, up to 80% of patients present with metastatic or unresectable disease [4]. Pancreatic cancer is immensely difficult to treat, largely due to the dense, fibrotic stroma that dominates much of the tumor microenvironment (TME) [5]. A significant portion of the stroma is composed of extracellular matrix (ECM) proteins deposited through a desmoplastic reaction [6]. Desmoplasia is a fibro-inflammatory process of the stroma that consists of immune cells, proliferative fibroblasts, and abundant deposition of ECM proteins such as collagens and fibronectin [7–10]. While several therapeutic strategies for PDAC exist, the fibrotic stroma is a significant barrier to drug efficacy. Recent investigations have implicated immune cells such as neutrophils and macrophages in their contributions to the PDAC fibrotic stroma (Figure 1). In this review, we outline how the PDAC TME is established and the significant contribution of fibrosis to the pathogenesis, therapeutic resistance, metabolic adaptation, and immuno-suppressive nature of PDAC. Moreover, we summarize the recent literature available on neutrophils and macrophages promoting PDAC fibrosis. Lastly, because of the therapeutic challenges presented by desmoplasia, we discuss current therapeutic strategies that target these immune cells with the aim of reducing PDAC fibrosis.

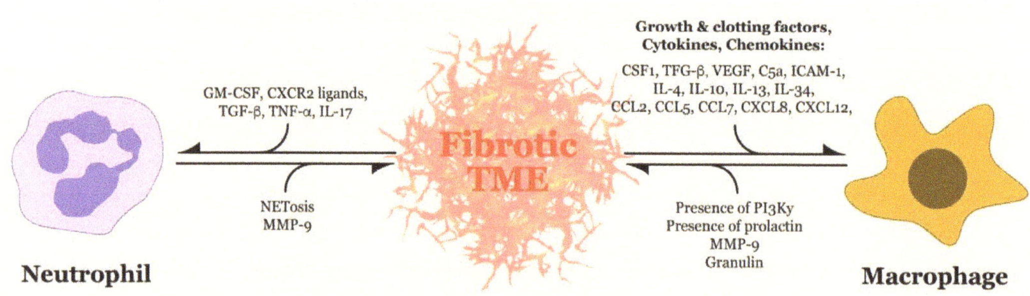

Figure 1. Crosstalk between the fibrotic tumor microenvironment (TME) and neutrophils and macrophages in pancreatic ductal adenocarcinoma. The fibrotic TME releases a variety of chemical mediators that recruit neutrophils and macrophages into the TME. In turn, neutrophils and macrophages possess characteristics and/or release their own factors that increase TME fibrosis. CSF1, colony stimulating factor 1; TGF-β, transforming growth factor-β; VEGF, vascular endothelial growth factor-A; C5a, complement component C5a; ICAM-1, intercellular adhesion molecule-1; IL, interleukin; CCL, chemokine (C-C motif) ligand; CXCL, chemokine (C-X-C motif) ligand; PI3Kγ, phosphatidylinositol 3-kinase gamma; MMP-9, matrix metalloproteinase 9; GM-CSF, granulocyte-macrophage colony-stimulating factor; TNF-α, tumor necrosis factor-alpha; NETosis, neutrophil extracellular trap release.

2. The Pancreatic Adenocarcinoma Tumor Microenvironment

The pancreatic TME consists of cellular and acellular components including ductal epithelial cells, fibroblasts, myofibroblasts, activated pancreatic stellate cells (PSCs), and a host of immune cells including regulatory T cells, myeloid-derived suppressor cells, tumor associated macrophages (TAMs), and tumor associated neutrophils (TANs) [11,12].

Quiescent fibroblasts are cells that comprise most of the stroma in various tissues. They are intimately involved in ECM modulation by secreting numerous ECM proteins such as collagens, elastin, and fibronectin [13,14]. Fibroblasts are typically recruited to an area of tissue insult by transforming growth factor-β (TGF-β), platelet-derived growth factor (PDGF), and fibroblast growth factor 2 (FGF2). After their recruitment, the fibroblasts are activated and promote the wound healing response through both cytoskeletal and ECM remodeling. After resolution of the injury, fibroblast activation is reversible through apoptosis. However, if the signals associated with tissue damage such as TGF-β, PDGF, and FGF2 are incessant, as is the situation in malignancy, the activated fibroblasts become hyper-proliferative and can become cancer-associated fibroblasts (CAFs) [15].

The PDAC stroma is imbued with a heterogeneous and plastic population of CAFs [6,11,16]. Fibroblasts inside the tumor mass differentiate into CAFs when exposed TGF-β, produced by PDAC cells, stromal cells, and TAMs [17,18]. The conversion of fibroblasts into CAFs is a positive feedback loop, as the formation of CAFs mechanically releases more TGF-β from its binding protein, latency-associated peptide (LAP) [19]. The CAFs have an active role in the TME, where they enable tumorigenic functions through the release of several pro-inflammatory cytokines, such as TGF-β, granulocyte-macrophage colony-stimulating factor (GM-CSF), colony stimulating factor 1 (CSF-1), and CCL2 [16,20,21]. Moreover, CAFs secrete multiple growth factors such as vascular endothelial growth factor-A (VEGF-A), hepatocyte growth factor (HGF), and platelet-derived growth factor (PDGF). These growth factors facilitate recruitment of immune cells and endothelial cells into the TME [19].

An important fibroblast population known as pancreatic stellate cells (PSC) are quiescent, star-shaped cells that reside in the basolateral portions of pancreatic acinar cells. Notable intracellular characteristics of PSCs include a large nucleus, limited mitochondria, and copious amounts of vitamin A- and albumin-containing fat droplets [22]. The role of quiescent PSCs has not yet been fully elucidated, however, it is thought that they are involved in structural support of the basement membrane by providing scaffolding [23]. During tumorigenesis, PSCs can become activated by the stimulating factors such as TGF-β, Interleukin-10 (IL-10), and PDGF that are released from PDAC cells and stromal cells [24]. Once activated, PSCs represent the most common subpopulation of CAFs [6,25]. In contrast to their quiescent counterparts, activated PSCs do not possess fat droplets. The mechanism by which fat droplets disappear or the impact of their absence on PDAC progression has not yet been answered [23,26]. The physiological hallmark of PSC activation is the expression of α-smooth muscle actin (α-SMA), a type of cytoskeletal protein [27]. Activated PSCs are a key contributor to the PDAC fibrotic stroma as they increasingly release ECM proteins such as collagen, periostin, fibronectin, matrix metalloproteinases (MMPs), and tissue inhibitors of matrix metalloproteinases (TIMPs) [25,28,29]. There have been conflicting results on the association of MMP type with patient survival, but increased MMP-7 has more consistently correlated with poor patient survival [30].

A recent proteomic analysis of the ECM in PDAC progression found that elevated levels of fibronectin and periostin were significantly associated with worse patient outcomes [31]. Increased deposition of collagen I and collagen IV correlate with reduced patient survival, whereas collagen III levels do not have a statistically significant association with patient survival [32]. Further, it was found that high circulating levels of collagen IV after PDAC resection correlated with reduced patient survival [33]. Interestingly, high alignment of collagens in PDAC tumors is associated with poor patient survival and correlates to stromal activation [34].

3. Contributions of Fibrosis in the TME to Pathogenesis of Pancreatic Cancer

There are multiple mechanisms through which the characteristic PDAC TME facilitates PDAC progression by enhancing tumor growth and promoting metastases (Figure 2).

3.1. Therapeutic Resistance through Limiting Penetration of Cytotoxic Agents

The overabundance and imbalance of released ECM proteins establishes a high interstitial pressure environment resulting in decreased perfusion of the tumor. This high pressure and diminished perfusion prevent infiltration of cytotoxic agents into the TME [35,36]. Several treatment strategies that target the stroma, outlined below, have been designed to enhance penetration of chemotherapy to the tumor, thereby increasing efficacy and treatment response.

3.2. Promotion of Hypoxia in the TME

Deposition of ECM proteins amplifies tissue tension and intra-tumoral pressure. These effects disrupt local blood circulation and oxygen diffusion in pancreatic tissue, leading to hypoxia [37,38]. In response to the hypoxic environment, hypoxia-inducible transcription factors (HIFs) are stabilized. The HIF-α/ARNT heterodimer, along with transcriptional coactivators, bind to hypoxia response elements (HREs) in gene promoters leading to transcription of genes in PDAC cells that facilitate glycolysis, tumorigenesis, and metastasis [39]. PSC exposure to HIF-α also increases expression of type I collagen, fibronectin, and periostin, thereby accentuating hypoxia secondary to ECM deposition and increased TME fibrosis [24,28,39,40].

The hypoxic TME also poses a significant challenge for radiotherapy interventions. Typically, radiation absorbed by the tissue requires oxygen to produce reactive oxygen species (ROS) that cause DNA damage to cancer cells, thereby shrinking the tumor. In the setting of hypoxia (under 10 mmHg), the efficacy of radiotherapy decreases and requires significantly increased dosage to reach the desired therapeutic effect [41]. Different strategies designed to alleviate hypoxia in the TME and improve radiotherapy penetration such as radiosensitizer drugs have been proposed [42,43] but will need to be tested against PDAC tumors.

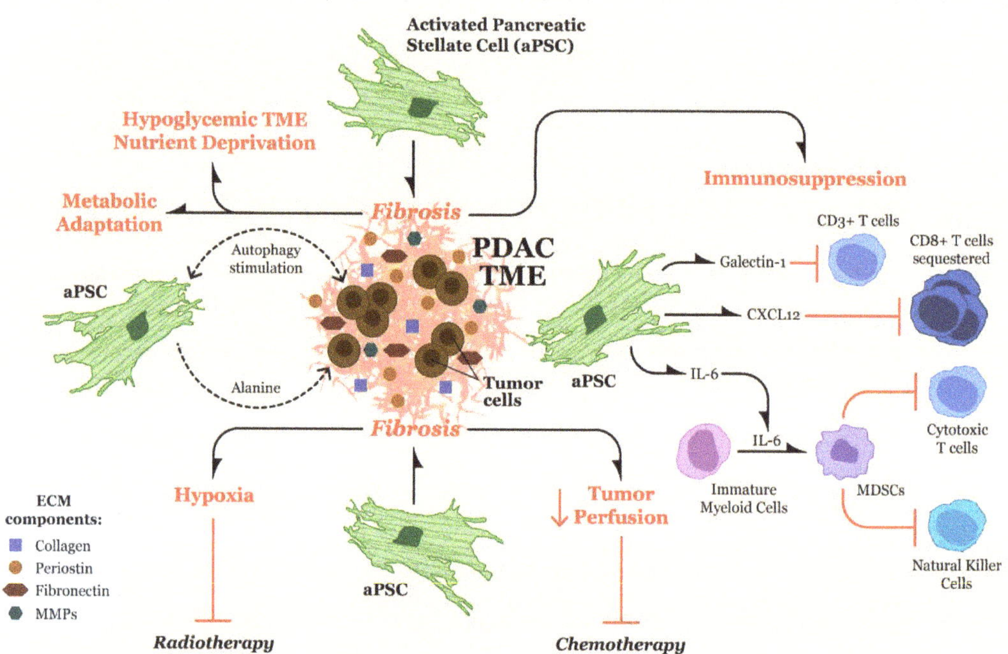

Figure 2. Contribution of extracellular matrix (ECM)/fibrosis to pancreatic ductal adenocarcinoma (PDAC) pathogenesis. Activated pancreatic stellate cells (aPSC) release significant quantities of ECM components, including collagen, periostin, fibronectin, and matrix metalloproteinases (MMPs) into the PDAC tumor microenvironment (TME) that contribute to fibrosis. The fibrotic PDAC TME results in several pathogenic effects (noted in red text). The abundance of fibrotic material in the PDAC TME can result in hypoxia and decreased tumor perfusion, which inhibit the therapeutic effects of radiotherapy and chemotherapy, respectively. Moreover, fibrosis can lead to hypoglycemia and nutrient deprivation in the TME. In response to nutrient deprivation, PDAC tumor cells metabolically adapt by stimulating autophagy in aPSCs, leading to release of alanine from aPSCs, which is then used for fuel by the PDAC tumor cells. The aPSCs may also stimulate autophagy in PDAC tumor cells. Immunosuppression in the TME is established in part by the release of galectin-1 and CXCL12 by aPSCs, inhibiting CD3[+] T cells and sequestering CD8[+] T cells, respectively. Additionally, release of interleukin-6 (IL-6) by aPSCs results in conversion of immature myeloid cells to myeloid-derived suppressor cells (MDSCs), which then inhibit infiltration by cytotoxic T cells and natural killer cells.

3.3. Altering Tumor Cell Metabolism

The reduction in perfusion associated with PDAC fibrosis also leads to nutrient deprivation of the tumor. To overcome this obstacle, PDAC cells maintain adequate nutrition by altering their metabolism to support tumor growth. The vast majority of PDAC cells possess the oncogenic KRAS mutation, enabling them to utilize glutamine as a nutritional source for cancer growth [44]. Aside from providing nutrition, glutamine is also used to promote hyaluronan production via the hexosamine biosynthesis pathway. A study done by Sherma et al. demonstrated that a small molecule glutamine analog (6-diazo-5-oxo-l-norleucine (DON)) was able to reduce the ECM deposition surrounding PDAC cells by inhibiting hexosamine biosynthesis [45]. KRAS mutant PDAC cells are also able to sustain themselves through macro-autophagy, a metabolic cell-survival process that relies on recycling of damaged organelles and proteins. Oncogenic KRAS positively regulates PDAC autophagy by promoting expression of vacuole membrane protein 1 (VMP1), a critical element of autophagosome formation [46,47]. There is also evidence that PDAC cells and activated PSCs engage in metabolic cross talk. Under hypoglycemic conditions, activated PSCs, induced into autophagy by PDAC cells, release alanine in the TME. The alanine is then internalized by PDAC cells and converted to pyruvate, which substitutes for glucose and glutamine in the tricarboxylic acid (TCA) cycle to maintain ATP generation [48,49].

3.4. Immunomodulation

In addition to secreting ECM proteins that provide a physical barrier to cytotoxic immune cells, activated PSCs release a variety of immunomodulatory factors that drive the PDAC TME into an immunosuppressive environment. One example is interleukin-6 (IL-6), which operates through the Janus kinase 2 (JAK2) and signal transducer and activator of transcription 3 (STAT3) signaling cascade. The activation of the JAK2/STAT3 signaling in immature myeloid cells results in their conversion to myeloid-derived suppressor cells (MDSCs). The MDSCs then release a variety of their own modulatory factors that suppress the actions of cytotoxic T cells and natural killer cells, thereby limiting the immune response against the PDAC tumor [50]. Another protein released by activated PSCs is galectin-1, which is a part of the β-galactoside-binding family. Galectin-1 is a contributor to tumor invasion and metastasis [51]. This concept was further evaluated in a study done by Tang et al. that examined the role of galectin-1 in PDAC. When co-culturing $CD3^+$ T cells with activated PSCs overexpressing galectin-1, the authors found that this led to significant apoptosis in the $CD3^+$ T cells. The authors also found that galectin-1-overexpressing PSCs shifted the Th1/Th2 cytokine balance towards Th2 cytokine release, which facilitates immune cell evasion [52,53]. Furthermore, secretion of C-X-C Motif Chemokine Ligand 12 (CXC12) by activated PSCs assists in sequestering $CD8^+$ T cells in the stroma distant from the tumor. The isolation of $CD8^+$ T cells in this distant compartment significantly reduces infiltration into the tumor, thereby establishing the immunosuppressive environment [54].

Interleukin-10 (IL-10) and TGF-β are potent immunosuppressive cytokines that are released into the TME by PDAC cells and immune cells during tumorigenesis [55,56]. These cytokines recruit regulatory T cells, which in turn also release IL-10 and TGF-β, inhibiting effector T cells and maintaining immunosuppression [57–59]. The presence of IL-10 and TGF-β in the PDAC TME also shift the Th1/Th2 cytokine balance towards Th2 cytokine release, thereby further enhancing the immunosuppressive TME [59,60].

Intra-pancreatic γδ T cells indirectly support PDAC pathogenesis by inhibition of αβ T cells using checkpoint receptor ligation [61]. Seifert et al. found that γδ T cells interacted with PSCs and stimulated their production of IL-6, which leads to increasing amounts of ECM deposition in the PDAC stroma. Therefore, the interaction of γδ T cells with PSCs contributes to immunosuppression by fortifying the fibrotic barrier environment [62].

4. Strategies for Targeting Fibrosis in PDAC

Given the significant role for the ECM in pancreatic cancer progression, several different strategies have been investigated that target the fibrotic TME. Approaches currently under evaluation seek to either reduce stromal ECM deposition to improve the delivery of cytotoxic agents, or target ECM proteins for direct delivery of therapeutics to the tumor, thereby limiting off-target effects.

Delivery of cytotoxic agents such as gemcitabine has been improved with the use of nab-paclitaxel [63,64]. Nab-paclitaxel/gemcitabine (AG) penetration was further improved in a phase 2 trial using pegvorhyaluronidase alfa (PEGPH20), but no substantial improvement was seen in a recent phase 3 clinical trial [65]. Treatment guidelines for borderline-resectable and locally advanced pancreatic cancers have recently shifted towards neoadjuvant therapy [66] and the use of combination therapies such as FOLFIRINOX [67,68]. Targeted delivery of FOLFIRINOX was improved using iontophoretic delivery [69]. Directed chemotherapy measures tend to have less associated side effects than systemic chemotherapy. Thus, these localized treatment strategies as a method to overcome ECM deposition warrant further investigation.

The literature describes a variety of strategies used to target CAFs in PDAC. In general, these methods include conversion of CAFs to their quiescent phenotypes, inhibition of CAF signaling cascades, depletion of CAFs, use of CAFs as a cellular vehicle for cyto-toxic agents, and targeting of CAF-derived ECM proteins [70–72]. Some previous studies targeting CAFs have unfortunately led to a more aggressive tumor [73] and/or severe side effects [74]. However, with the discovery of CAF heterogeneity, specific targeting of CAFs could lead to improved therapeutic benefit [75,76]. Regarding clinical trials, several ongoing PDAC trials are investigating therapeutics that disrupt CAF signaling or reprogram CAFs to quiescence (Table 1). These clinical trials should be closely followed to determine if these stromal interventions increase chemotherapy efficacy.

Table 1. Clinical Trials Targeting Fibroblasts in PDAC.

Strategy	Therapeutic	Trial Phase	Trial Status	NCT Number
	Tocilizumab	1b/2	Recruiting	NCT03193190
	Tocilizumab	2	Recruiting	NCT02767557
	Tocilizumab	2	Active	NCT04258150
Disrupt CAF Signaling	Siltuximab	1,2	Recruiting	NCT04191421
	Canakinumab	1	Recruiting	NCT04581343
	Plerixafor	2	Recruiting	NCT04177810
	Plerixafor	1	Completed	NCT02179970
	BL-8040	2	Active	NCT02826486
	ATRA [1]	1	Completed	NCT03307148
	ATRA [1]	2	Not yet recruiting	NCT04241276
	Vitamin D3	3	Recruiting	NCT03472833
Reprogramming to Quiescence	Paricalcitrol	2	Completed	NCT03331562
	Paricalcitrol	1	Recruiting	NCT03519308
	Paricalcitrol	2	Recruiting	NCT04617067
	Paricalcitrol	1	Active	NCT03883919
	Paricalcitrol	2	Recruiting	NCT04524702

Source: Clinicaltrials.gov. [1] All-trans retinoic acid.

Activated PSCs have a prominent role in PDAC ECM deposition. As a result, several investigations have been performed to elucidate therapeutic strategies against PSCs, hypothesizing that inhibition of these fibroblasts would lead to reduced fibrosis and, therefore, enhanced cytotoxic efficacy. For example, epidermal growth factor receptor (EGFR) can activate PSCs, and it was found that inhibition of EGFR reduced fibrosis [77]. PSCs can also be activated to increase desmoplasia by the Sonic Hedgehog (SHH) protein as part of the hedgehog pathway [78]. Studies impeding SHH signaling have led to mixed results [36,79]. Targeting the renin-angiotensin system, which has been demonstrated to activate PSCs [80], using olmesartan, an angiotensin II type-1 receptor blocker, decreased collagen deposition of PSCs, in vitro [81]. More recent PSC-targeting strategies include the use of phytochemicals such as curcumin, which can hinder the gene expression of type I and III collagen, thereby decreasing fibrotic production [82]. Future PSC inhibition studies will likely make use of more directed treatment strategies such as nanotechnology [83]. These strategies aim to reduce PSC activation to decrease ECM deposition and improve efficacy of treatments that are limited by substantial fibrosis in the TME.

Another strategy may resurrect the century old concept of therapeutic infection, as introduced by the sarcoma surgeon Dr. William Coley with "Coley's toxins" [84], using the natural properties of infectious agents such as bacteria to target the TME and modulate anti-cancer immune responses. Group A Streptococcus Streptococcal collagen-like protein 1 (Scl1), is a major GAS adhesin, which exhibits selective binding to ECM proteins [85]. Scl1 binds to tumor-associated isoforms of cellular fibronectin (cFn) containing type IIII repeats, extra domain A and/or B (EDA/EDB/cFn) also known as oncofetal Fn [86–88]. Binding to EDA and EDB is mediated through conserved structural determinants present within the Scl1 globular V domain and facilitates GAS adherence and biofilm formation in the host [89–91]. In vitro, Scl1 mediates biofilm formation on matrices deposited by cancer-associated fibroblasts (CAFs) and osteosarcoma (Saos-2) cells containing EDA/EDB/cFn isoforms [87,89]. Importantly, oncofetal cFn is expressed in many cancers [92], including pancreatic tumors [93], suggesting a potential for bacterial targeting of tumors by Scl1 after injection [94]. Indeed, EDB expression in pancreatic tumors has been leveraged to develop imaging probes for EDB fibronectin to visualize pancreatic tumors [95,96].

While CAF-targeting and therapeutic infection are promising areas of continued research, studies thus far have not yet fully revealed definitive clinical benefit. Therefore, it is important to simultaneously explore other therapeutic options for a highly aggressive malignancy such as PDAC. Another exciting field of treatment focuses on targeting immune cells, especially neutrophils and macrophages, due to their significant pro-fibrotic effects in the PDAC TME. Successful targeting of these immune cells has the potential to mitigate both immunosuppression and fibrosis.

5. Immune Characterization of the TME and Impact on the ECM

As mentioned previously, the TME harbors a heterogenous population of immune cells such as macrophages, neutrophils, dendritic cells, natural killer cells, effector T lymphocytes, regulatory T lymphocytes, MDSCs, and B lymphocytes [97]. In general, immune cells modulate the TME through direct interactions with the tumor or indirectly by releasing a variety of chemical mediators. These cellular communications can both facilitate and hinder the effectiveness of therapeutics in the TME [98]. Given the significant fibrotic barrier in the PDAC TME that obstructs therapeutic delivery, it is important to explore the contribution of immune cells to fibrosis. Specifically, neutrophils and macrophages have been implicated for their role in ECM deposition. Neutrophils and macrophages in physiologic settings contribute to the natural wound healing process in injured tissue without fibrosis. However, pathological disruptions in this homeostatic mechanism can result in a fibrotic phenotype [99]. Therefore, in this section we will examine both key immune cells to elucidate their mechanisms of ECM modulation.

5.1. Neutrophils

Neutrophils, also known as polymorphonuclear leukocytes, are the most common circulating leukocyte and play a key role in microbial defense [100,101]. Classic effector immune responses of neutrophils include phagocytosis and secretion of hydrolytic enzymes, granule-derived myeloperoxidase, and antimicrobial proteins/peptides. Additionally, neutrophils further participate in the immune response by releasing lipid mediators, cytokines, chemokines, and extracellular vesicles [102].

Recruitment signals into the TME for neutrophils include the ligands that bind to CXCR2, such as CXCL1 and CXCL2 [103]. It is also likely that tumor-derived GM-CSF recruits neutrophils into the TME, as this mechanism has been implicated in other cancers including gastric adenocarcinoma [104]. Like macrophages, neutrophils in the TME are capable of polarizing into different phenotypes: N1 and N2. Although the N1/N2 terminology facilitates discussion of these phenotypes, these cells function on a spectrum, therefore, our preference is to describe them as N1-like and N2-like. Conversion into either phenotype designates the neutrophil as a tumor-associated neutrophil (TAN). The N1-like phenotype is considered anti-tumorigenic as it releases reactive oxygen species (ROS), Fas, intercellular adhesion molecule (ICAM)-1, and tumor necrosis factor (TNF-α). These products are cytotoxic towards the tumor and hinder immunosuppression of the TME. The N2-like phenotype appears to promote tumorigenesis by remodeling the ECM and supporting angiogenesis of the tumor. This is accomplished through secretion of arginase, MMP-9, VEGF, and a variety of chemokines [105,106]. ECM remodeling by MMP-9 facilitates release and subsequent activation of VEGF from the ECM, thereby increasing vascularization of the tumor [107].

Neutrophils can also neutralize bacteria and other pathogens through formation of neutrophil extracellular traps (NETs) [108]. In this process, neutrophils release decondensed DNA, histones, high mobility group box 1 protein (HMGB1), ROS, and granules that ensnare and kill bacteria [108,109]. Typically, the expulsion of intracellular contents is a slow process that occurs as the neutrophil is dying. However, an alternative NET mechanism can occur that is independent of cell death and results in expedited degranulation [110,111]. In the unstimulated neutrophil, the DNA is tightly wrapped around histones and stored as heterochromatin. Upon exposure to the pathogen, the heterochromatin is decondensed by peptidyl arginine deiminase 4 (PAD4), which catalyzes the citrullination of histones [109,112]. Decondensation of histones is also facilitated from the interaction between histones and neutrophil elastase (NE) after NE translocation into the cell nucleus [110].

Although NETs are beneficial for protection against microbes, recent studies have shown that they contribute to pathogenesis of sterile inflammatory diseases including PDAC [113]. Neutrophils from PDAC are primed for NET formation and NETs are increased in both the circulation and TME during PDAC progression [114–116]. NETs promote the pathogenesis through multiple mechanisms including stimulating primary pancreatic tumor growth [114], driving cancer-associated hypercoagulability [117–120], promoting formation of metastatic disease [121–123], and supporting immunosuppression [124,125]. Several studies have also implicated NETs in activating PSCs to modulate the TME. PSCs transform into the activated state upon binding extracellular DNA [126], which is a predominant component of NETs. Indeed, interactions between the receptor for advanced glycation end products (RAGE) on quiescent PSCs and the DNA released from neutrophils during NETosis result in activation of PSCs [114]. As mentioned previously, activated PSCs heavily contribute to the fibrotic stroma of PDAC through deposition of ECM proteins, providing a mechanism through which neutrophils and NETs promote the fibrotic TME.

5.2. Factors Promoting Neutrophil Extracellular Traps in PDAC

Given the expansive role for NETs in pancreatic pathogenesis, identifying the signals that trigger NETosis is critical to targeting this cancer-promoting phenomenon. Numerous potential targets have been identified. NET formation is dependent on RAGE, as mice with RAGE knockout resulted in significantly decreased extracellular DNA [115]. A recent study done by Zhang et al. explored the role of IL-17 in PDAC tumorigenesis and immunosuppression. Using an in vitro NET formation assay, the authors discovered that when KPC cells from spontaneous PDAC mice preconditioned with IL-17 were used as conditioned media for neutrophils, NET formation was significantly higher than control neutrophils exposed to IL-17 alone. The authors also found a similar significant result for TNFα [124]. Thus, it is likely that IL-17 and TNFα are crucial factors involved in the recruitment of TANs and NETs.

A recent study reported that amyloid fibrils, insoluble fibers resistant to degradation, can trigger neutrophils into NET activation [127]. To determine if amyloid fibrils contributed to NET activation in the pancreatic tumor microenvironment, Munir et al. used mass spectrometry to investigate the presence of amyloid proteins. They found that the Amyloid β A4 protein (APP) was highly expressed in CAFs. Interestingly, they found that *APP* mRNA was also found in PSCs, though to a lesser extent. The authors showed that CAFs induce NET formation, but by inhibiting secretion of APP, found that CAFs were unable to stimulate NETs. Moreover, through blocking the potential APP receptor, CD11b, the authors noted that neutrophils were no longer stimulated into NETosis [128], implicating several potential targets for NET formation in PDAC.

Research into the stimuli for NETs in PDAC is relatively new, so it is also important to explore how NETs are promoted in other forms of cancer. A recent study done by Li et al. examined the function of Sciellin (SCEL), a precursor to the cornified envelope, which is a protective barrier in the upper epidermis [129], in gallbladder cancer progression. Using a co-culture experiment, the authors found that SCEL induced expression of NETs and citrullinated-histone 3, which is a critical marker of NET formation [130]. Similar to gallbladder cancer, SCEL is markedly elevated in pancreatic cancer [131]. Therefore, it would be integral for future studies to examine if SCEL participates in neutrophil/NET recruitment in PDAC.

5.3. Therapeutic Strategies for Targeting Neutrophils and NETs

Research over the past few years has investigated potential therapeutic strategies for NETs (Figure 3). In addition to promoting inflammation, NETs can also lead to coagulation as the expelled intracellular contents create a scaffold for thrombus formation [132]. A recent study done by Kajioka et al. found that NETs can capture PDAC cells and influence their migration and invasion capabilities. The authors also explored the possibility of targeting NETs in pancreatic cancer cells using recombinant thrombomodulin (rTM), a type of endothelial cell surface protein. The authors found that, when treating pancreatic cancer cells with rTM, HMGB1 released from NETs was degraded. As a result, the capture/migration of PDAC tumor cells controlled by NETs was inhibited thereby reducing metastasis to the liver [121]. TM has also been tested against NETs in situations other than pancreatic cancer. For example, it was determined that rTM reduced histone-induced NET release via citrullinated histone 3 staining in kidney sections of rTM-treated rats [133]. Another study examined the impact of rTM treatment on NETs in septic shock rat models as this condition leads to intravascular coagulation. The authors discovered that rTM treatment reduced systemic NETs in septic shock rat models compared to control, as determined by examining levels of citrullinated histone H3 and DNA, or NE and DNA [134]. The reduction in NETs seen in these studies show promising results for rTM, and this treatment strategy should continue to be thoroughly evaluated in additional pancreatic cancer studies.

A second therapeutic strategy for NETs is targeting the DNA that is expelled from neutrophils during this process. Deoxyribonuclease I (DNAse I) treatment of murine acute lung injury models induces degradation of NETs structure [135]. In another study related to lung injury, when methicillin-resistant *Staphylococcus aureus* (MRSA)-infected mice were treated with DNase I, it was found that neutrophil elastase-DNA (NE-DNA) ELISA measurements were reduced in bronchoalveolar lavage blood [136]. As of this writing, the severe acute respiratory syndrome coronavirus 2/coronavirus disease 2019 (SARS-CoV-2/COVID-19) pandemic is still ongoing. SARS-CoV-2 has been associated with excessive NETs and coagulation [137]. A study done by Park et al. treated severe SARS-CoV-2 patients with either free DNase I or DNase-I-coated melanin-like nanospheres (DNase-I pMNSs). With both treatments, the authors noted significant decreases in extracellular DNA, NET percentage, myeloperoxidase activity, and NE, along with a significant increase in relative plasma DNase activity. Interestingly, treatment with DNase-I pMNSs was more effective than free DNase I at reducing cytokine secretion from neutrophils such as NF-κβ and TNF-α [138]. As a result of these beneficial effects, it is worth exploring the use of DNase-I nanotechnology in the setting of PDAC NETs. Although many DNase I studies have focused on lung injury, a study done by Xia et al. examined the impact of DNase I in the setting of colorectal cancer metastasis to the liver. Due to the short biological half-life of DNase I, the investigators used adeno-associated virus (AAV) as the vector for long-term expression of the enzyme in the liver. The authors found decreased levels of citrullinated histone H3 and NETs in the tumors of mice treated with AAV-DNase I, as compared to the tumors in mice that were treated with AAV-null [139].

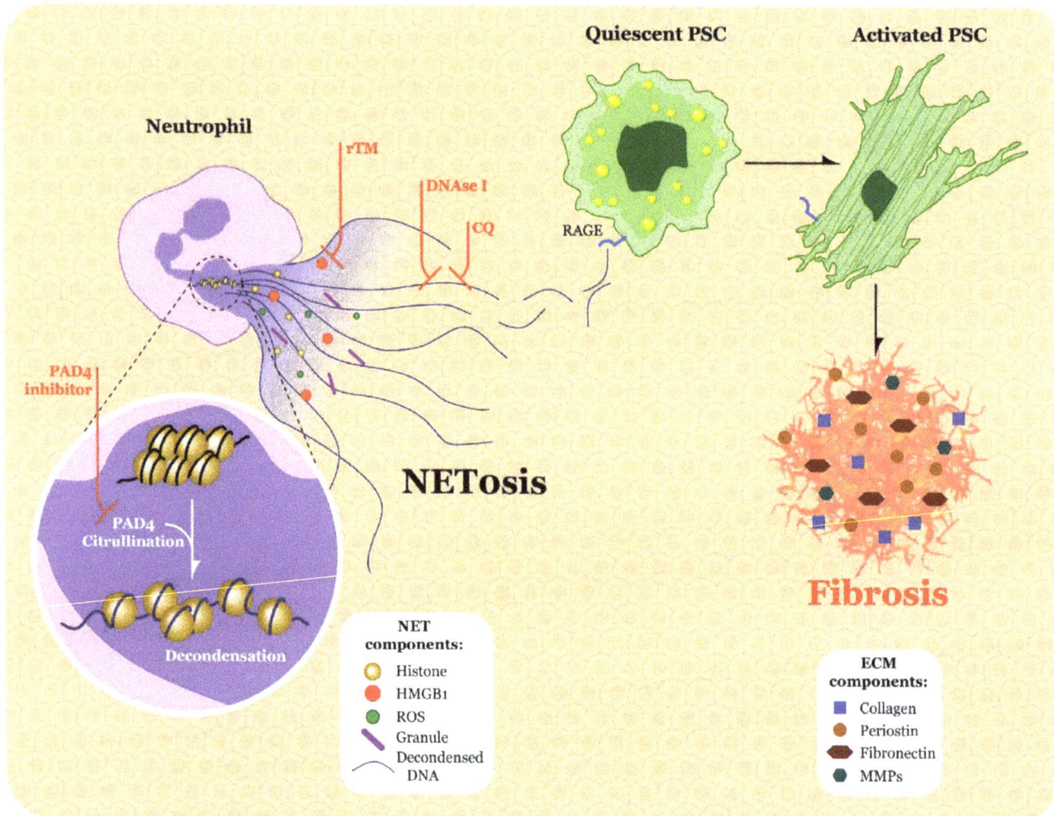

Figure 3. Proposed neutrophil extracellular trap (NET) targeting strategies. Peptidyl arginine deiminase 4 (PAD4) inhibitors block PAD4-mediated citrullination of histones, thereby preventing heterochromatin decondensation and subsequent NET release. Recombinant thrombomodulin (rTM) can inhibit high mobility group box 1 protein (HMGB1), reducing capture and migration of tumor cells by NETs. DNAse I and chloroquine (CQ) degrade and decrease secretion of decondensed DNA, respectively, thereby reducing interactions of decondensed DNA with the receptor for advanced glycation end products (RAGE), preventing pancreatic stellate cell (PSC) activation and subsequent extracellular matrix (ECM) deposition. ROS, reactive oxygen species; MMPs, matrix metalloproteinases.

Another therapeutic approach to NET inhibition involves the use of chloroquine (CQ). In a study examining the potential effect of NETs on high density lipoprotein (HDL) in systemic lupus erythematosus (SLE), Smith et al. found that CQ hindered NET formation in both the control neutrophils and in a type of peripheral blood lupus neutrophils called low-density granulocytes [140]. Treatment with chloroquine resulted in a decrease in serum DNA in the Kras pancreatic cancer mouse model. Moreover, of 15 PDAC patients who were treated with neoadjuvant gemcitabine plus hydroxychloroquine, 12 patients had a significant reduction in circulating DNA levels [115]. Inhibition of NETs by CQ also reverses the hypercoagulable state seen in PDAC [119]. It is also important to note that PAD4 deficiency has been shown to reduce NET formation, and, therefore, PAD4 inhibitor treatments should also be thoroughly investigated as a potential therapeutic target [119,141].

There are a few clinical trials relevant to the treatment of NETs, although they are outside the context of PDAC. Table 2 delineates current NET targeting strategies in clinical trials. One clinical trial (NCT03250689) examined the effect of Danirixin, a selective CXCR2 antagonist, on NETs in chronic obstructive pulmonary disease (COPD) patients. The study was eventually terminated due to changes in the benefit risk profile of Danirixin that was determined in another clinical trial (NCT03034967). A clinical trial at McGill University Health Center, DISCONNECT-1 (NCT04409925) is currently recruiting to evaluate the safety of inhaled rhDNase I and its impact on NETs in severe SARS-CoV-2 patients. Another clinical trial (NCT03368092) at University Hospital in Strasbourg, France, is also evaluating the effect of inhaled rhDNase I on NET-induced lung injury.

Table 2. Clinical Trials Targeting NETs in Various Diseases.

Therapeutic	Trial Phase	Trail Status	Context	Trial ID
rhDNAse I	1	Recruiting	Severe SARS CoV-2 [1]	NCT04409925
	3	Recruiting	Moderate to Severe ARDS [2]	NCT03368092
Danirixin	2	Terminated	COPD [3]	NCT03250689
NucleoCapture Device	N/A	Recruiting	SA-AKI [4]	NCT04749238

Source: Clinicaltrials.gov. [1] Severe Acute Respiratory Syndrome Coronavirus 2. [2] Acute Respiratory Distress Syndrome. [3] Chronic Obstructive Pulmonary Disease. [4] Sepsis-Associated Acute Kidney Injury.

5.4. Tumor Associated Macrophages

Macrophages are a group of immune cells that possess heterogenous function and serve as the first line of immune protection in nearly every tissue [142]. An over-simplified view of macrophage differentiation is that macrophages undergo polarization into different phenotypes depending on the cytokine exposure. The classical activation pathway, in the presence of Th1-derived cytokines such as IFN-γ, colony stimulating factor 2 (CSF2), or toll-like receptor (TLR) activation, gives rise to the M1-like macrophage phenotype considered more protective against cancer cells. The alternatively activated pathway, in the presence of Th2-derived cytokines such as IL-4, IL-10, IL-13, TGF-β, prostaglandin E2, or colony stimulating factor 1 (CSF1), gives rise to the M2-like macrophage phenotype, which typically facilitates tumor progression [143,144]. However, it is now understood that macrophage polarization extends beyond the dichotomy of M1/M2 phenotypes and is better defined as a spectrum [145,146]. Tumor-associated macrophages (TAMs) are considered to have an M1-like phenotype during the early process of tumorigenesis, and then eventually switch to an M2-like phenotype [147]. In the TME, TAMs contribute to PDAC pathogenesis through their promotion of inflammation, tumor angiogenesis, metastasis, immune evasion, and ECM modulation [148]. In this section, we will focus on the various mechanisms by which TAMs alter the ECM.

5.5. Effect of TAMs on ECM

There are several investigations in the literature that implicate TAMs for their role in enhancing the ECM deposition in PDAC. Co-culturing quiescent pancreatic stellate cells with macrophage cell lines in the presence of Heparin-binding EGF (HB-EGF) activates stellate cells and promotes α-SMA expression [149,150]. Activation of these pancreatic stellate cells likely lead to increased deposition of ECM proteins in the tumor stroma. When comparing human pancreatic tissue samples possessing both PDAC lesions and adjacent unaffected tissue, Zhu et al. found a positive correlation between amount of tissue fibrosis and number of macrophages. This group also found, through analysis of gene ontology, that embryonically-derived macrophages expressed higher levels of ECM remodeling genes as compared to monocyte-derived macrophages. For example, qPCR demonstrated higher expression levels of the ECM-producing enzymes hyaluronan synthases 2 and 3 [151]. Activated M2-like macrophages participate in ECM remodeling by secreting MMPs, which exert digestive effects on the ECM [152]. A recent study done by Tekin et al. analyzed mRNA expression of various proteases by quiescent macrophages in the TME. The authors found that MMP9 was significantly produced and led to cleavage of protease-activated receptor-1 (PAR1), a G protein-coupled receptor linked to tumor growth [153]. Because MMPs participate in ECM remodeling, they play a significant role in the levels of fibrosis in the TME.

Tissue resident macrophages, the predominate macrophage subsets in the pancreatic TME, express the prolactin receptor [154] and prolactin has been reported to contribute to the fibrosis of the TME. One of the downstream effectors of prolactin receptor signaling is focal adhesion kinase 1 (FAK1). Treatment with a FAK1 inhibitor significantly decreases fibrosis in transgenic models of murine pancreatic cancer. Therefore, activation of tissue-resident macrophages by prolactin could regulate collagen deposition in the TME.

Incubation of murine fibroblasts with macrophages containing the lipid kinase PI3Kγ increased collagen mRNA in those fibroblasts as compared to murine fibroblasts incubated with PI3Kγ-deficient macrophages. Additionally, it was found that pancreata from both PI3Kγ-deficient KPC mice and KPC mice treated with a PI3Kγ inhibitor displayed significantly less fibrosis as compared to controls. Reduced collagen protein and gene expression was also observed in orthotopic LMP tumors treated with a PI3Kγ inhibitor [155]. As these PI3Kγ-macrophages are present in PDAC, their pro-fibrotic effects and potential as therapeutic targets necessitate further investigation.

Macrophages are also involved in establishing a pre-metastatic niche that promotes PDAC metastasis to the liver, suggesting a role for macrophages modulating fibrosis in the TME. In a study done by Nielsen et al., it was determined that, after exposure to a variety of cancer cell derived factors, M2-like macrophages and metastasis-associated macrophages release granulin. Not only does the secretion of granulin itself likely contribute to the fibrotic stroma, it also activates resident hepatic stellate cells promoting their differentiation into myofibroblasts. These myofibroblasts then release a number of proteins related to ECM remodeling. In particular, the myofibroblasts release high levels of periostin, which contribute to the fibrotic stroma in the TME and facilitate pancreatic tumor growth and invasion into the liver [156].

In addition to PDAC, macrophages are known to contribute to fibrosis in other disease models. For example, it has been recently shown that macrophages expressing the AP-1 transcription factor Fra-2 contributes to the ECM deposition in idiopathic pulmonary fibrosis by releasing type VI collagen. It is unclear if there are Fra-2-expressing macrophages present in the PDAC TME. Therefore, an interesting future investigation would be to search for the existence of these specific macrophages in the PDAC TME as it would be another factor leading to the desmoplastic reaction [157].

Tunica Interna endothelial cell kinase (Tie2)-expressing macrophages (TEMs) are a distinct subtype of macrophages considered to be highly pro-angiogenic and immunosuppressive in the TME [158–160]. Tie-2 expressing macrophages have been associated with poor survival in gastric cancer patients [161] and in PDAC patients when M2-like TAMs are also present [162]. Tie2 is a receptor tyrosine kinase that binds to angiopoietin 1 (ANG-1) and angiopoietin 2 (ANG-2). In circulation, TEMs highly express the pro-angiogenic genes *MMP-9*, *VEGFA*, *COX-2*, and *WNT5A*. In the TME, ANG-2 is secreted by endothelial cells and sometimes tumor cells. ANG-2 levels are typically higher than ANG-1 in the TME. Binding of ANG-2 in the TME leads to upregulation of two other pro-angiogenic genes cathepsin-B (*CTSB*) and thymidine phosphorylase (*TP*) as well as the highly immunosuppressive *IL-10* [163]. To our knowledge, there are currently no studies available on TEM-mediated fibrosis in the TME. Given the substantial modulatory functions of TEMs on TME angiogenesis and immunosuppression, an analysis of their potential role in PDAC fibrosis would further contribute to their recent growing importance as a therapeutic target.

5.6. Recruitment of Macrophages

Various chemokines and cytokines released by the tumor promote macrophage recruitment into the TME. IL-4, IL-10, IL-13, IL-34, TGF-β, and complement component

C5a all have been implicated in this macrophage recruitment. Colony-stimulating factor (CSF)-1 leads to myeloid progenitor differentiation into monocytes and macrophages. Moreover, it has been demonstrated that CSF-1 is involved in generating the M2-like phenotype of macrophages [164–166]. As pancreatic tumorigenesis proceeds, C-C motif chemokine ligand 2 (CCL2) is released by the tumor cells, leading to substantial attraction of circulating monocytes to the TME. Further, the imbalanced release of various chemical mediators such as CCL5, CCL7, CXCL8, CXCL12, and VEGF also serve as chemoattractants for macrophages and facilitate conversion into the M2-like phenotype [152]. Additionally, a recent study done by Tekin et al. demonstrates a positive correlation between number of macrophages in the TME and PAR1 expression in pancreatic tumor tissues [153].

Pancreatic acinar cells with *KRAS* mutation express intercellular adhesion molecule-1 (ICAM-1), which can recruit macrophages to the TME. The infiltration of macrophages facilitates the conversion of acinar to ductal phenotype, which is an integral early component of pancreas carcinogenesis [149]. Two subsets of macrophages are present within the pancreatic inter-acinar stroma. One population is derived from primitive hematopoiesis, whereas the other population is derived from definitive hematopoiesis and substitutions with circulating myeloid cells [151,167].

5.7. Strategies to Target Macrophages

There are several different therapeutic strategies described in the literature that seek to reduce the impact of TAMs in the TME (Figure 4). In general, these therapeutic options target different properties of TAMs such as their survival, polarization, recruitment, phagocytosis, and angiogenesis [168]. Different pharmacological techniques for TAMs such as targeting chemokine-chemokine receptors and tyrosine kinases, as well as the use of bisphosphonates and nanotechnology have been evaluated [169]. Table 3 lists the clinical trials of therapeutics being tested against TAMs in pancreatic cancer. In this section, we will discuss some of the more recent and novel pharmacological approaches to attenuate the influences of TAMs on ECM production.

Table 3. Clinical Trials Targeting Tumor-Associated Macrophages in Pancreatic Cancer.

Target	Therapeutic	Trial Phase	Trial Status	Additional Interventions	Trial ID
CSF1-R	IMC-CS4 (LY3022855)	1	Recruiting	Cyclophosphamide, GVAX, Pembrolizumab	NCT03153410
	Cabiralizumab (FPA008)	1a/1b	Completed	Nivolumab	NCT02526017
	Cabiralizumab (FPA008)	2	Completed	Nivolumab +/− Chemotherapy	NCT03336216
	Pexidartinib	1	Completed	Durvalumab	NCT02777710
CSF1	MCS110	1b/2	Completed	PDR001	NCT02807844
CCR2	PF-04136309	1	Completed	FOLFIRINOX	NCT01413022
	PF-04136309	1b	Completed	nab-paclitaxel and gemcitabine	NCT02732938
	CCX872-B	1	Active	FOLFIRINOX	NCT02345408
CXCR4	BL-8040	2b	Active	Pembrolizumab	NCT02907099

Source: Clinicaltrials.gov.

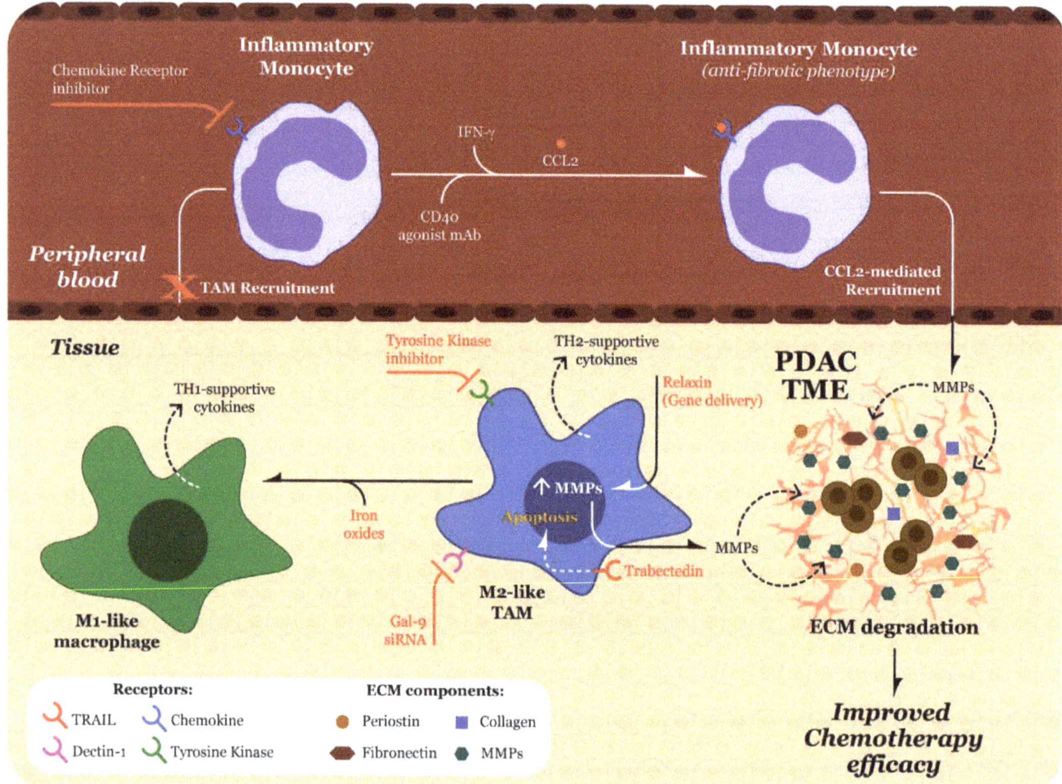

Figure 4. Potential tumor associated macrophage (TAM) targeting strategies. Strategies have been developed that target macrophages in the peripheral blood (top panel) and the PDAC TME (bottom panel). Inhibition of the chemokine receptor may decrease recruitment of inflammatory monocytes into the tissue, where they can polarize into M2-like TAMs. Treatment with a CD40 agonist monoclonal antibody (mAb) can increase systemic levels of interferon-γ (IFN-γ), thereby polarizing the inflammatory monocyte into an anti-fibrotic phenotype. The binding of CCL2 then recruits the inflammatory monocyte into the tissue, where it releases various matrix metalloproteinases (MMPs) that can degrade the abundant extracellular matrix (ECM) of the pancreatic ductal adenocarcinoma tumor microenvironment (PDAC TME), improving chemotherapy efficacy. In general, M1-like macrophages tend to release TH1-supportive cytokines, which are considered more protective against cancer cells. M2-like TAMs usually release TH2-supportive cytokines, which tend to support cancer progression. The literature describes a variety of methods (noted in red text) to target M2-like TAMs. Inhibition of the tyrosine kinase receptor may reduce survival and proliferation of M2-like TAMs. Both iron oxides and targeting of the dectin-1 receptor by galectin-9 small interfering RNA (siRNA) can repolarize M2-like TAMs into the M1-like phenotype. Trabectedin operates through the TNF-related apoptosis-inducing ligand (TRAIL) receptor, thereby leading to apoptosis of the M2-like TAM or switching their secretion profile to that of an inflammatory phenotype. Gene delivery of relaxin (RLN) can upregulate MMP9 and MMP13 genes, thereby leading to increased release of MMPs into the PDAC TME that can degrade fibrosis and improve chemotherapy efficacy.

Targeting chemokine-chemokine signaling represents a promising strategy to limit macrophage infiltration. Targeting the CCL2/CCR2 axis in recruitment of CCR2$^+$ inflammatory monocytes to the PDAC TME, where they can differentiate into macrophages, is one strategy that has been examined. The small molecule CCR2 inhibitor, PF-04136309, decreases levels of circulating inflammatory monocytes in tumor bearing mice, effectively blocking TAM recruitment to the PDAC TME [170]. The safety and efficacy of PF-04136309 along with FOLFIRINOX on TAMs in PDAC was tested in a recent phase Ib clinical trial

(NCT01413022). Of 47 enrolled patients, 39 received the combination therapy, whereas eight patients received FOLFIRINOX alone. Using flow cytometry on six post combination therapy treatment tumor biopsies, the authors found a mean reduction in TAMs from 9.0% to 2.4%. Further, they found a significant decrease in peripheral blood CCR2$^+$ monocytes with the combination therapy as compared to FOLFIRINOX alone, which indicates blockage of TAM recruitment by the PDAC TME [171].

Another strategy in treating TAMs involves the targeting of tyrosine kinases. As mentioned previously, CSF-1 is a cytokine involved in the polarization of macrophages into a tumor-supportive phenotype. In general, efforts to target the CSF-1/CSF-1R interaction are CSF-1/CSF-1R antibodies and CSF-1R kinase inhibitors. A randomized phase 2b clinical trial (NCT03336216) tested the efficacy of cabiralizumab, an antibody that blocks CSF-1R, in combination with nivolumab, with or without chemotherapy, in patients with advanced pancreatic cancer. Unfortunately, the clinical trial sponsor reported that the combination therapy with and without chemotherapy was not beneficial compared to standard chemotherapy [172]. A recent phase 1b trial (NCT02713529) was completed that tested the safety and efficacy of AMG 820, an anti-CSF1R monoclonal antibody, in combination with anti-PD-1 antibody pembrolizumab in adults with advanced pancreatic cancer, colorectal cancer (CRC), or non-small cell lung cancer (NSCLC). Although AMG 820 plus pembrolizumab was shown to have an adequate safety profile, none of the pancreatic cancer patients met the pre-determined threshold for efficacy. There were two CRC patients and one NSCLC patient who achieved a response of immune-related partial response [173]. Blocking tyrosine kinases is also being evaluated for several other types of gastrointestinal cancers, hopefully leading to more progress in this therapeutic strategy.

A third strategy recently evaluated for targeting macrophage polarization is the use of nanotechnology. Several types of nanoparticles that are used to target TAMs have been evaluated [174]. Iron oxides have been demonstrated to switch the polarization of M2-like macrophages into the M1-like phenotype. Additionally, iron oxides increase ROS production and induce apoptosis in cancer cells [175]. In a study done by Zhao et al., the authors developed a tumor-derived antigenic microparticle (T-MP) that contained nano-iron oxide. The surface of the T-MP was tethered with adjuvant CPG oligodeoxynucleotides-loaded liposomes. Once this combination vaccine was delivered into the TME, it was determined that repolarization of M2-like macrophages to M1-like had occurred [176]. Interference with the galectin-9/dectin axis, which has been previously implicated in the conversion of macrophages to the M2-like phenotype, is another strategy targeting the polarization of macrophages in the PDAC TME. A nanoscale delivery system composed of bone marrow mesenchymal stem cell exosomes that were electropermeabilization-loaded with galectin-9 siRNA has been evaluated [177]. Additionally, these exosomes contained oxaliplatin to induce death in tumor cells. After co-delivery of the siRNA and oxaliplatin into orthotopic pancreatic tumor-bearing C57BL/6J mice, the authors found significant re-polarization of TAMs into the M1-like phenotype via flow cytometry and immunofluorescence staining of tumor sections, using CD206 as a marker for M2-like macrophages and CD 16/32 for the M1-like phenotype [177]. Given the many potential benefits of nanotechnology, such as increased stability and decreased side effects [178], they are certainly worth continued exploration for application in targeting TAMs in the PDAC TME.

Macrophage polarization has also been targeted using trabectedin, an isoquinoline cytotoxic agent that was initially isolated from a Caribbean tunicate [179]. Trabectedin can operate through the monocyte specific TNF-related apoptosis-inducing ligand (TRAIL) receptors 1 and 2, resulting in the extrinsic apoptotic pathway through activation of caspase-8 [180]. A study using a patient-derived orthotopic mouse model of gemcitabine-resistant PDAC reported that treatment with trabectedin inhibited but did not regress PDAC tumor growth [181]. In a PDAC mouse model, it was shown that depletion of TAMs by trabectedin significantly increased infiltration of CD4 and CD8 T cells into the TME. Notably, in trabectedin treated mice, both infiltrating CD4 and CD8 T cells produced lower levels of the immunosuppressive cytokine IL-10. CD4 T cells also produced increased levels of IFN-γ. Lastly, it was demonstrated there was higher levels of TAM secretion of inflammatory mediators such as IL2, IL12, IL17, and TNFα, suggesting a switch to the inflammatory M1-like phenotype of TAMs [182]. It has not yet been explored whether this possible re-polarization by trabectedin could lead to reduced fibrosis in the PDAC TME. Trabectedin was approved in 2015 by the Food and Drug Administration for treatment of liposarcomas and leiomyosarcomas and, therefore, has not yet been extensively evaluated for use in PDAC [183]. A phase II clinical trial tested single agent trabectedin in patients with gemcitabine resistant metastatic PDAC. Unfortunately, the primary endpoint measure of progression-free survival at six months from treatment was not met [184]. An interesting future investigation should analyze the impact of trabectedin on PDAC fibrosis and determine if combination with other cytotoxic agents improves clinical outcomes.

Some investigators have found success by reprogramming TAMs to deplete fibrosis. Treatment with an agonist CD40 monoclonal antibody increased the systemic release of IFN-γ, leading to polarization of CCR2$^+$ monocytes into an anti-fibrotic phenotype. These inflammatory monocytes are then recruited into the PDAC TME via CCL2 release. Once in the TME, monocytes differentiate to inflammatory macrophages that are able to release various MMPs that deplete ECM proteins such as fibronectin and type I collagen, thereby reducing fibrosis and increasing the efficacy of cytotoxic agents in the PDAC TME [185]. Previous studies have shown the beneficial effect of the anti-fibrotic hormone relaxin (RLN) in reducing fibrosis in PDAC and liver metastasis from various cancers [186,187]. A study done by Zhou et al. found that more than 70% of cells in both macrophage and fibroblast populations expressed the relaxin family peptide receptor type 1 (RXFP1). After *RLN* gene delivery, the authors observed significant increases *MMP9* and *MMP13* mRNAs in the PDAC TME compared to the PBS-treated group, thereby leading to ECM degradation [188].

Regarding treatment strategies for TEMs, disruption of the ANG-2/Tie2 signaling pathway in vivo has been shown to inhibit tumor growth and reduce tumor microvasculature using monoclonal antibodies [189,190] and peptides [191]. A recent study using rebastinib, a selective inhibitor of Tie2, decreased both Tie2-expressing macrophage infiltration and TME vasculature density in a mouse model of mammary cancer, but reduced only Tie2-expressing macrophage infiltration in a pancreatic neuroendocrine tumor model [192]. Rebastinib is currently being evaluated in clinical trials in combination with chemotherapy for treatment of metastatic breast cancer (NCT02824575) and other advanced solid tumors (NCT03717415 and NCT03601897). To our knowledge, there have been no studies or clinical trials targeting Tie2-expressing macrophages in the context of PDAC.

5.8. Beyond Neutrophils and Macrophages

Neutrophils and macrophages are the most extensively investigated immune cell types regarding fibrotic production. Generally, the involvement of immune cells appears to contribute to fibrosis in many disease contexts. Some immune cells, such as regulatory T cells (Tregs) and natural killer T (NKT) cells, have conflicting roles in fibrosis [193]. For example, in the TME, factors released from CAFs such as TGF-β cause Tregs to release their own TGF-β, which influences the conversion of quiescent fibroblasts into CAFs, likely promoting ECM deposition [194]. Although, in a study examining human immunodeficiency virus type 1 (HIV-1) infection in a humanized mouse model, the presence of Tregs mitigates liver fibrosis [195]. NKT cells have been shown to reduce collagen in the liver by selectively removing hepatic stellate cells after treatment with IL-30 [196]. In contrast, NKT cells have also been implicated in fibrosis production following liver injury through a CXCR6-dependent mechanism [197].

Dendritic cells release MMP9, which can have modulatory effects on the ECM, but more studies are required to clarify their relationship with fibrogenesis [198]. There is evidence that both T helper 2 and T helper 17 cells activate hepatic stellate cells, which in turn secrete collagen [199]. T helper 17 cells can release IL-17, which can promote hepatic stellate cell expression of collagen I and influence their conversion into fibrogenic myofibroblasts [200]. As mentioned earlier, $\gamma\delta$ T cells have been demonstrated to contribute to both the immunosuppressive and fibrotic TME in PDAC [62]. Overall, crosstalk between various immune cells and CAFs are evident [201], but several more studies are needed to investigate the potential pro-fibrotic or anti-fibrotic roles of the various immune cells in PDAC specifically.

6. Conclusions and Future Perspectives

The desmoplastic reaction heavily contributes to the poor prognosis of PDAC. The overabundance of ECM proteins establishes a fibrotic stroma and TME that are highly refractory to cytotoxic chemotherapy, immunotherapy, and radiotherapy. Targeting the stroma directly in pre-clinical studies has unfortunately led to inconsistent results and in some instances, a more aggressive disease [202]. Thus, various other therapeutic options such as targeting immune cell modulation of the ECM should be explored. Although it has been known for some time that NETs can contribute to various pathologies, their effect on PSC activation is a relatively new discovery. Thus, the laboratory investigations that target NETs in PDAC models are also quite new and limited in number yet promising. To our knowledge, there are currently no active clinical trials targeting NETs in the context of PDAC. With regard to TAMs in PDAC, several more laboratory studies and clinical trials of therapeutic strategies have been published. Further pre-clinical studies using NET-targeting therapies in combination with neoadjuvant and/or adjuvant cytotoxic agents is warranted. This strategy is already being evaluated for TAMs in clinical trials. Targeting both NETs and TAMs could deplete some of the fibrosis surrounding the tumor, thereby enabling better penetration of cytotoxic agents into the TME. While current strategies have focused on either macrophage or neutrophil targeting, limited efforts have been made to target both immune cells [103], which may be critical for efficacy. With the recent advancements in chemotherapy such as FOLFORINOX and innovations in more directed cytotoxic delivery, the addition of immune cell-targeting agents could be the extra boost needed to win the battle against this devastating disease.

Author Contributions: Conceptualization, B.A.B., R.S.A., T.D.E. and S.L.; writing—original draft preparation, R.S.A. and B.A.B.; writing—review and editing, B.A.B., T.D.E. and S.L.; visualization, T.D.E. and R.S.A.; supervision, B.A.B. All authors have read and agreed to the published version of the manuscript.

Funding: Research reported in this publication was supported by the National Institutes of Health under award number 5U54GM104942-04 (B.A.B.), CA192064 and CA194013 (T.D.E.). The content is solely the responsibility of the authors and does not necessarily represent the official views of the National Institutes of Health.

Institutional Review Board Statement: Not applicable.

Informed Consent Statement: Not applicable.

Data Availability Statement: The data presented in this study are openly available.

Conflicts of Interest: The authors declare no conflict of interest.

References

1. Siegel, R.L.; Miller, K.D.; Jemal, A. Cancer statistics, 2020. *CA Cancer J. Clin.* **2020**, *70*, 7–30. [CrossRef]
2. Rawla, P.; Sunkara, T.; Gaduputi, V. Epidemiology of pancreatic cancer: Global trends, etiology and risk factors. *World J. Oncol.* **2019**, *10*, 10–27. [CrossRef]
3. Tavakkoli, A.; Singal, A.G.; Waljee, A.K.; Elmunzer, B.J.; Pruitt, S.L.; McKey, T.; Rubenstein, J.H.; Scheiman, J.M.; Murphy, C.C. Racial disparities and trends in pancreatic cancer incidence and mortality in the united states. *Clin. Gastroenterol. Hepatol.* **2020**, *18*, 171–178.e110. [CrossRef]
4. Mizrahi, J.D.; Surana, R.; Valle, J.W.; Shroff, R.T. Pancreatic cancer. *Lancet* **2020**, *395*, 2008–2020. [CrossRef]
5. Melstrom, L.G.; Salazar, M.D.; Diamond, D.J. The pancreatic cancer microenvironment: A true double agent. *J. Surg. Oncol.* **2017**, *116*, 7–15. [CrossRef] [PubMed]
6. Weniger, M.; Honselmann, K.C.; Liss, A.S. The extracellular matrix and pancreatic cancer: A complex relationship. *Cancers* **2018**, *10*, 316. [CrossRef]

7.   DeClerck, Y.A. Desmoplasia: A response or a niche? *Cancer Discov.* **2012**, *2*, 772–774. [CrossRef] [PubMed]
8.   Hingorani, S.R.; Petricoin, E.F.; Maitra, A.; Rajapakse, V.; King, C.; Jacobetz, M.A.; Ross, S.; Conrads, T.P.; Veenstra, T.D.; Hitt, B.A.; et al. Preinvasive and invasive ductal pancreatic cancer and its early detection in the mouse. *Cancer Cell* **2003**, *4*, 437–450. [CrossRef]
9.   Pandol, S.; Edderkaoui, M.; Gukovsky, I.; Lugea, A.; Gukovskaya, A. Desmoplasia of pancreatic ductal adenocarcinoma. *Clin. Gastroenterol. Hepatol.* **2009**, *7*, S44–S47. [CrossRef]
10. Suklabaidya, S.; Dash, P.; Das, B.; Suresh, V.; Sasmal, P.K.; Senapati, S. Experimental models of pancreatic cancer desmoplasia. *Lab. Investig.* **2018**, *98*, 27–40. [CrossRef]
11. von Ahrens, D.; Bhagat, T.D.; Nagrath, D.; Maitra, A.; Verma, A. The role of stromal cancer-associated fibroblasts in pancreatic cancer. *J. Hematol. Oncol.* **2017**, *10*, 76. [CrossRef]
12. Ren, B.; Cui, M.; Yang, G.; Wang, H.; Feng, M.; You, L.; Zhao, Y. Tumor microenvironment participates in metastasis of pancreatic cancer. *Mol. Cancer* **2018**, *17*, 108. [CrossRef] [PubMed]
13. Soundararajan, M.; Kannan, S. Fibroblasts and mesenchymal stem cells: Two sides of the same coin? *J. Cell Physiol.* **2018**, *233*, 9099–9109. [CrossRef] [PubMed]
14. LeBleu, V.S.; Neilson, E.G. Origin and functional heterogeneity of fibroblasts. *FASEB J.* **2020**, *34*, 3519–3536. [CrossRef] [PubMed]
15. Kalluri, R. The biology and function of fibroblasts in cancer. *Nat. Rev. Cancer* **2016**, *16*, 582–598. [CrossRef] [PubMed]
16. Pereira, B.A.; Vennin, C.; Papanicolaou, M.; Chambers, C.R.; Herrmann, D.; Morton, J.P.; Cox, T.R.; Timpson, P. Caf subpopulations: A new reservoir of stromal targets in pancreatic cancer. *Trends Cancer* **2019**, *5*, 724–741. [CrossRef]
17. Massague, J. Tgfbeta in cancer. *Cell* **2008**, *134*, 215–230. [CrossRef]
18. Kim, J.; Bae, J.S. Tumor-associated macrophages and neutrophils in tumor microenvironment. *Mediators Inflamm.* **2016**, *2016*, 6058147. [CrossRef]
19. Eble, J.A.; Niland, S. The extracellular matrix in tumor progression and metastasis. *Clin. Exp. Metastasis* **2019**, *36*, 171–198. [CrossRef]
20. Vennin, C.; Murphy, K.J.; Morton, J.P.; Cox, T.R.; Pajic, M.; Timpson, P. Reshaping the tumor stroma for treatment of pancreatic cancer. *Gastroenterology* **2018**, *154*, 820–838. [CrossRef] [PubMed]
21. Pan, B.; Liao, Q.; Niu, Z.; Zhou, L.; Zhao, Y. Cancer-associated fibroblasts in pancreatic adenocarcinoma. *Future Oncol.* **2015**, *11*, 2603–2610. [CrossRef]
22. Allam, A.; Thomsen, A.R.; Gothwal, M.; Saha, D.; Maurer, J.; Brunner, T.B. Pancreatic stellate cells in pancreatic cancer: In focus. *Pancreatology* **2017**, *17*, 514–522. [CrossRef]
23. Zhou, Y.; Sun, B.; Li, W.; Zhou, J.; Gao, F.; Wang, X.; Cai, M.; Sun, Z. Pancreatic stellate cells: A rising translational physiology star as a potential stem cell type for beta cell neogenesis. *Front. Physiol.* **2019**, *10*, 218. [CrossRef]
24. Jin, G.; Hong, W.; Guo, Y.; Bai, Y.; Chen, B. Molecular mechanism of pancreatic stellate cells activation in chronic pancreatitis and pancreatic cancer. *J. Cancer* **2020**, *11*, 1505–1515. [CrossRef]
25. Apte, M.V.; Park, S.; Phillips, P.A.; Santucci, N.; Goldstein, D.; Kumar, R.K.; Ramm, G.A.; Buchler, M.; Friess, H.; McCarroll, J.A.; et al. Desmoplastic reaction in pancreatic cancer: Role of pancreatic stellate cells. *Pancreas* **2004**, *29*, 179–187. [CrossRef] [PubMed]
26. Apte, M.V.; Pirola, R.C.; Wilson, J.S. Pancreatic stellate cells: A starring role in normal and diseased pancreas. *Front. Physiol.* **2012**, *3*, 344. [CrossRef] [PubMed]
27. Yang, Y.; Kim, J.W.; Park, H.S.; Lee, E.Y.; Yoon, K.H. Pancreatic stellate cells in the islets as a novel target to preserve the pancreatic beta-cell mass and function. *J. Diabetes Investig.* **2020**, *11*, 268–280. [CrossRef] [PubMed]
28. Bynigeri, R.R.; Jakkampudi, A.; Jangala, R.; Subramanyam, C.; Sasikala, M.; Rao, G.V.; Reddy, D.N.; Talukdar, R. Pancreatic stellate cell: Pandora's box for pancreatic disease biology. *World J. Gastroenterol.* **2017**, *23*, 382–405. [CrossRef] [PubMed]
29. Phillips, P.A.; McCarroll, J.A.; Park, S.; Wu, M.J.; Pirola, R.; Korsten, M.; Wilson, J.S.; Apte, M.V. Rat pancreatic stellate cells secrete matrix metalloproteinases: Implications for extracellular matrix turnover. *Gut* **2003**, *52*, 275–282. [CrossRef] [PubMed]
30. Garcea, G.; Neal, C.P.; Pattenden, C.J.; Steward, W.P.; Berry, D.P. Molecular prognostic markers in pancreatic cancer: A systematic review. *Eur J. Cancer* **2005**, *41*, 2213–2236. [CrossRef]
31. Tian, C.; Clauser, K.R.; Ohlund, D.; Rickelt, S.; Huang, Y.; Gupta, M.; Mani, D.R.; Carr, S.A.; Tuveson, D.A.; Hynes, R.O. Proteomic analyses of ecm during pancreatic ductal adenocarcinoma progression reveal different contributions by tumor and stromal cells. *Proc. Natl. Acad. Sci. USA* **2019**, *116*, 19609–19618. [CrossRef] [PubMed]
32. Whatcott, C.J.; Diep, C.H.; Jiang, P.; Watanabe, A.; LoBello, J.; Sima, C.; Hostetter, G.; Shepard, H.M.; Von Hoff, D.D.; Han, H. Desmoplasia in primary tumors and metastatic lesions of pancreatic cancer. *Clin. Cancer Res.* **2015**, *21*, 3561–3568. [CrossRef] [PubMed]
33. Ohlund, D.; Lundin, C.; Ardnor, B.; Oman, M.; Naredi, P.; Sund, M. Type iv collagen is a tumour stroma-derived biomarker for pancreas cancer. *Br. J. Cancer* **2009**, *101*, 91–97. [CrossRef] [PubMed]
34. Drifka, C.R.; Loeffler, A.G.; Mathewson, K.; Keikhosravi, A.; Eickhoff, J.C.; Liu, Y.; Weber, S.M.; Kao, W.J.; Eliceiri, K.W. Highly aligned stromal collagen is a negative prognostic factor following pancreatic ductal adenocarcinoma resection. *Oncotarget* **2016**, *7*, 76197–76213. [CrossRef]
35. Biancur, D.E.; Kimmelman, A.C. The plasticity of pancreatic cancer metabolism in tumor progression and therapeutic resistance. *Biochim. Biophys. Acta Rev. Cancer* **2018**, *1870*, 67–75. [CrossRef]

36. Olive, K.P.; Jacobetz, M.A.; Davidson, C.J.; Gopinathan, A.; McIntyre, D.; Honess, D.; Madhu, B.; Goldgraben, M.A.; Caldwell, M.E.; Allard, D.; et al. Inhibition of hedgehog signaling enhances delivery of chemotherapy in a mouse model of pancreatic cancer. *Science* **2009**, *324*, 1457–1461. [CrossRef]

37. Erkan, M.; Hausmann, S.; Michalski, C.W.; Fingerle, A.A.; Dobritz, M.; Kleeff, J.; Friess, H. The role of stroma in pancreatic cancer: Diagnostic and therapeutic implications. *Nat. Rev. Gastroenterol. Hepatol.* **2012**, *9*, 454–467. [CrossRef]

38. Shah, V.M.; Sheppard, B.C.; Sears, R.C.; Alani, A.W. Hypoxia: Friend or foe for drug delivery in pancreatic cancer. *Cancer Lett.* **2020**, *492*, 63–70. [CrossRef] [PubMed]

39. Tao, J.; Yang, G.; Zhou, W.; Qiu, J.; Chen, G.; Luo, W.; Zhao, F.; You, L.; Zheng, L.; Zhang, T.; et al. Targeting hypoxic tumor microenvironment in pancreatic cancer. *J. Hematol. Oncol.* **2021**, *14*, 14. [CrossRef]

40. Daniel, S.K.; Sullivan, K.M.; Labadie, K.P.; Pillarisetty, V.G. Hypoxia as a barrier to immunotherapy in pancreatic adenocarcinoma. *Clin. Transl. Med.* **2019**, *8*, 10. [CrossRef]

41. Sorensen, B.S.; Horsman, M.R. Tumor hypoxia: Impact on radiation therapy and molecular pathways. *Front. Oncol.* **2020**, *10*, 562. [CrossRef] [PubMed]

42. Graham, K.; Unger, E. Overcoming tumor hypoxia as a barrier to radiotherapy, chemotherapy and immunotherapy in cancer treatment. *Int. J. Nanomed.* **2018**, *13*, 6049–6058. [CrossRef]

43. Riess, J.G. Understanding the fundamentals of perfluorocarbons and perfluorocarbon emulsions relevant to in vivo oxygen delivery. *Artif. Cells Blood Substit. Immobil. Biotechnol.* **2005**, *33*, 47–63. [CrossRef]

44. Son, J.; Lyssiotis, C.A.; Ying, H.; Wang, X.; Hua, S.; Ligorio, M.; Perera, R.M.; Ferrone, C.R.; Mullarky, E.; Shyh-Chang, N.; et al. Glutamine supports pancreatic cancer growth through a kras-regulated metabolic pathway. *Nature* **2013**, *496*, 101–105. [CrossRef]

45. Sharma, N.S.; Gupta, V.K.; Garrido, V.T.; Hadad, R.; Durden, B.C.; Kesh, K.; Giri, B.; Ferrantella, A.; Dudeja, V.; Saluja, A.; et al. Targeting tumor-intrinsic hexosamine biosynthesis sensitizes pancreatic cancer to anti-pd1 therapy. *J. Clin. Investig.* **2020**, *130*, 451–465. [CrossRef]

46. New, M.; Tooze, S. The role of autophagy in pancreatic cancer-recent advances. *Biology* **2019**, *9*, 7. [CrossRef]

47. Lo Re, A.E.; Fernandez-Barrena, M.G.; Almada, L.L.; Mills, L.D.; Elsawa, S.F.; Lund, G.; Ropolo, A.; Molejon, M.I.; Vaccaro, M.I.; Fernandez-Zapico, M.E. Novel akt1-gli3-vmp1 pathway mediates kras oncogene-induced autophagy in cancer cells. *J. Biol. Chem.* **2012**, *287*, 25325–25334. [CrossRef] [PubMed]

48. Gorgulu, K.; Diakopoulos, K.N.; Kaya-Aksoy, E.; Ciecielski, K.J.; Ai, J.; Lesina, M.; Algul, H. The role of autophagy in pancreatic cancer: From bench to the dark bedside. *Cells* **2020**, *9*, 1063. [CrossRef]

49. Sousa, C.M.; Biancur, D.E.; Wang, X.; Halbrook, C.J.; Sherman, M.H.; Zhang, L.; Kremer, D.; Hwang, R.F.; Witkiewicz, A.K.; Ying, H.; et al. Pancreatic stellate cells support tumour metabolism through autophagic alanine secretion. *Nature* **2016**, *536*, 479–483. [CrossRef]

50. Mace, T.A.; Bloomston, M.; Lesinski, G.B. Pancreatic cancer-associated stellate cells: A viable target for reducing immunosuppression in the tumor microenvironment. *Oncoimmunology* **2013**, *2*, e24891. [CrossRef] [PubMed]

51. Wu, M.H.; Hong, T.M.; Cheng, H.W.; Pan, S.H.; Liang, Y.R.; Hong, H.C.; Chiang, W.F.; Wong, T.Y.; Shieh, D.B.; Shiau, A.L.; et al. Galectin-1-mediated tumor invasion and metastasis, up-regulated matrix metalloproteinase expression, and reorganized actin cytoskeletons. *Mol. Cancer Res.* **2009**, *7*, 311–318. [CrossRef] [PubMed]

52. Tang, D.; Yuan, Z.; Xue, X.; Lu, Z.; Zhang, Y.; Wang, H.; Chen, M.; An, Y.; Wei, J.; Zhu, Y.; et al. High expression of galectin-1 in pancreatic stellate cells plays a role in the development and maintenance of an immunosuppressive microenvironment in pancreatic cancer. *Int. J. Cancer* **2012**, *130*, 2337–2348. [CrossRef]

53. Looi, C.K.; Chung, F.F.; Leong, C.O.; Wong, S.F.; Rosli, R.; Mai, C.W. Therapeutic challenges and current immunomodulatory strategies in targeting the immunosuppressive pancreatic tumor microenvironment. *J. Exp. Clin. Cancer Res.* **2019**, *38*, 162. [CrossRef]

54. Ene-Obong, A.; Clear, A.J.; Watt, J.; Wang, J.; Fatah, R.; Riches, J.C.; Marshall, J.F.; Chin-Aleong, J.; Chelala, C.; Gribben, J.G.; et al. Activated pancreatic stellate cells sequester cd8$^+$ t cells to reduce their infiltration of the juxtatumoral compartment of pancreatic ductal adenocarcinoma. *Gastroenterology* **2013**, *145*, 1121–1132. [CrossRef] [PubMed]

55. Padoan, A.; Plebani, M.; Basso, D. Inflammation and pancreatic cancer: Focus on metabolism, cytokines, and immunity. *Int. J. Mol. Sci.* **2019**, *20*, 676. [CrossRef]

56. Tantau, A.; Leucuta, D.C.; Tantau, M.; Botan, E.; Zaharie, R.; Mandrutiu, A.; Tomuleasa, I.C. Inflammation, tumoral markers and interleukin-17, -10, and -6 profiles in pancreatic adenocarcinoma and chronic pancreatitis. *Dig. Dis. Sci.* **2020**.

57. Yao, W.; Maitra, A.; Ying, H. Recent insights into the biology of pancreatic cancer. *EBioMedicine* **2020**, *53*, 102655. [CrossRef] [PubMed]

58. Saka, D.; Gokalp, M.; Piyade, B.; Cevik, N.C.; Arik Sever, E.; Unutmaz, D.; Ceyhan, G.O.; Demir, I.E.; Asimgil, H. Mechanisms of t-cell exhaustion in pancreatic cancer. *Cancers* **2020**, *12*, 2274. [CrossRef]

59. Roshani, R.; McCarthy, F.; Hagemann, T. Inflammatory cytokines in human pancreatic cancer. *Cancer Lett.* **2014**, *345*, 157–163. [CrossRef]

60. Farajzadeh Valilou, S.; Keshavarz-Fathi, M.; Silvestris, N.; Argentiero, A.; Rezaei, N. The role of inflammatory cytokines and tumor associated macrophages (tams) in microenvironment of pancreatic cancer. *Cytokine Growth Factor Rev.* **2018**, *39*, 46–61. [CrossRef] [PubMed]

61. Daley, D.; Zambirinis, C.P.; Seifert, L.; Akkad, N.; Mohan, N.; Werba, G.; Barilla, R.; Torres-Hernandez, A.; Hundeyin, M.; Mani, V.R.K.; et al. Gammadelta t cells support pancreatic oncogenesis by restraining alphabeta t cell activation. *Cell* **2016**, *166*, 1485–1499.e1415. [CrossRef]

62. Seifert, A.M.; List, J.; Heiduk, M.; Decker, R.; von Renesse, J.; Meinecke, A.C.; Aust, D.E.; Welsch, T.; Weitz, J.; Seifert, L. Gamma-delta t cells stimulate il-6 production by pancreatic stellate cells in pancreatic ductal adenocarcinoma. *J. Cancer Res. Clin. Oncol.* **2020**, *146*, 3233–3240. [CrossRef] [PubMed]

63. Von Hoff, D.D.; Ervin, T.; Arena, F.P.; Chiorean, E.G.; Infante, J.; Moore, M.; Seay, T.; Tjulandin, S.A.; Ma, W.W.; Saleh, M.N.; et al. Increased survival in pancreatic cancer with nab-paclitaxel plus gemcitabine. *N. Engl. J. Med.* **2013**, *369*, 1691–1703. [CrossRef]

64. Frese, K.K.; Neesse, A.; Cook, N.; Bapiro, T.E.; Lolkema, M.P.; Jodrell, D.I.; Tuveson, D.A. nab-Paclitaxel potentiates gemcitabine activity by reducing cytidine deaminase levels in a mouse model of pancreatic cancer. *Cancer Discov.* **2012**, *2*, 260–269. [CrossRef]

65. Van Cutsem, E.; Tempero, M.A.; Sigal, D.; Oh, D.Y.; Fazio, N.; Macarulla, T.; Hitre, E.; Hammel, P.; Hendifar, A.E.; Bates, S.E.; et al. Randomized Phase III Trial of Pegvorhyaluronidase Alfa with Nab-Paclitaxel Plus Gemcitabine for Patients with Hyaluronan-High Metastatic Pancreatic Adenocarcinoma. *J. Clin. Oncol.* **2020**, *38*, 3185–3194. [CrossRef] [PubMed]

66. Chawla, A.; Ferrone, C.R. Neoadjuvant Therapy for Resectable Pancreatic Cancer: An Evolving Paradigm Shift. *Front. Oncol.* **2019**, *9*, 1085. [CrossRef] [PubMed]

67. Gupta, R.; Amanam, I.; Chung, V. Current and future therapies for advanced pancreatic cancer. *J. Surg. Oncol.* **2017**, *116*, 25–34. [CrossRef] [PubMed]

68. Suker, M.; Beumer, B.R.; Sadot, E.; Marthey, L.; Faris, J.E.; Mellon, E.A.; El-Rayes, B.F.; Wang-Gillam, A.; Lacy, J.; Hosein, P.J.; et al. FOLFIRINOX for locally advanced pancreatic cancer: A systematic review and patient-level meta-analysis. *Lancet Oncol.* **2016**, *17*, 801–810. [CrossRef]

69. Byrne, J.D.; Jajja, M.R.N.; O'Neill, A.T.; Schorzman, A.N.; Keeler, A.W.; Luft, J.C.; Zamboni, W.C.; DeSimone, J.M.; Yeh, J.J. Impact of formulation on the iontophoretic delivery of the FOLFIRINOX regimen for the treatment of pancreatic cancer. *Cancer Chemother. Pharmacol.* **2018**, *81*, 991–998. [CrossRef]

70. Chen, X.; Song, E. Turning foes to friends: Targeting cancer-associated fibroblasts. *Nat. Rev. Drug Discov.* **2019**, *18*, 99–115. [CrossRef]

71. Biffi, G.; Tuveson, D.A. Diversity and biology of cancer-associated fibroblasts. *Physiol. Rev.* **2021**, *101*, 147–176. [CrossRef] [PubMed]

72. Sunami, Y.; Boker, V.; Kleeff, J. Targeting and reprograming cancer-associated fibroblasts and the tumor microenvironment in pancreatic cancer. *Cancers* **2021**, *13*, 697. [CrossRef] [PubMed]

73. Ozdemir, B.C.; Pentcheva-Hoang, T.; Carstens, J.L.; Zheng, X.; Wu, C.C.; Simpson, T.R.; Laklai, H.; Sugimoto, H.; Kahlert, C.; Novitskiy, S.V.; et al. Depletion of carcinoma-associated fibroblasts and fibrosis induces immunosuppression and accelerates pancreas cancer with reduced survival. *Cancer Cell* **2014**, *25*, 719–734. [CrossRef] [PubMed]

74. Piersma, B.; Hayward, M.K.; Weaver, V.M. Fibrosis and cancer: A strained relationship. *Biochim Biophys Acta Rev. Cancer* **2020**, *1873*, 188356. [CrossRef]

75. Norton, J.; Foster, D.; Chinta, M.; Titan, A.; Longaker, M. Pancreatic cancer associated fibroblasts (caf): Under-explored target for pancreatic cancer treatment. *Cancers* **2020**, *12*, 1347. [CrossRef]

76. Huang, H.; Brekken, R.A. Recent advances in understanding cancer-associated fibroblasts in pancreatic cancer. *Am. J. Physiol. Cell Physiol.* **2020**, *319*, C233–c243. [CrossRef]

77. Blaine, S.A.; Ray, K.C.; Branch, K.M.; Robinson, P.S.; Whitehead, R.H.; Means, A.L. Epidermal growth factor receptor regulates pancreatic fibrosis. *Am. J. Physiol. Gastrointest Liver Physiol.* **2009**, *297*, G434–G441. [CrossRef]

78. Bailey, J.M.; Swanson, B.J.; Hamada, T.; Eggers, J.P.; Singh, P.K.; Caffery, T.; Ouellette, M.M.; Hollingsworth, M.A. Sonic hedgehog promotes desmoplasia in pancreatic cancer. *Clin. Cancer Res.* **2008**, *14*, 5995–6004. [CrossRef]

79. Pitarresi, J.R.; Liu, X.; Avendano, A.; Thies, K.A.; Sizemore, G.M.; Hammer, A.M.; Hildreth, B.E., 3rd; Wang, D.J.; Steck, S.A.; Donohue, S.; et al. Disruption of stromal hedgehog signaling initiates rnf5-mediated proteasomal degradation of pten and accelerates pancreatic tumor growth. *Life Sci. Alliance* **2018**, *1*, e201800190. [CrossRef]

80. Reinehr, R.; Zoller, S.; Klonowski-Stumpe, H.; Kordes, C.; Haussinger, D. Effects of angiotensin ii on rat pancreatic stellate cells. *Pancreas* **2004**, *28*, 129–137. [CrossRef]

81. Masamune, A.; Hamada, S.; Kikuta, K.; Takikawa, T.; Miura, S.; Nakano, E.; Shimosegawa, T. The angiotensin ii type i receptor blocker olmesartan inhibits the growth of pancreatic cancer by targeting stellate cell activities in mice. *Scand. J. Gastroenterol.* **2013**, *48*, 602–609. [CrossRef] [PubMed]

82. Ramakrishnan, P.; Loh, W.M.; Gopinath, S.C.B.; Bonam, S.R.; Fareez, I.M.; Mac Guad, R.; Sim, M.S.; Wu, Y.S. Selective phytochemicals targeting pancreatic stellate cells as new anti- fibrotic agents for chronic pancreatitis and pancreatic cancer. *Acta Pharm. Sin. B* **2020**, *10*, 399–413. [CrossRef]

83. Elechalawar, C.K.; Hossen, M.N.; Shankarappa, P.; Peer, C.J.; Figg, W.D.; Robertson, J.D.; Bhattacharya, R.; Mukherjee, P. Targeting pancreatic cancer cells and stellate cells using designer nanotherapeutics in vitro. *Int. J. Nanomed.* **2020**, *15*, 991–1003. [CrossRef]

84. McCarthy, E.F. The toxins of william b. Coley and the treatment of bone and soft-tissue sarcomas. *Iowa. Orthop. J.* **2006**, *26*, 154–158. [PubMed]

85. Caswell, C.C.; Oliver-Kozup, H.; Han, R.; Lukomska, E.; Lukomski, S. Scl1, the multifunctional adhesin of group a streptococcus, selectively binds cellular fibronectin and laminin, and mediates pathogen internalization by human cells. *FEMS Microbiol. Lett.* **2010**, *303*, 61–68. [CrossRef]

86. McNitt, D.H.; Choi, S.J.; Keene, D.R.; Van De Water, L.; Squeglia, F.; Berisio, R.; Lukomski, S. Surface-exposed loops and an acidic patch in the scl1 protein of group a streptococcus enable scl1 binding to wound-associated fibronectin. *J. Biol. Chem.* **2018**, *293*, 7796–7810. [CrossRef] [PubMed]

87. McNitt, D.H.; Choi, S.J.; Allen, J.L.; Hames, R.A.; Weed, S.A.; Van De Water, L.; Berisio, R.; Lukomski, S. Adaptation of the group a streptococcus adhesin scl1 to bind fibronectin type iii repeats within wound-associated extracellular matrix: Implications for cancer therapy. *Mol. Microbiol.* **2019**, *112*, 800–819. [CrossRef] [PubMed]

88. McNitt, D.H.; Van De Water, L.; Marasco, D.; Berisio, R.; Lukomski, S. Streptococcal collagen-like protein 1 binds wound fibronectin: Implications in pathogen targeting. *Curr. Med. Chem.* **2019**, *26*, 1933–1945. [CrossRef]

89. Oliver-Kozup, H.; Martin, K.H.; Schwegler-Berry, D.; Green, B.J.; Betts, C.; Shinde, A.V.; Van De Water, L.; Lukomski, S. The group a streptococcal collagen-like protein-1, scl1, mediates biofilm formation by targeting the extra domain a-containing variant of cellular fibronectin expressed in wounded tissue. *Mol. Microbiol.* **2013**, *87*, 672–689. [CrossRef]

90. Oliver-Kozup, H.A.; Elliott, M.; Bachert, B.A.; Martin, K.H.; Reid, S.D.; Schwegler-Berry, D.E.; Green, B.J.; Lukomski, S. The streptococcal collagen-like protein-1 (scl1) is a significant determinant for biofilm formation by group a streptococcus. *BMC Microbiol.* **2011**, *11*, 262. [CrossRef] [PubMed]

91. Bachert, B.A.; Choi, S.J.; LaSala, P.R.; Harper, T.I.; McNitt, D.H.; Boehm, D.T.; Caswell, C.C.; Ciborowski, P.; Keene, D.R.; Flores, A.R.; et al. Unique footprint in the scl1.3 locus affects adhesion and biofilm formation of the invasive m3-type group a streptococcus. *Front. Cell Infect. Microbiol.* **2016**, *6*, 90. [CrossRef] [PubMed]

92. Gopal, S.; Veracini, L.; Grall, D.; Butori, C.; Schaub, S.; Audebert, S.; Camoin, L.; Baudelet, E.; Radwanska, A.; Beghelli-de la Forest Divonne, S.; et al. Fibronectin-guided migration of carcinoma collectives. *Nat. Commun.* **2017**, *8*, 14105. [CrossRef]

93. Astrof, S.; Crowley, D.; George, E.L.; Fukuda, T.; Sekiguchi, K.; Hanahan, D.; Hynes, R.O. Direct test of potential roles of eiiia and eiiib alternatively spliced segments of fibronectin in physiological and tumor angiogenesis. *Mol. Cell Biol.* **2004**, *24*, 8662–8670. [CrossRef] [PubMed]

94. Kumra, H.; Reinhardt, D.P. Fibronectin-targeted drug delivery in cancer. *Adv. Drug Deliv. Rev.* **2016**, *97*, 101–110. [CrossRef]

95. Han, Z.; Zhang, S.; Fujiwara, K.; Zhang, J.; Li, Y.; Liu, J.; van Zijl, P.C.M.; Lu, Z.R.; Zheng, L.; Liu, G. Extradomain-b fibronectin-targeted dextran-based chemical exchange saturation transfer magnetic resonance imaging probe for detecting pancreatic cancer. *Bioconjug. Chem.* **2019**, *30*, 1425–1433. [CrossRef] [PubMed]

96. Qiao, P.; Ayat, N.R.; Vaidya, A.; Gao, S.; Sun, W.; Chou, S.; Han, Z.; Gilmore, H.; Winter, J.M.; Lu, Z.R. Magnetic resonance molecular imaging of extradomain b fibronectin improves imaging of pancreatic cancer tumor xenografts. *Front. Oncol.* **2020**, *10*, 586727. [CrossRef]

97. Lei, X.; Lei, Y.; Li, J.K.; Du, W.X.; Li, R.G.; Yang, J.; Li, J.; Li, F.; Tan, H.B. Immune cells within the tumor microenvironment: Biological functions and roles in cancer immunotherapy. *Cancer Lett.* **2020**, *470*, 126–133. [CrossRef]

98. Wu, T.; Dai, Y. Tumor microenvironment and therapeutic response. *Cancer Lett.* **2017**, *387*, 61–68. [CrossRef]

99. Diegelmann, R.F.; Evans, M.C. Wound healing: An overview of acute, fibrotic and delayed healing. *Front. Biosci.* **2004**, *9*, 283–289. [CrossRef]

100. Tazzyman, S.; Lewis, C.E.; Murdoch, C. Neutrophils: Key mediators of tumour angiogenesis. *Int. J. Exp. Pathol.* **2009**, *90*, 222–231. [CrossRef]

101. Malech, H.L.; Deleo, F.R.; Quinn, M.T. The role of neutrophils in the immune system: An overview. *Methods Mol. Biol.* **2014**, *1124*, 3–10.

102. Nemeth, T.; Sperandio, M.; Mocsai, A. Neutrophils as emerging therapeutic targets. *Nat. Rev. Drug Discov.* **2020**, *19*, 253–275. [CrossRef]

103. Nywening, T.M.; Belt, B.A.; Cullinan, D.R.; Panni, R.Z.; Han, B.J.; Sanford, D.E.; Jacobs, R.C.; Ye, J.; Patel, A.A.; Gillanders, W.E.; et al. Targeting both tumour-associated cxcr2(+) neutrophils and ccr2(+) macrophages disrupts myeloid recruitment and improves chemotherapeutic responses in pancreatic ductal adenocarcinoma. *Gut* **2018**, *67*, 1112–1123. [CrossRef]

104. Wang, T.T.; Zhao, Y.L.; Peng, L.S.; Chen, N.; Chen, W.; Lv, Y.P.; Mao, F.Y.; Zhang, J.Y.; Cheng, P.; Teng, Y.S.; et al. Tumour-activated neutrophils in gastric cancer foster immune suppression and disease progression through gm-csf-pd-l1 pathway. *Gut* **2017**, *66*, 1900–1911. [CrossRef]

105. Mizuno, R.; Kawada, K.; Itatani, Y.; Ogawa, R.; Kiyasu, Y.; Sakai, Y. The role of tumor-associated neutrophils in colorectal cancer. *Int. J. Mol. Sci.* **2019**, *20*, 529. [CrossRef]

106. Masucci, M.T.; Minopoli, M.; Carriero, M.V. Tumor associated neutrophils. Their role in tumorigenesis, metastasis, prognosis and therapy. *Front. Oncol.* **2019**, *9*, 1146. [CrossRef] [PubMed]

107. Liang, W.; Ferrara, N. The complex role of neutrophils in tumor angiogenesis and metastasis. *Cancer Immunol. Res.* **2016**, *4*, 83–91. [CrossRef]

108. Brinkmann, V.; Reichard, U.; Goosmann, C.; Fauler, B.; Uhlemann, Y.; Weiss, D.S.; Weinrauch, Y.; Zychlinsky, A. Neutrophil extracellular traps kill bacteria. *Science* **2004**, *303*, 1532–1535. [CrossRef] [PubMed]

109. Sorensen, O.E.; Borregaard, N. Neutrophil extracellular traps-the dark side of neutrophils. *J. Clin. Investig.* **2016**, *126*, 1612–1620. [CrossRef] [PubMed]

110. Papayannopoulos, V. Neutrophil extracellular traps in immunity and disease. *Nat. Rev. Immunol* **2018**, *18*, 134–147. [CrossRef] [PubMed]
111. Pilsczek, F.H.; Salina, D.; Poon, K.K.; Fahey, C.; Yipp, B.G.; Sibley, C.D.; Robbins, S.M.; Green, F.H.; Surette, M.G.; Sugai, M.; et al. A novel mechanism of rapid nuclear neutrophil extracellular trap formation in response to staphylococcus aureus. *J. Immunol.* **2010**, *185*, 7413–7425. [CrossRef] [PubMed]
112. Wang, S.; Wang, Y. Peptidylarginine deiminases in citrullination, gene regulation, health and pathogenesis. *Biochim. Biophys. Acta* **2013**, *1829*, 1126–1135. [CrossRef]
113. Takesue, S.; Ohuchida, K.; Shinkawa, T.; Otsubo, Y.; Matsumoto, S.; Sagara, A.; Yonenaga, A.; Ando, Y.; Kibe, S.; Nakayama, H.; et al. Neutrophil extracellular traps promote liver micrometastasis in pancreatic ductal adenocarcinoma via the activation of cancerassociated fibroblasts. *Int. J. Oncol.* **2020**, *56*, 596–605.
114. Miller-Ocuin, J.L.; Liang, X.; Boone, B.A.; Doerfler, W.R.; Singhi, A.D.; Tang, D.; Kang, R.; Lotze, M.T.; Zeh, H.J., 3rd. DNA released from neutrophil extracellular traps (nets) activates pancreatic stellate cells and enhances pancreatic tumor growth. *Oncoimmunology* **2019**, *8*, e1605822. [CrossRef]
115. Boone, B.A.; Orlichenko, L.; Schapiro, N.E.; Loughran, P.; Gianfrate, G.C.; Ellis, J.T.; Singhi, A.D.; Kang, R.; Tang, D.; Lotze, M.T.; et al. The receptor for advanced glycation end products (rage) enhances autophagy and neutrophil extracellular traps in pancreatic cancer. *Cancer Gene Ther.* **2015**, *22*, 326–334. [CrossRef]
116. Jin, W.; Xu, H.X.; Zhang, S.R.; Li, H.; Wang, W.Q.; Gao, H.L.; Wu, C.T.; Xu, J.Z.; Qi, Z.H.; Li, S.; et al. Tumor-infiltrating nets predict postsurgical survival in patients with pancreatic ductal adenocarcinoma. *Ann. Surg. Oncol.* **2019**, *26*, 635–643. [CrossRef]
117. Jung, H.S.; Gu, J.; Kim, J.E.; Nam, Y.; Song, J.W.; Kim, H.K. Cancer cell-induced neutrophil extracellular traps promote both hypercoagulability and cancer progression. *PLoS ONE* **2019**, *14*, e0216055. [CrossRef]
118. Hisada, Y.; Grover, S.P.; Maqsood, A.; Houston, R.; Ay, C.; Noubouossie, D.F.; Cooley, B.C.; Wallen, H.; Key, N.S.; Thalin, C.; et al. Neutrophils and neutrophil extracellular traps enhance venous thrombosis in mice bearing human pancreatic tumors. *Haematologica* **2020**, *105*, 218–225. [CrossRef]
119. Boone, B.A.; Murthy, P.; Miller-Ocuin, J.; Doerfler, W.R.; Ellis, J.T.; Liang, X.; Ross, M.A.; Wallace, C.T.; Sperry, J.L.; Lotze, M.T.; et al. Chloroquine reduces hypercoagulability in pancreatic cancer through inhibition of neutrophil extracellular traps. *BMC Cancer* **2018**, *18*, 678. [CrossRef]
120. Abdol Razak, N.; Elaskalani, O.; Metharom, P. Pancreatic cancer-induced neutrophil extracellular traps: A potential contributor to cancer-associated thrombosis. *Int. J. Mol. Sci.* **2017**, *18*, 487. [CrossRef] [PubMed]
121. Kajioka, H.; Kagawa, S.; Ito, A.; Yoshimoto, M.; Sakamoto, S.; Kikuchi, S.; Kuroda, S.; Yoshida, R.; Umeda, Y.; Noma, K.; et al. Targeting neutrophil extracellular traps with thrombomodulin prevents pancreatic cancer metastasis. *Cancer Lett.* **2021**, *497*, 1–13. [CrossRef]
122. Yang, L.; Liu, Q.; Zhang, X.; Liu, X.; Zhou, B.; Chen, J.; Huang, D.; Li, J.; Li, H.; Chen, F.; et al. DNA of neutrophil extracellular traps promotes cancer metastasis via ccdc25. *Nature* **2020**, *583*, 133–138. [CrossRef]
123. Albrengues, J.; Shields, M.A.; Ng, D.; Park, C.G.; Ambrico, A.; Poindexter, M.E.; Upadhyay, P.; Uyeminami, D.L.; Pommier, A.; Kuttner, V.; et al. Neutrophil extracellular traps produced during inflammation awaken dormant cancer cells in mice. *Science* **2018**, *361*, eaao4227. [CrossRef]
124. Zhang, Y.; Chandra, V.; Riquelme Sanchez, E.; Dutta, P.; Quesada, P.R.; Rakoski, A.; Zoltan, M.; Arora, N.; Baydogan, S.; Horne, W.; et al. Interleukin-17-induced neutrophil extracellular traps mediate resistance to checkpoint blockade in pancreatic cancer. *J. Exp. Med.* **2020**, *217*, e20190354. [CrossRef]
125. Teijeira, A.; Garasa, S.; Gato, M.; Alfaro, C.; Migueliz, I.; Cirella, A.; de Andrea, C.; Ochoa, M.C.; Otano, I.; Etxeberria, I.; et al. Cxcr1 and cxcr2 chemokine receptor agonists produced by tumors induce neutrophil extracellular traps that interfere with immune cytotoxicity. *Immunity* **2020**, *52*, 856–871.e858. [CrossRef]
126. Zambirinis, C.P.; Levie, E.; Nguy, S.; Avanzi, A.; Barilla, R.; Xu, Y.; Seifert, L.; Daley, D.; Greco, S.H.; Deutsch, M.; et al. Tlr9 ligation in pancreatic stellate cells promotes tumorigenesis. *J. Exp. Med.* **2015**, *212*, 2077–2094. [CrossRef] [PubMed]
127. Azevedo, E.P.; Guimaraes-Costa, A.B.; Torezani, G.S.; Braga, C.A.; Palhano, F.L.; Kelly, J.W.; Saraiva, E.M.; Foguel, D. Amyloid fibrils trigger the release of neutrophil extracellular traps (nets), causing fibril fragmentation by net-associated elastase. *J. Biol Chem.* **2012**, *287*, 37206–37218. [CrossRef] [PubMed]
128. Munir, H.; Jones, J.O.; Janowitz, T.; Hoffmann, M.; Euler, M.; Martins, C.P.; Welsh, S.J.; Shields, J.D. Stromal-driven and amyloid beta-dependent induction of neutrophil extracellular traps modulates tumor growth. *Nat. Commun.* **2021**, *12*, 683. [CrossRef] [PubMed]
129. Schafer, M.; Werner, S. The cornified envelope: A first line of defense against reactive oxygen species. *J. Investig. Dermatol.* **2011**, *131*, 1409–1411. [CrossRef]
130. Li, Y.; Yuan, R.; Ren, T.; Yang, B.; Miao, H.; Liu, L.; Cai, C.; Yang, Y.; Hu, Y.; Jiang, C.; et al. Role of sciellin in gallbladder cancer proliferation and formation of neutrophil extracellular traps. *Cell Death Dis.* **2021**, *12*, 30. [CrossRef]
131. Cheng, Y.; Wang, K.; Geng, L.; Sun, J.; Xu, W.; Liu, D.; Gong, S.; Zhu, Y. Identification of candidate diagnostic and prognostic biomarkers for pancreatic carcinoma. *EBioMedicine* **2019**, *40*, 382–393. [CrossRef]
132. de Bont, C.M.; Boelens, W.C.; Pruijn, G.J.M. Netosis, complement, and coagulation: A triangular relationship. *Cell Mol. Immunol.* **2019**, *16*, 19–27. [CrossRef]

133. Shrestha, B.; Ito, T.; Kakuuchi, M.; Totoki, T.; Nagasato, T.; Yamamoto, M.; Maruyama, I. Recombinant thrombomodulin suppresses histone-induced neutrophil extracellular trap formation. *Front. Immunol.* **2019**, *10*, 2535. [CrossRef]

134. Helms, J.; Clere-Jehl, R.; Bianchini, E.; Le Borgne, P.; Burban, M.; Zobairi, F.; Diehl, J.L.; Grunebaum, L.; Toti, F.; Meziani, F.; et al. Thrombomodulin favors leukocyte microvesicle fibrinolytic activity, reduces netosis and prevents septic shock-induced coagulopathy in rats. *Ann. Intensive. Care* **2017**, *7*, 118. [CrossRef]

135. Liu, S.; Su, X.; Pan, P.; Zhang, L.; Hu, Y.; Tan, H.; Wu, D.; Liu, B.; Li, H.; Li, Y.; et al. Neutrophil extracellular traps are indirectly triggered by lipopolysaccharide and contribute to acute lung injury. *Sci. Rep.* **2016**, *6*, 37252. [CrossRef]

136. Lefrancais, E.; Mallavia, B.; Zhuo, H.; Calfee, C.S.; Looney, M.R. Maladaptive role of neutrophil extracellular traps in pathogen-induced lung injury. *JCI Insight* **2018**, *3*, e98178. [CrossRef]

137. Allegra, A.; Innao, V.; Allegra, A.G.; Musolino, C. Coagulopathy and thromboembolic events in patients with sars-cov-2 infection: Pathogenesis and management strategies. *Ann. Hematol.* **2020**, *99*, 1953–1965. [CrossRef]

138. Park, H.H.; Park, W.; Lee, Y.Y.; Kim, H.; Seo, H.S.; Choi, D.W.; Kwon, H.K.; Na, D.H.; Kim, T.H.; Choy, Y.B.; et al. Bioinspired dnase-i-coated melanin-like nanospheres for modulation of infection-associated netosis dysregulation. *Adv. Sci. (Weinh.)* **2020**, *7*, 2001940. [CrossRef]

139. Xia, Y.; He, J.; Zhang, H.; Wang, H.; Tetz, G.; Maguire, C.A.; Wang, Y.; Onuma, A.; Genkin, D.; Tetz, V.; et al. Aav-mediated gene transfer of dnase i in the liver of mice with colorectal cancer reduces liver metastasis and restores local innate and adaptive immune response. *Mol. Oncol.* **2020**, *14*, 2920–2935. [CrossRef]

140. Smith, C.K.; Vivekanandan-Giri, A.; Tang, C.; Knight, J.S.; Mathew, A.; Padilla, R.L.; Gillespie, B.W.; Carmona-Rivera, C.; Liu, X.; Subramanian, V.; et al. Neutrophil extracellular trap-derived enzymes oxidize high-density lipoprotein: An additional proatherogenic mechanism in systemic lupus erythematosus. *Arthritis. Rheumatol.* **2014**, *66*, 2532–2544. [CrossRef]

141. Suzuki, M.; Ikari, J.; Anazawa, R.; Tanaka, N.; Katsumata, Y.; Shimada, A.; Suzuki, E.; Tatsumi, K. Pad4 deficiency improves bleomycin-induced neutrophil extracellular traps and fibrosis in mouse lung. *Am. J. Respir. Cell Mol. Biol.* **2020**, *63*, 806–818. [CrossRef] [PubMed]

142. Franken, L.; Schiwon, M.; Kurts, C. Macrophages: Sentinels and regulators of the immune system. *Cell Microbiol.* **2016**, *18*, 475–487. [CrossRef]

143. van Dalen, F.J.; van Stevendaal, M.; Fennemann, F.L.; Verdoes, M.; Ilina, O. Molecular repolarisation of tumour-associated macrophages. *Molecules* **2018**, *24*, 9. [CrossRef] [PubMed]

144. Najafi, M.; Hashemi Goradel, N.; Farhood, B.; Salehi, E.; Nashtaei, M.S.; Khanlarkhani, N.; Khezri, Z.; Majidpoor, J.; Abouzaripour, M.; Habibi, M.; et al. Macrophage polarity in cancer: A review. *J. Cell Biochem.* **2019**, *120*, 2756–2765. [CrossRef]

145. Xue, J.; Schmidt, S.V.; Sander, J.; Draffehn, A.; Krebs, W.; Quester, I.; De Nardo, D.; Gohel, T.D.; Emde, M.; Schmidleithner, L.; et al. Transcriptome-based network analysis reveals a spectrum model of human macrophage activation. *Immunity* **2014**, *40*, 274–288. [CrossRef] [PubMed]

146. Chambers, M.; Rees, A.; Cronin, J.G.; Nair, M.; Jones, N.; Thornton, C.A. Macrophage plasticity in reproduction and environmental influences on their function. *Front. Immunol.* **2020**, *11*, 607328. [CrossRef] [PubMed]

147. Wang, J.; Li, D.; Cang, H.; Guo, B. Crosstalk between cancer and immune cells: Role of tumor-associated macrophages in the tumor microenvironment. *Cancer Med.* **2019**, *8*, 4709–4721. [CrossRef]

148. Malekghasemi, S.; Majidi, J.; Baghbanzadeh, A.; Abdolalizadeh, J.; Baradaran, B.; Aghebati-Maleki, L. Tumor-associated macrophages: Protumoral macrophages in inflammatory tumor microenvironment. *Adv. Pharm. Bull.* **2020**, *10*, 556–565. [CrossRef]

149. Pandol, S.J.; Edderkaoui, M. What are the macrophages and stellate cells doing in pancreatic adenocarcinoma? *Front. Physiol.* **2015**, *6*, 125. [CrossRef]

150. Shi, C.; Washington, M.K.; Chaturvedi, R.; Drosos, Y.; Revetta, F.L.; Weaver, C.J.; Buzhardt, E.; Yull, F.E.; Blackwell, T.S.; Sosa-Pineda, B.; et al. Fibrogenesis in pancreatic cancer is a dynamic process regulated by macrophage-stellate cell interaction. *Lab. Investig.* **2014**, *94*, 409–421. [CrossRef]

151. Zhu, Y.; Herndon, J.M.; Sojka, D.K.; Kim, K.W.; Knolhoff, B.L.; Zuo, C.; Cullinan, D.R.; Luo, J.; Bearden, A.R.; Lavine, K.J.; et al. Tissue-resident macrophages in pancreatic ductal adenocarcinoma originate from embryonic hematopoiesis and promote tumor progression. *Immunity* **2017**, *47*, 323–338.e326. [CrossRef]

152. Lankadasari, M.B.; Mukhopadhyay, P.; Mohammed, S.; Harikumar, K.B. Taming pancreatic cancer: Combat with a double edged sword. *Mol. Cancer* **2019**, *18*, 48. [CrossRef]

153. Tekin, C.; Aberson, H.L.; Waasdorp, C.; Hooijer, G.K.J.; de Boer, O.J.; Dijk, F.; Bijlsma, M.F.; Spek, C.A. Macrophage-secreted mmp9 induces mesenchymal transition in pancreatic cancer cells via par1 activation. *Cell Oncol. (Dordr.)* **2020**, *43*, 1161–1174. [CrossRef]

154. Tandon, M.; Coudriet, G.M.; Criscimanna, A.; Socorro, M.; Eliliwi, M.; Singhi, A.D.; Cruz-Monserrate, Z.; Bailey, P.; Lotze, M.T.; Zeh, H.; et al. Prolactin promotes fibrosis and pancreatic cancer progression. *Cancer Res.* **2019**, *79*, 5316–5327. [CrossRef]

155. Kaneda, M.M.; Cappello, P.; Nguyen, A.V.; Ralainirina, N.; Hardamon, C.R.; Foubert, P.; Schmid, M.C.; Sun, P.; Mose, E.; Bouvet, M.; et al. Macrophage pi3kgamma drives pancreatic ductal adenocarcinoma progression. *Cancer Discov.* **2016**, *6*, 870–885. [CrossRef]

156. Nielsen, S.R.; Quaranta, V.; Linford, A.; Emeagi, P.; Rainer, C.; Santos, A.; Ireland, L.; Sakai, T.; Sakai, K.; Kim, Y.S.; et al. Macrophage-secreted granulin supports pancreatic cancer metastasis by inducing liver fibrosis. *Nat. Cell Biol.* **2016**, *18*, 549–560. [CrossRef] [PubMed]

157. Ucero, A.C.; Bakiri, L.; Roediger, B.; Suzuki, M.; Jimenez, M.; Mandal, P.; Braghetta, P.; Bonaldo, P.; Paz-Ares, L.; Fustero-Torre, C.; et al. Fra-2-expressing macrophages promote lung fibrosis in mice. *J. Clin. Investig.* **2019**, *129*, 3293–3309. [CrossRef]

158. Venneri, M.A.; De Palma, M.; Ponzoni, M.; Pucci, F.; Scielzo, C.; Zonari, E.; Mazzieri, R.; Doglioni, C.; Naldini, L. Identification of proangiogenic tie2-expressing monocytes (tems) in human peripheral blood and cancer. *Blood* **2007**, *109*, 5276–5285. [CrossRef]

159. Pucci, F.; Venneri, M.A.; Biziato, D.; Nonis, A.; Moi, D.; Sica, A.; Di Serio, C.; Naldini, L.; De Palma, M. A distinguishing gene signature shared by tumor-infiltrating tie2-expressing monocytes, blood "resident" monocytes, and embryonic macrophages suggests common functions and developmental relationships. *Blood* **2009**, *114*, 901–914. [CrossRef] [PubMed]

160. Lewis, C.E.; De Palma, M.; Naldini, L. Tie2-expressing monocytes and tumor angiogenesis: Regulation by hypoxia and angiopoietin-2. *Cancer Res.* **2007**, *67*, 8429–8432. [CrossRef] [PubMed]

161. Yang, W.J.; Hao, Y.X.; Yang, X.; Fu, X.L.; Shi, Y.; Yue, H.L.; Yin, P.; Dong, H.L.; Yu, P.W. Overexpression of tie2 is associated with poor prognosis in patients with gastric cancer. *Oncol. Lett.* **2018**, *15*, 8027–8033. [CrossRef]

162. Atanasov, G.; Potner, C.; Aust, G.; Schierle, K.; Dietel, C.; Benzing, C.; Krenzien, F.; Bartels, M.; Eichfeld, U.; Schmelzle, M.; et al. Tie2-expressing monocytes and m2-polarized macrophages impact survival and correlate with angiogenesis in adenocarcinoma of the pancreas. *Oncotarget* **2018**, *9*, 29715–29726. [CrossRef] [PubMed]

163. Coffelt, S.B.; Tal, A.O.; Scholz, A.; De Palma, M.; Patel, S.; Urbich, C.; Biswas, S.K.; Murdoch, C.; Plate, K.H.; Reiss, Y.; et al. Angiopoietin-2 regulates gene expression in tie2-expressing monocytes and augments their inherent proangiogenic functions. *Cancer Res.* **2010**, *70*, 5270–5280. [CrossRef] [PubMed]

164. Habtezion, A.; Edderkaoui, M.; Pandol, S.J. Macrophages and pancreatic ductal adenocarcinoma. *Cancer Lett.* **2016**, *381*, 211–216. [CrossRef] [PubMed]

165. Svensson, J.; Jenmalm, M.C.; Matussek, A.; Geffers, R.; Berg, G.; Ernerudh, J. Macrophages at the fetal-maternal interface express markers of alternative activation and are induced by m-csf and il-10. *J. Immunol.* **2011**, *187*, 3671–3682. [CrossRef] [PubMed]

166. Mantovani, A.; Marchesi, F.; Malesci, A.; Laghi, L.; Allavena, P. Tumour-associated macrophages as treatment targets in oncology. *Nat. Rev. Clin. Oncol.* **2017**, *14*, 399–416. [CrossRef]

167. Calderon, B.; Carrero, J.A.; Ferris, S.T.; Sojka, D.K.; Moore, L.; Epelman, S.; Murphy, K.M.; Yokoyama, W.M.; Randolph, G.J.; Unanue, E.R. The pancreas anatomy conditions the origin and properties of resident macrophages. *J. Exp. Med.* **2015**, *212*, 1497–1512. [CrossRef]

168. Pathria, P.; Louis, T.L.; Varner, J.A. Targeting tumor-associated macrophages in cancer. *Trends Immunol.* **2019**, *40*, 310–327. [CrossRef]

169. Wang, N.; Wang, S.; Wang, X.; Zheng, Y.; Yang, B.; Zhang, J.; Pan, B.; Gao, J.; Wang, Z. Research trends in pharmacological modulation of tumor-associated macrophages. *Clin. Transl. Med.* **2021**, *11*, e288.

170. Sanford, D.E.; Belt, B.A.; Panni, R.Z.; Mayer, A.; Deshpande, A.D.; Carpenter, D.; Mitchem, J.B.; Plambeck-Suess, S.M.; Worley, L.A.; Goetz, B.D.; et al. Inflammatory monocyte mobilization decreases patient survival in pancreatic cancer: A role for targeting the ccl2/ccr2 axis. *Clin. Cancer Res.* **2013**, *19*, 3404–3415. [CrossRef] [PubMed]

171. Nywening, T.M.; Wang-Gillam, A.; Sanford, D.E.; Belt, B.A.; Panni, R.Z.; Cusworth, B.M.; Toriola, A.T.; Nieman, R.K.; Worley, L.A.; Yano, M.; et al. Targeting tumour-associated macrophages with ccr2 inhibition in combination with folfirinox in patients with borderline resectable and locally advanced pancreatic cancer: A single-centre, open-label, dose- finding, non-randomised, phase 1b trial. *Lancet Oncol.* **2016**, *17*, 651–662. [CrossRef]

172. Five Prime Therapeutics Provides Update on Phase 2 Trial of Cabiralizumab Combined with Opdivo®in Pancreatic Cancer. 2020. Available online: https://www.businesswire.com/news/home/20200218005144/en/Five-Prime-Therapeutics-Provides-Update-on-Phase-2-Trial-of-Cabiralizumab-Combined-with-Opdivo%C2%AE-in-Pancreatic-Cancer (accessed on 16 June 2021).

173. Razak, A.R.; Cleary, J.M.; Moreno, V.; Boyer, M.; Calvo Aller, E.; Edenfield, W.; Tie, J.; Harvey, R.D.; Rutten, A.; Shah, M.A.; et al. Safety and efficacy of amg 820, an anti-colony-stimulating factor 1 receptor antibody, in combination with pembrolizumab in adults with advanced solid tumors. *J. Immunother Cancer* **2020**, *8*, e001006. [CrossRef] [PubMed]

174. Yang, M.; Li, J.; Gu, P.; Fan, X. The application of nanoparticles in cancer immunotherapy: Targeting tumor microenvironment. *Bioact. Mater.* **2021**, *6*, 1973–1987. [CrossRef]

175. Zanganeh, S.; Hutter, G.; Spitler, R.; Lenkov, O.; Mahmoudi, M.; Shaw, A.; Pajarinen, J.S.; Nejadnik, H.; Goodman, S.; Moseley, M.; et al. Iron oxide nanoparticles inhibit tumour growth by inducing pro- inflammatory macrophage polarization in tumour tissues. *Nat. Nanotechnol.* **2016**, *11*, 986–994. [CrossRef]

176. Zhao, H.; Zhao, B.; Wu, L.; Xiao, H.; Ding, K.; Zheng, C.; Song, Q.; Sun, L.; Wang, L.; Zhang, Z. Amplified cancer immunotherapy of a surface-engineered antigenic microparticle vaccine by synergistically modulating tumor microenvironment. *ACS Nano* **2019**, *13*, 12553–12566. [CrossRef] [PubMed]

177. Zhou, W.; Zhou, Y.; Chen, X.; Ning, T.; Chen, H.; Guo, Q.; Zhang, Y.; Liu, P.; Li, C.; Chu, Y.; et al. Pancreatic cancer-targeting exosomes for enhancing immunotherapy and reprogramming tumor microenvironment. *Biomaterials* **2021**, *268*, 120546. [CrossRef] [PubMed]

178. Yao, Y.; Zhou, Y.; Liu, L.; Xu, Y.; Chen, Q.; Wang, Y.; Wu, S.; Deng, Y.; Zhang, J.; Shao, A. Nanoparticle-based drug delivery in cancer therapy and its role in overcoming drug resistance. *Front. Mol. Biosci.* **2020**, *7*, 193. [CrossRef]

179. Brodowicz, T. Trabectedin in soft tissue sarcomas. *Future Oncol.* **2014**, *10*, s1–s5. [CrossRef]
180. Germano, G.; Frapolli, R.; Belgiovine, C.; Anselmo, A.; Pesce, S.; Liguori, M.; Erba, E.; Uboldi, S.; Zucchetti, M.; Pasqualini, F.; et al. Role of macrophage targeting in the antitumor activity of trabectedin. *Cancer Cell* **2013**, *23*, 249–262. [CrossRef]
181. Kawaguchi, K.; Igarashi, K.; Murakami, T.; Kiyuna, T.; Lwin, T.M.; Hwang, H.K.; Delong, J.C.; Clary, B.M.; Bouvet, M.; Unno, M.; et al. Mek inhibitors cobimetinib and trametinib, regressed a gemcitabine- resistant pancreatic-cancer patient-derived orthotopic xenograft (pdox). *Oncotarget* **2017**, *8*, 47490–47496. [CrossRef]
182. Borgoni, S.; Iannello, A.; Cutrupi, S.; Allavena, P.; D'Incalci, M.; Novelli, F.; Cappello, P. Depletion of tumor-associated macrophages switches the epigenetic profile of pancreatic cancer infiltrating t cells and restores their anti-tumor phenotype. *Oncoimmunology* **2018**, *7*, e1393596. [CrossRef] [PubMed]
183. Ratan, R.; Patel, S.R. Chemotherapy for soft tissue sarcoma. *Cancer* **2016**, *122*, 2952–2960. [CrossRef] [PubMed]
184. Belli, C.; Piemonti, L.; D'Incalci, M.; Zucchetti, M.; Porcu, L.; Cappio, S.; Doglioni, C.; Allavena, P.; Ceraulo, D.; Maggiora, P.; et al. Phase ii trial of salvage therapy with trabectedin in metastatic pancreatic adenocarcinoma. *Cancer Chemother. Pharmacol.* **2016**, *77*, 477–484. [CrossRef]
185. Long, K.B.; Gladney, W.L.; Tooker, G.M.; Graham, K.; Fraietta, J.A.; Beatty, G.L. Ifny and ccl2 cooperate to redirect tumor-infiltrating monocytes to degrade fibrosis and enhance chemotherapy efficacy in pancreatic carcinoma. *Cancer Discov.* **2016**, *6*, 400–413. [CrossRef]
186. Mardhian, D.F.; Storm, G.; Bansal, R.; Prakash, J. Nano-targeted relaxin impairs fibrosis and tumor growth in pancreatic cancer and improves the efficacy of gemcitabine in vivo. *J. Control. Release* **2018**, *290*, 1–10. [CrossRef]
187. Hu, M.; Wang, Y.; Xu, L.; An, S.; Tang, Y.; Zhou, X.; Li, J.; Liu, R.; Huang, L. Relaxin gene delivery mitigates liver metastasis and synergizes with check point therapy. *Nat. Commun.* **2019**, *10*, 2993. [CrossRef]
188. Zhou, X.; Liu, Y.; Hu, M.; Wang, M.; Liu, X.; Huang, L. Relaxin gene delivery modulates macrophages to resolve cancer fibrosis and synergizes with immune checkpoint blockade therapy. *Sci. Adv.* **2021**, *7*, eabb6596. [CrossRef] [PubMed]
189. Mazzieri, R.; Pucci, F.; Moi, D.; Zonari, E.; Ranghetti, A.; Berti, A.; Politi, L.S.; Gentner, B.; Brown, J.L.; Naldini, L.; et al. Targeting the ang2/tie2 axis inhibits tumor growth and metastasis by impairing angiogenesis and disabling rebounds of proangiogenic myeloid cells. *Cancer Cell* **2011**, *19*, 512–526. [CrossRef] [PubMed]
190. Daly, C.; Eichten, A.; Castanaro, C.; Pasnikowski, E.; Adler, A.; Lalani, A.S.; Papadopoulos, N.; Kyle, A.H.; Minchinton, A.I.; Yancopoulos, G.D.; et al. Angiopoietin-2 functions as a tie2 agonist in tumor models, where it limits the effects of vegf inhibition. *Cancer Res.* **2013**, *73*, 108–118. [CrossRef]
191. Huang, H.; Lai, J.Y.; Do, J.; Liu, D.; Li, L.; Del Rosario, J.; Doppalapudi, V.R.; Pirie-Shepherd, S.; Levin, N.; Bradshaw, C.; et al. Specifically targeting angiopoietin-2 inhibits angiogenesis, tie2-expressing monocyte infiltration, and tumor growth. *Clin. Cancer Res.* **2011**, *17*, 1001–1011. [CrossRef]
192. Harney, A.S.; Karagiannis, G.S.; Pignatelli, J.; Smith, B.D.; Kadioglu, E.; Wise, S.C.; Hood, M.M.; Kaufman, M.D.; Leary, C.B.; Lu, W.P.; et al. The selective tie2 inhibitor rebastinib blocks recruitment and function of tie2(hi) macrophages in breast cancer and pancreatic neuroendocrine tumors. *Mol. Cancer Ther.* **2017**, *16*, 2486–2501. [CrossRef]
193. Zhang, M.; Zhang, S. T cells in fibrosis and fibrotic diseases. *Front. Immunol.* **2020**, *11*, 1142. [CrossRef]
194. Najafi, M.; Farhood, B.; Mortezaee, K. Contribution of regulatory t cells to cancer: A review. *J. Cell Physiol* **2019**, *234*, 7983–7993. [CrossRef] [PubMed]
195. Nunoya, J.; Washburn, M.L.; Kovalev, G.I.; Su, L. Regulatory t cells prevent liver fibrosis during hiv type 1 infection in a humanized mouse model. *J. Infect. Dis.* **2014**, *209*, 1039–1044. [CrossRef] [PubMed]
196. Mitra, A.; Satelli, A.; Yan, J.; Xueqing, X.; Gagea, M.; Hunter, C.A.; Mishra, L.; Li, S. Il-30 (il27p28) attenuates liver fibrosis through inducing nkg2d-rae1 interaction between nkt and activated hepatic stellate cells in mice. *Hepatology* **2014**, *60*, 2027–2039. [CrossRef] [PubMed]
197. Wehr, A.; Baeck, C.; Heymann, F.; Niemietz, P.M.; Hammerich, L.; Martin, C.; Zimmermann, H.W.; Pack, O.; Gassler, N.; Hittatiya, K.; et al. Chemokine receptor cxcr6-dependent hepatic nk t cell accumulation promotes inflammation and liver fibrosis. *J. Immunol.* **2013**, *190*, 5226–5236. [CrossRef]
198. Rahman, A.H.; Aloman, C. Dendritic cells and liver fibrosis. *Biochim. Biophys. Acta* **2013**, *1832*, 998–1004. [CrossRef]
199. Pellicoro, A.; Ramachandran, P.; Iredale, J.P.; Fallowfield, J.A. Liver fibrosis and repair: Immune regulation of wound healing in a solid organ. *Nat. Rev. Immunol.* **2014**, *14*, 181–194. [CrossRef]
200. Koyama, Y.; Brenner, D.A. Liver inflammation and fibrosis. *J. Clin. Investig.* **2017**, *127*, 55–64. [CrossRef]
201. An, Y.; Liu, F.; Chen, Y.; Yang, Q. Crosstalk between cancer-associated fibroblasts and immune cells in cancer. *J. Cell Mol. Med.* **2020**, *24*, 13–24. [CrossRef]
202. Jiang, B.; Zhou, L.; Lu, J.; Wang, Y.; Liu, C.; You, L.; Guo, J. Stroma-targeting therapy in pancreatic cancer: One coin with two sides? *Front. Oncol.* **2020**, *10*, 576399. [CrossRef] [PubMed]

Article

Diagnostic and Prognostic Utility of the Extracellular Vesicles Subpopulations Present in Pleural Effusion

Joman Javadi [1,*], André Görgens [2], Hanna Vanky [1], Dhanu Gupta [2], Anders Hjerpe [1], Samir EL-Andaloussi [2], Daniel Hagey [2,†] and Katalin Dobra [1,†]

[1] Division of Pathology, Department of Laboratory Medicine, Karolinska Institutet, 141 52 Stockholm, Sweden; hanna.hjerpe.vanky@stud.ki.se (H.V.); anders.hjerpe@ki.se (A.H.); katalin.dobra@ki.se (K.D.)

[2] Division of BCM, Department of Laboratory Medicine, Karolinska Institutet, 141 52 Stockholm, Sweden; andre.gorgens@ki.se (A.G.); dhanu.gupta@ki.se (D.G.); Samir.el-andaloussi@ki.se (S.E.-A.); daniel.hagey@ki.se (D.H.)

* Correspondence: joman.javadi@ki.se; Tel.: +46-76-261-5122

† Shared last authors.

Citation: Javadi, J.; Görgens, A.; Vanky, H.; Gupta, D.; Hjerpe, A.; EL-Andaloussi, S.; Hagey, D.; Dobra, K. Diagnostic and Prognostic Utility of the Extracellular Vesicles Subpopulations Present in Pleural Effusion. *Biomolecules* **2021**, *11*, 1606. https://doi.org/10.3390/biom11111606

Academic Editors: George Tzanakakis and Dragana Nikitovic

Received: 8 September 2021
Accepted: 20 October 2021
Published: 29 October 2021

Publisher's Note: MDPI stays neutral with regard to jurisdictional claims in published maps and institutional affiliations.

Abstract: Extracellular vesicles (EVs), comprising exosomes, microvesicles, and apoptotic bodies, are released by all cells into the extracellular matrix and body fluids, where they play important roles in intercellular communication and matrix remodeling in various pathological conditions. Malignant pleural mesothelioma (MPM) is a primary tumor of mesothelial origin, predominantly related to asbestos exposure. The detection of MPM at an early stage and distinguishing it from benign conditions and metastatic adenocarcinomas (AD) is sometimes challenging. Pleural effusion is often the first available biological material and an ideal source for characterizing diagnostic and prognostic factors. Specific proteins have previously been identified as diagnostic markers in effusion, but it is not currently known whether these are associated with vesicles or released in soluble form. Here, we study and characterize tumor heterogeneity and extracellular vesicle diversity in pleural effusion as diagnostic or prognostic markers for MPM. We analyzed extracellular vesicles and soluble proteins from 27 pleural effusions, which were collected and processed at the department of pathology and cytology at Karolinska University Hospital, representing three different patient groups, MPM ($n = 9$), benign ($n = 6$), and AD ($n = 12$). The vesicles were fractionated into apoptotic bodies, microvesicles, and exosomes by differential centrifugation and characterized by nanoparticle tracking analysis and Western blotting. Multiplex bead-based flow cytometry analysis showed that exosomal markers were expressed differently on EVs present in different fractions. Further characterization of exosomes by a multiplex immunoassay (Luminex) showed that all soluble proteins studied were also present in exosomes, though the ratio of protein concentration present in supernatant versus exosomes varied. The proportion of Angiopoietin-1 present in exosomes was generally higher in benign compared to malignant samples. The corresponding ratios of Mesothelin, Galectin-1, Osteopontin, and VEGF were higher in MPM effusions compared to those in the benign group. These findings demonstrate that relevant diagnostic markers can be recovered from exosomes.

Keywords: malignant pleural mesothelioma; pleural effusion; extracellular vesicles; biomarkers

1. Introduction

Malignant pleural mesothelioma (MPM) is an aggressive mesenchymal tumor arising from mesothelial cells of pleura, characterized by the production of hyaluronan and a spectrum of cell surface and matrix proteoglycans, some of which are useful biomarkers facilitating early clinical diagnosis, the monitoring of tumor burden, and effect of therapy. Early diagnosis from pleural effusion is challenging but essential to improving patient survival [1,2]. As pleural effusion is the first available clinical material, cytological examination provides the earliest diagnostic evaluation [3,4] by combining cytomorphology, immunocytochemistry, fluorescence in situ hybridization, biomarker analyses, and electron

157

microscopy [3,5,6]. In addition to early diagnosis, distinguishing MPM from metastatic adenocarcinoma (AD) is another challenge in patients with malignant pleural effusion. CEA, HBME1, TTF1, CK7, and Calretinin are the most useful immunohistochemistry markers for AD [3,7].

The tumor tissue microenvironment is an important and dynamic regulator of tumor progression and development of metastases [8]. This is also true for the malignant effusion, which contains various soluble factors facilitating or inhibiting this process. Cells communicate with their surroundings, thereby regulating the behavior of benign cells and the extracellular matrix [9]. The secretion of factors for paracrine stimulation is one way to achieve such regulation, while the formation extracellular vesicles is another way to transfer signals to neighboring cells [10].

Extracellular vesicles (EVs), comprising exosomes, microvesicles, and apoptotic bodies, are heterogeneous nanoparticles ranging from 30 nm to 4 µm. They are released into the extracellular matrix and body fluids by all cell types but arise via distinct biological processes. For instance, apoptotic bodies are the largest (1–4 µm), as they are formed during programmed cell death. In contrast, all cells release exosomes, which are the smallest (50–100 nm) and formed during the maturation of endosomes into multi-vesicular bodies by inward vesiculation of the endosomal membrane, and microvesicles, which are between 200 nm and 1 µm and formed by outward budding of the plasma membrane [8,11–16]. Although there are no definitive markers to completely separate these vesicle populations, apoptotic bodies are known to contain apoptotic material, such as cleaved caspase 9. It is currently impossible to separate microvesicles and exosomes. However, microvesicles are enriched in the cell surface tetraspanin CD9, while exosomes are enriched in CD81 [17–19].

The released EVs carry functional factors, such as lipids, proteins, and nucleic acids, to recipient cells, which highlights their importance as mediators of cell-to-cell communication [15,20,21]. Vesicles secreted into the extracellular fluid can be taken up by target cells via direct fusion with the plasma membrane or by endocytosis. This transfer of their bioactive content can then regulate intracellular signaling pathways or gene expression in the recipient cells [22]. On the other hand, this EV cargo can also be utilized a biomarker of disease [23].

Pleural effusions contain different cells, including macrophages, lymphocytes, neutrophils, mesothelial cells, as well as, in malignant conditions, tumor cells of various origin [24,25]. All these cells may secrete EVs into the pleura, and bear heterogenous molecular surface markers, such as CD81, CD63, and CD9, proteins, RNAs, or DNA, which can be used as biomarkers [26–28]. We have previously shown the diagnostic applicability of a series of biomarkers and optimized a Luminex based multiparameter battery to assess them [29]. Some of these factors are biologically active, regulating cell growth and angiogenesis. Their appearance in the effusion fluid may reflect the tumor promoting effects of the tumor cells themselves or the reaction of the surrounding benign tissues. Since these factors are potential targets for individualized treatment, it is interesting to understand how the expression of such factors varies from case to case and if the factors are transported in association with different vesicles or released directly into the effusion supernatant.

The aim of the present investigation is to study the nature and variability of these factors in effusion caused by adenocarcinoma, malignant mesothelioma, and benign reactivity. To achieve this, we have used differential centrifugation to separate the cells and different classes of EVs from the soluble components within pleural effusion. We then analyze these components using Western blotting, nanoparticle tracking (NTA), multiplex bead-based flow cytometric EV surface protein profiling, and Luminex antigen detection. A better understanding of the factors related to vesicle mediated intercellular communication may give us tools to monitor cancer progression and identify novel therapeutic targets.

2. Materials and Methods

2.1. Sample Collection and Study Design

Pleural effusions from patients with malignant pleural mesothelioma (MPM; $n = 9$), metastatic lung adenocarcinoma (AD; $n = 12$), and benign reactive mesothelial proliferations (BE; $n = 6$) were evaluated. Pleural effusions were collected at different time point at the Department of Pathology and Cytology, Karolinska University Hospital by thoracocentesis under ethical permit number 2009/1138-31/3. After initial centrifugation, the cell pellets were taken for diagnostic cytology (Figure S1) and clinically established biomarker analysis, while the remaining effusions were kept at 4 °C until processing.

2.2. Fractionation of Extracellular Vesicles

All samples were centrifuged at $300\times g$ for 10 min to collect any remaining cells. Supernatants were centrifuged further at $2000\times g$ for 10 min to isolate apoptotic bodies (ABs) and $10,000\times g$ for 10 min to isolate microvesicles. Further centrifugation at $100,000\times g$ for 90 min sedimented exosomes, while dissolved free proteins remained in the supernatant. The exosome pellets were resuspended in filtered PBS and kept at −80 °C for further analysis. All ultracentrifugation steps were performed using the Beckman Coulter (Brea, CA, USA) Type 70 Ti rotor at 4 °C. Furthermore, supernatant free soluble proteins were concentrated by centrifugal filters (Amicon Ultra-15, REF UFC901024) (Figure 1A).

(A)

Differential centrifugation

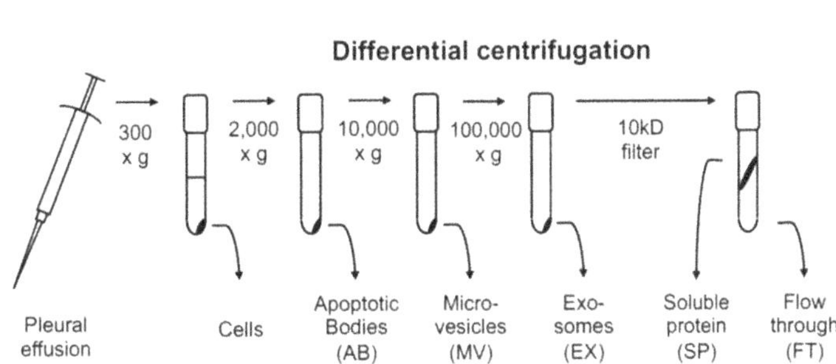

(B)

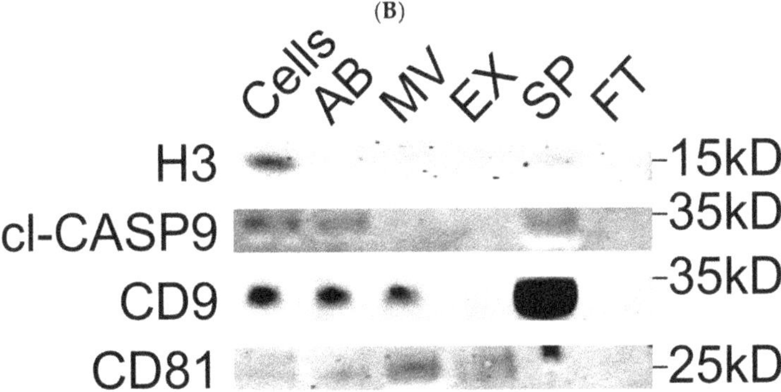

Figure 1. (**A**) Schematic of differential centrifugation protocol. (**B**) Western blot analysis of differential centrifugation fractions. The abundance of common markers (Histone H3 (H3), cleaved caspase-9 (cl-CASP9), CD9 and CD81) in different fractions was determined using Western blot. The molecular weight of each protein is shown on the right.

2.3. NTA Analysis

To ascertain the average concentrations and sizes of exosomes in the samples, nanoparticle tracking analysis (NanoSight Techniques, LM, Malvern, UK) was used according to manufacturer's instructions characterizing nanoparticles from 10–1000 nm in solution. The exosomes were diluted in PBS (1:2500) and applied directly to the NanoSight LM 10.

2.4. Luminex Assay with Human Premixed Multi-Analyte Kit

Two human premixed multi-analyte kits from R&D system were used to assess the levels of 10 different biomarkers in exosomes and soluble protein derived from pleural effusion. The first kit (cat: LXSAHM-09, Minneapolis, MN, USA) was used for analyzing Angiopoietin-1, HGF, MMP-7, Osteopontin, TIMP-1, Galectin, Mesothelin, NRG1-b1, and Syndecan-1 simultaneously. The second kit (cat: LXSAHM-01, Minneapolis, MN, USA) was used for analyzing VEGF. In total, we analyzed 27 pleural effusions, of which 9 were from MPM patients, 12 from AD patients, and 6 were benign effusions. Effusions were diluted 5-fold using the dilution buffer included in the kit. All standards and samples were assayed in duplicate according to the manufacturer's instructions.

2.5. Multiplex Bead-Based EV Flow Cytometry Assay

Different fractions of pleural effusion from MPM, AD, and BE patients (centrifuged at $2000\times g$, $10,000\times g$, and $120,000\times g$) were analyzed by bead-based multiplex EV flow cytometry assays (MACSPlex Exosome Kit, human, Miltenyi Biotec Corston, UK), following the manufacturer's instructions with slight modifications [30]. The MACSPlex Exosome Kit allows detection of 37 surface epitopes (CD1c, CD2, CD3, CD4, CD8, CD9, CD11c, CD14, CD19, CD20, CD24, CD25, CD29, CD31, CD40, CD41b, CD42a, CD44, CD45, CD49e, CD56, CD62P, CD63, CD69, CD81, CD86, CD105, CD133/1, CD142, CD146, CD209, CD326, HLA-ABC, HLA-DRDPDQ, MCSP, ROR1, and SSEA-4), plus two internal isotype controls (mIgG1 and REA). Briefly, samples were incubated with the antibody coated MACSPlex Exosome Capture Beads. Subsequently, EVs bound to the MACSPlex Exosome Capture Beads were labeled with the MACSPlex Pan-Exosome Detection Reagents (APC-conjugated antiCD9, anti-CD63, and anti-CD81 detection antibody). Consequently, these complexes were analyzed by flow cytometry (MACSPlex Analyzer 10 Bergisch Gladbach, Germany) based on the fluorescence characteristics of both beads and the detection reagent, as described previously [30]. Positive signals indicate the abundance of the respective surface epitope on EVs within the sample.

2.6. Western Blot Analysis

To verify the content of the differential centrifugation fractions, Western blotting was performed with antibodies directed against well described cellular components. Albumin and IgG were removed using SpinTrap columns (28-9480-20, GE Healthcare, Buckinghamshire, UK) according to the manufacturer's instructions. Sample fractions were prepared in 0.5M DTT, 8% SDS, 0.4 M Sodium carbonate, and 10% Glycerol and boiled for 5 min at 95 °C. Then, 20 μL of the samples were loaded on a NuPAGE Novex 4–12% (w/v) Bis-Tris pre-cast SDS-PAGE gel (NP0321BOX, Invitrogen, New York, NY, USA). Gels were run for 90 min at 120 V. Gels were transferred to a nitrocellulose membrane using the iBlot system (LC2009, Invitrogen, New York, NY, USA) according to the manufacturer's instructions and then incubated at room temperature for 60 min in blocking buffer (927-70001, Li-Cor, Lincoln, NE, USA). After this point, the membrane was incubated overnight at 4 °C with primary anti-bodies in 1:1 blocking buffer: PBST. We used the following antibodies: CD9 (Abcam ab-92726, 1:2000, Cambridge, UK), cleaved-caspase 9 (Cell Signaling 9505S, 1:1000, Stockholm, Sweden), CD81 (Santa Cruz sc-9158, 1:200, Danvers, MA, USA), and histone H3 (Santa Cruz FL-136, 1:250, Danvers, MA, USA). After washing 3×15 min in PBST, freshly prepared secondary antibody in PBST (Li-Cor 926-68020, 926-152 68071, 926-68079, 1:15,000) was added to the membrane and incubated for 60 min at room temperature followed by 3×15 min washes with PBST. Membranes were developed using both

700- and 800-nm channels on the LI-COR Odyssey Imager and exported from the Li-Cor Image Studio 5.2 software.

3. Results

In order to comprehensively separate the vesicles and soluble proteins associated with BE, MPM, and AD pleural effusion, we collected samples from 27 patients and subjected them to differential centrifugation [31]. Due to the distinct sizes and densities of these components, this allowed us to separate the cells, apoptotic bodies, microvesicles, and exosomes from the soluble proteins within the effusion.

3.1. Validation of Differential Centrifugation Fraction of Pleural Effusion

To verify that the fractions isolated from pleural effusions contained the relevant hypothesized components, we performed Western blot against proteins known to be enriched in each. Although we detected Histone H3 specifically in the cell fraction, cleaved Caspase-9 was found in cells, apoptotic bodies, and soluble protein. As a primarily cell surface tetraspanin, CD9 was detected in cells, apoptotic bodies, microvesicles, and soluble protein. In contrast, CD81 was primarily found in association with the microvesicle and exosome fractions (Figure 1B). These results validated that differential centrifugation was successful in separating the hypothesized components within pleural effusion.

3.2. Measurement of Exosome Particle Size and Concentration

The concentration and size distribution of particles in the exosome fraction were assessed by NTA in 14 patients: five benign, five AD, and four MPM. The size distribution ranged from 30 to 600 nm in diameter and the vesicular concentration varied between individuals (0.6×10^6 to 8.74×10^6 particles/mL) (Figure 2A–C). The mean exosome concentration was 3.8×10^6, 3.1×10^6, and 3.5×10^6 particles/mL for MPM, AD, and benign patients, respectively.

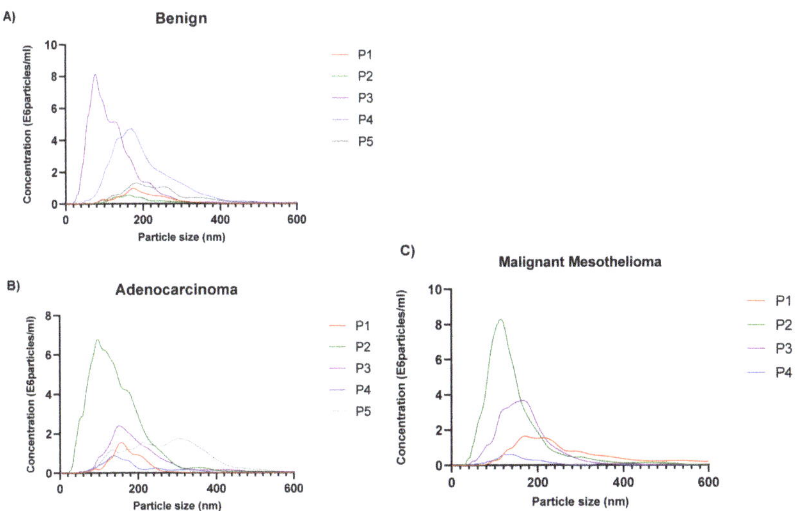

Figure 2. Exosome's size and concentration measured by NTA. Size distribution plots from five benign pleural effusions (0.7×10^6 to 8.7×10^6 particles/mL) (**A**), five adenocarcinoma patients (0.7×10^6 to 7.4×10^6 particles/mL) (**B**) and four malignant mesothelioma patients (0.6×10^6 to 8.7×10^6 particles/mL) (**C**).

3.3. Detection of Surface Protein Markers

To characterize the surface markers present on EVs within the different fractions derived from MPM, AD, and benign patient pleural effusions, we performed a multiplex bead-based flow cytometry assay which was extensively optimized previously [30]. When comparing EV surface protein profiles derived from the different EV subtypes to one another, the highest median fluorescence intensity values were consistently found in the exosome fraction, with the tetraspanins CD63 and CD81 showing the greatest levels. In general, the ratio of CD9 to CD81, as well as the levels of all non-tetraspanin markers decreased during differential centrifugation (Figures 3–5). Although there were still significant CD9, CD63, and CD81 signals in terms of their abundance on EVs, this was not as high as implied by our Western blotting results (Figures 1, 3D, 4D and 5D).

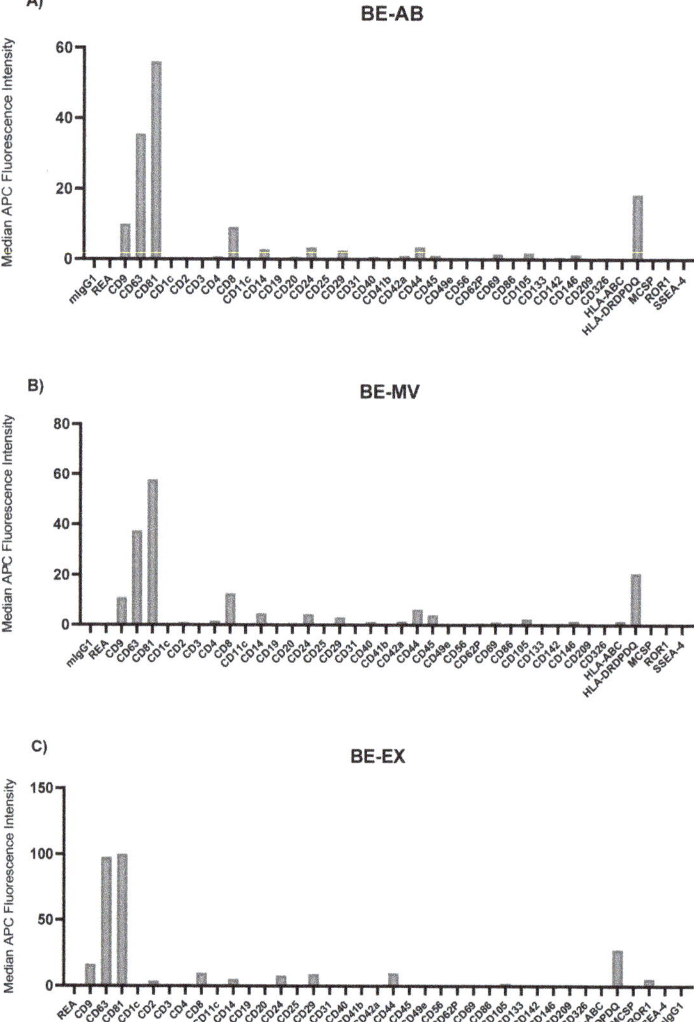

Figure 3. *Cont.*

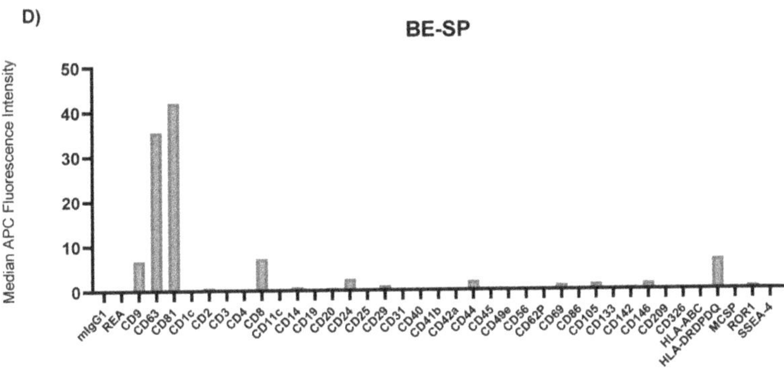

Figure 3. Median fluorescence intensity EV surface protein profiles of benign (BE) pleural effusion differential centrifugation fractions: apoptotic bodies (AB) (**A**), microvesicles (MV) (**B**), exosomes (EX) (**C**) and soluble protein in supernatant (SP) (**D**). mIgG1 and REA indicate isotype control and represent negative markers. Each bar corresponds to one surface epitope measurement, as indicated on the *x*-axis.

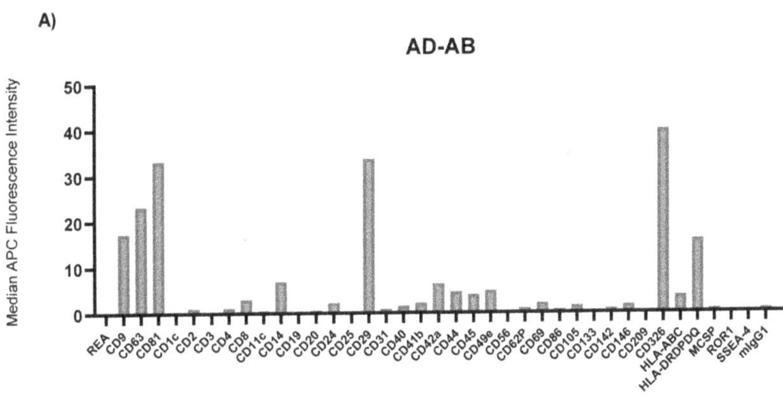

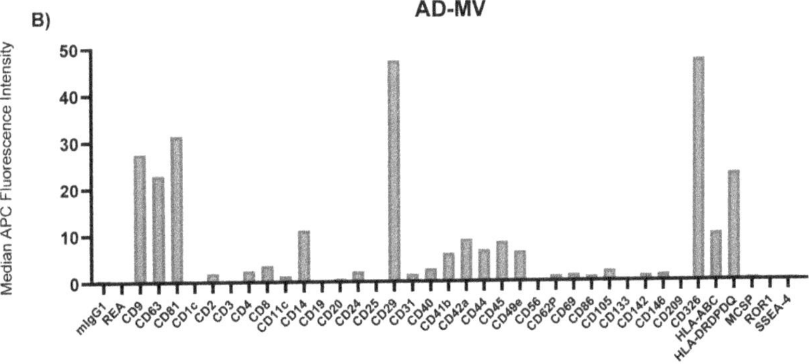

Figure 4. *Cont.*

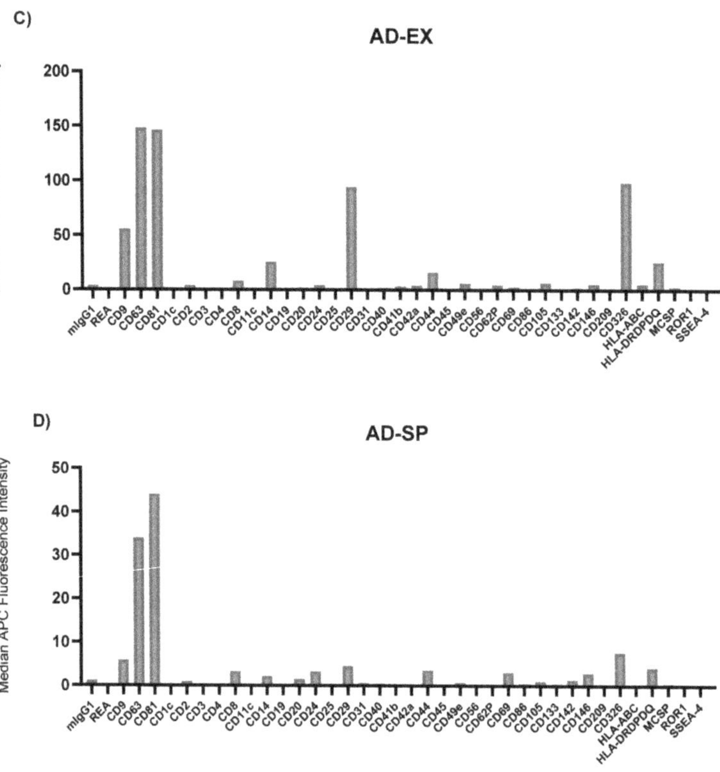

Figure 4. Median fluorescence intensity EV surface protein profiles of adenocarcinoma (AD) pleural effusion differential centrifugation fractions: apoptotic bodies (AB) (**A**), microvesicles (MV) (**B**), exosomes (EX) (**C**) and soluble protein in supernatant (SP) (**D**). mIgG1 and REA indicate isotype control and represent negative markers. Each bar corresponds to one surface epitope measurement, as indicated on the *x*-axis.

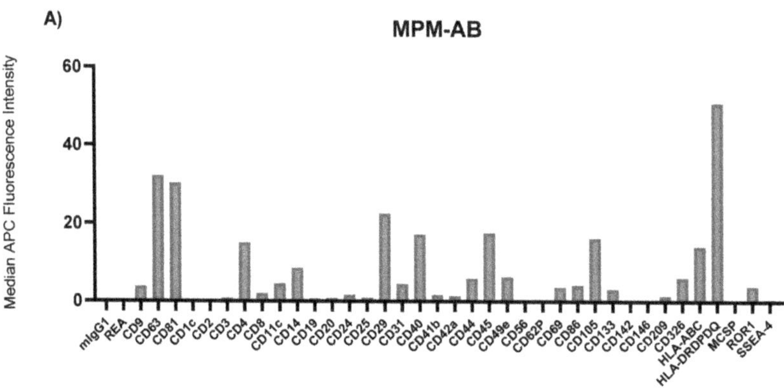

Figure 5. *Cont.*

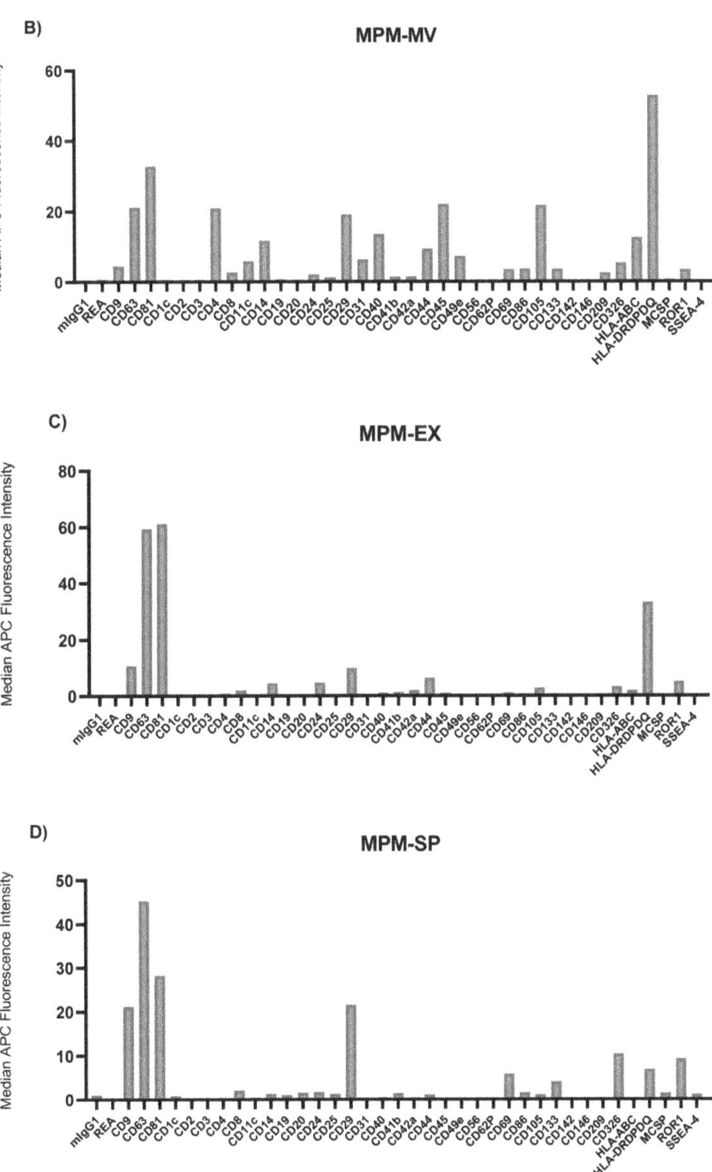

Figure 5. Median fluorescence intensity EV surface protein profiles of malignant pleural mesothelioma (MPM) effusion differential centrifugation fractions: apoptotic bodies (AB) (**A**), microvesicles (MV) (**B**), exosomes (EX) (**C**) and soluble protein in supernatant (SP) (**D**). mIgG1 and REA indicate isotype control and represent negative markers. Each bar corresponds to one surface epitope measurement, as indicated on the *x*-axis.

There were also large differences between the EV surface profiles detected in benign, AD, and MPM pleural effusions. The most notable trend was that the benign effusion was much less complex than those of AD or MPM, such that only the classic EV tetraspanins, CD9, CD63, CD81, as well as HLA-DRDPDQ, were detected above MFI 15 (Figure 3). The AD effusion showed the next most complex profile in addition to the highest absolute levels of EV tetraspanins, and particularly CD9, of any sample. In addition to these, this sample showed very high levels of CD29 and CD326. Although lower, the AD effusion also contained CD14, CD42a, CD44, CD49e, HLA-ABC, and HLA-DRDPDQ above MFI 5 in multiple EV populations (Figure 4). By direct comparison, MPM effusion contained all of the proteins expressed in AD, except CD42a, in addition to CD4, CD40, CD45, and CD105 (Figure 5). Thus, the MPM effusion was the most complex of the samples studied, though it showed the lowest levels of the EV tetraspanins.

3.4. Quantification of Biomarkers in Different Pleural Effusion Fractions

Our previous study showed the diagnostic and prognostic value of 10 different angiogenesis proteins in pleural effusion from MPM, AD, and benign patients. Using Luminex assays, we checked the expression levels of these angiogenesis proteins in microvesicles and exosomes derived from pleural effusion from these three patient groups. We detected that the expression level of Galectin-1, Mesothelin, Osteopontin, VEGF, MMP-7, and HGF were significantly lower in exosomes compared with the supernatant of MPM, AD, and benign conditions. Additionally, there were no significant differences in the expression levels of NRG1-β1, Angiopoietin-1, TIMP-1, and Syndecan-1 in the exosomes compared to supernatant (Figure 6).

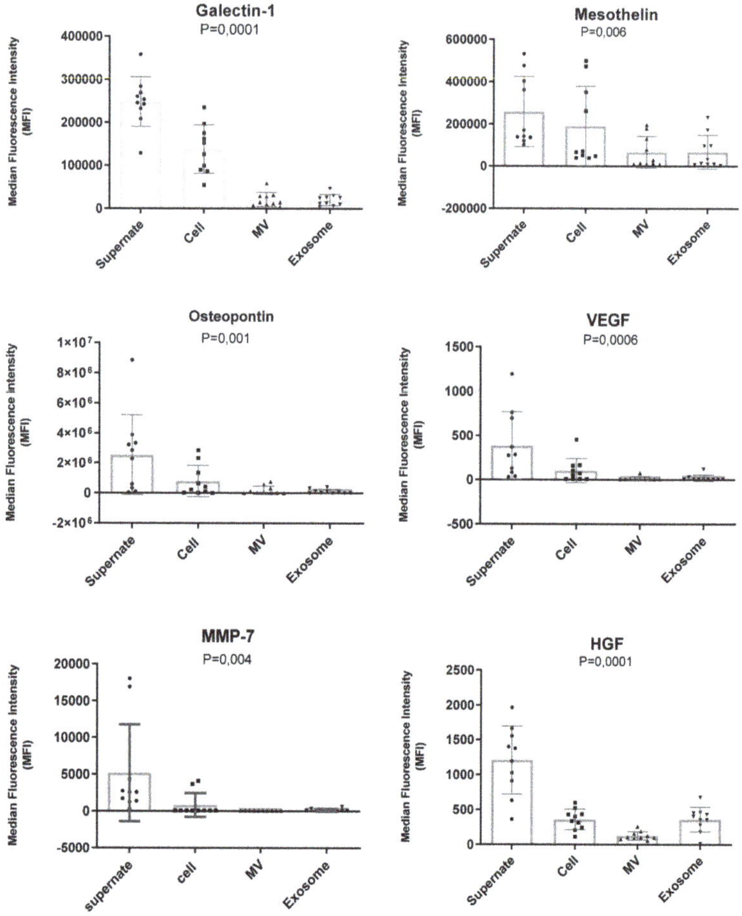

Figure 6. *Cont.*

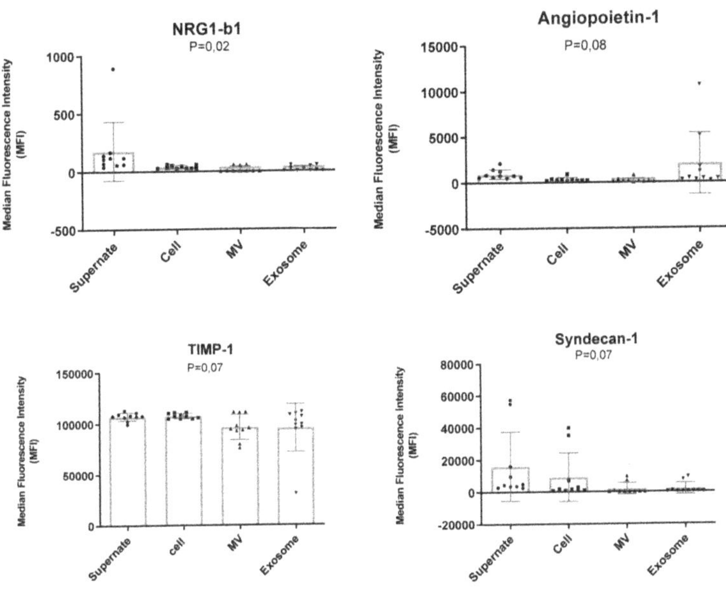

Figure 6. Level of angiogenesis proteins in differential centrifugation fractions of pleural effusion. Levels of Galectin-1, Mesothelin, Osteopontin, VEGF, MMP-7 and HGF were significantly higher in soluble protein when compared with their levels in exosomes. Significance was assessed by two-tailed *t*-test at $p \leq 0.05$.

3.5. Presence of Proteins in Exosomes vs. Supernatant

Next, we focused on whether angiogenesis proteins show different vesicular associations dependent on the disease studied. The results showed that Angiopoietin-1 and TIMP-1 are preferentially transported in exosomes, whereas other proteins occur mainly dissolved in the supernatant with very little in vesicular structures. Osteopontin, Galectin-1, Mesothelin, and VEGF were higher, whereas HGF and SDC-1 were lower in exosomes derived from MPM patients compared to AD and benign patients (Table 1). Interestingly, SDC-1 was represented more in exosomes from AD effusions compared to MPM, while VEGF was represented more in exosomes derived from MPM effusion compared to that from AD patients.

Table 1. Comparison of protein concentration in exosomes and supernatant of specific diseases.

Patient Group	Angiopoietin-1		HGF		Osteopontin		TIMP-1		Galectin-1	
	Exosomes	Supernatant	Exosomes	Supernatant	Exosomes	Supernatant	Exosomes	Supernatant	Exosomes	Supernatant
AD	49.60%	50.40%	32.50%	67.50%	1.70%	98.30%	48.20%	51.80%	6.70%	93.30%
MPM	33.50%	66.50%	17.20%	82.80%	5.50%	94.50%	49.50%	50.50%	9.30%	90.70%
Benign	81%	19%	32.60%	67.40%	2.10%	97.90%	41.90%	58.10%	5.10%	94.90%

Patient Group	Mesothelin		NRG1-b1		SDC-1		VEGF		MMP-7	
	Exosomes	Supernatant	Exosomes	Supernatant	Exosomes	Supernatant	Exosomes	Supernatant	Exosomes	Supernatant
AD	7.50%	92.50%	6.50%	93.50%	12.20%	87.80%	0.00%	100%	2.90%	96.90%
MPM	25.40%	74.60%	10.50%	89.50%	0.00%	100%	12.30%	87.70%	2.90%	93.10%
Benign	10.30%	89.70%	8.10%	91.90%	0.00%	100%	0.00%	100%	8.90%	91%

4. Discussion

Malignant mesothelioma is an aggressive malignancy with limited therapeutic options. In this research project, we studied exosomes isolated from fresh pleural effusion from patients with malignant pleural mesothelioma, metastatic adenocarcinoma, and benign mesothelial proliferations.

Pleural effusion contains a variety of cells among them mesothelial cells, macrophages, lymphocytes, leukocytes, as well as malignant cells when caused by cancer. Cell-to cell communication between malignant cells and malignant to non-malignant cells in effusion may play critical role for cancer progression.

We show that the obtained size and concentration of the exosomes were in accordance with previous studies [32–34]. There were differences in their total concentration. The concentrations of exosomes varied among individuals. In the malignant mesothelioma group, three of the effusions came from the same patient but at different time points. The first effusion showed a higher exosomes concentration, and in the subsequent effusions the amount of the exosomes gradually declined. This can depend on many factors, including tumor burden in the pleural cavity and the rate at which vesicles are formed and the fluid volume is replaced after thoracocentesis. The cellular origin of the extracellular vesicles, including the exosomes, can be identified by the surface proteins that reflect the membrane proteins in the original cells [9]. In this study, we demonstrated that the presence of 15 surface proteins (CD9, CD63, CD81, CD2, CD8, CD14, CD29, CD44, CD49e, CD62p, CD105, CD146, CD326, HLA-ABC, and MCSP) were higher, whereas HLA-DRDPDQ and ROR1 were lower on the exosome fraction derived from metastatic adenocarcinoma patient compared to MPM patient suggesting that these markers can be associated with MPM (Table S1).

These extracellular vesicles carried molecules that regulate a variety of cellular processes, including cell–cell adhesion (CD2, CD8, CD9, and CD146), immune regulation (CD24, CD40), extracellular matrix regulation (CD44), TGF-β receptor activation (CD105), as well as cell growth, differentiation, and migration. The results indicate that all cells present communicate with the pleura as vesicles. The content of these vesicles thus reflects the different regulatory signals present in the effusion.

Additionally, we performed Luminex analysis on differential centrifugation fractions including, apoptotic bodies, microvesicles, exosomes, and supernatant, to measure the concentration of 10 different proteins (among which some were angiogenic-related proteins). In our analysis of extracellular vesicular proteins, Osteopontin, Galectin-1, Mesothelin, and VEGF had higher concentrations in exosomes isolated from MPM patients. In our previous study, we showed that Galectin-1, Mesothelin, Osteopontin, and VEGF have higher levels in malignant pleural mesothelioma compared with the benign patients and Galectin-1 and Mesothelin have higher levels in malignant pleural mesothelioma compared with the metastatic adenocarcinoma patients [29]. These results may explain the higher level of Osteopontin, Galectin-1, Mesothelin, and VEGF in exosomes derived from pleural effusion from MPM patients.

These proteins are biologically active and can regulate signaling pathways in the recipient cells which leads to alteration of their phenotype [35]. For example, mesothelin can inhibit apoptosis by activating PI3K/Akt and MAPK/ERK signaling pathways or increase cell proliferation by activating the Stat3 signaling pathway. The interaction of mesothelin with CA125 (cell surface glycoprotein) facilitates tumor invasion and metastasis. Osteopontin, Galectin-1, and VEGF regulate many cellular processes, including cell proliferation, adhesion, tumor formation, migration, and angiogenesis [36–40].

In addition to this, we showed that SDC-1 and HGF had higher concentrations in exosomes derived from pleural effusion from metastatic adenocarcinoma patients. Exosomes are rich in proteins and previous studies have shown that enzymatic activities and proteins regulate exosomes secretion and mechanisms of proteins loading in exosomes. High heparanase expression influences exosome secretion and regulates their protein contents, among others. SDC-1 and VEGF increase its concentration in exosomes upon heparanase overexpression [41].

The presence of these biologically active proteins may be more stable in exosomes and can actively participate in cell–cell communication by transferring their cargo to recipient cells, eliciting effects related to angiogenesis and therapy resistance. This makes exosomes suitable candidates for future targeting and carriers of precise drug delivery systems. In case these biologically active proteins will be objects for future targeted therapies, it may be important to clarify their possible presence inside vesicles.

5. Conclusions

Exosomes derived from pleural effusion from different patient groups represent different proteins according to their different cell types. Exosomes isolated from pleural effusion from MPM patients have lower levels of CD9, CD63, CD81, CD2, CD8, CD14, CD29, CD44, CD49e, CD62p, CD105, CD146, CD326, HLA-ABC, and MCSP and higher levels of HLA-DRDPDQ and ROR1 surface proteins compared to AD patients. Additionally, Galectin-1, Mesothelin, Osteopontin, and VEGF have higher levels, whereas Angiopoitien-1 has a lower level in exosomes derived from MPM patients compared to the benign patients. Therefore, these proteins can be diagnostic markers for MPM patients.

Supplementary Materials: The following are available online at https://www.mdpi.com/article/10.3390/biom11111606/s1, Figure S1: The diagnostic picture in MPM (a–e), lung adenocarcinoma (f–j) and reactive, but benign, mesothelium (k–n). Immunohistochemical stainings used are Papanicolaou (a,f,k), May Grünewald Giemsa (b,g,l), Calretinin (red)/BerEp4 (brown) (c,h,m), EMA (red)/Desmin (brown) (d,i,n), Mesothelin (e) and TFF-1 (j). In addition to the tumor cells there is benign mesothelium, white blood cells and plenty of macrophages. The mor tumor specific markers (d,e,i,j) show that the tumor cells only make out part of the cell populations. Diagnoses are based on standardized immunocytochemical reaction patterns [42], Table S1: Median Fluorescence Intensity (MFI) values of EV's surface proteins.

Author Contributions: Conceptualization, K.D. and D.H.; methodology, K.D., A.H., D.H., H.V., D.G. and J.J.; software, J.J. and D.H.; validation, J.J. and D.H.; formal analysis, J.J.; investigation, J.J.; writing—original draft preparation, J.J.; writing—review and editing, J.J., K.D., D.H. and A.G.; visualization, J.J. and K.D.; supervision, K.D. and D.H.; project administration, K.D.; funding acquisition, S.E.-A. and K.D. All authors have read and agreed to the published version of the manuscript.

Funding: This research was funded by The Swedish Cancer Society, grant number CAN 2018/653; Stockholm County Council, grant number LS 2015-1198; and The Cancer Society in Stockholm, grant number 174073.

Institutional Review Board Statement: The study was conducted according to the guidelines of the Declaration of Helsinki and approved by the Institutional Ethics Committee of Karolinska Institute (2009/1138-31/3, 2020/10/11).

Informed Consent Statement: Informed consent was obtained from all subjects involved in the study. Written informed consent has been obtained from the patient(s) to publish this paper.

Conflicts of Interest: The authors declare no conflict of interest. The funders had no role in the design of the study; in the collection, analyses, or interpretation of data; in the writing of the manuscript, or in the decision to publish the results.

References

1. Xu, L.L.; Yang, Y.; Wang, Z.; Wang, X.J.; Tong, Z.H.; Shi, H.Z. Malignant pleural mesothelioma: Diagnostic value of medical thoracoscopy and long-term prognostic analysis. *BMC Pulm. Med.* **2018**, *18*, 56. [CrossRef] [PubMed]
2. Kondola, S.; Manners, D.; Nowak, A.K. Malignant pleural mesothelioma: An update on diagnosis and treatment options. *Ther. Adv. Respir. Dis.* **2016**, *10*, 275–288. [CrossRef] [PubMed]
3. Hjerpe, A.; Abd-Own, S.; Dobra, K. Cytopathologic Diagnosis of Epithelioid and Mixed-Type Malignant Mesothelioma: Ten Years of Clinical Experience in Relation to International Guidelines. *Arch. Pathol. Lab. Med.* **2018**, *142*, 893–901. [CrossRef] [PubMed]
4. Porcel, J.M. Diagnosis and characterization of malignant effusions through pleural fluid cytological examination. *Curr. Opin. Pulm. Med.* **2019**, *25*, 362–368. [CrossRef] [PubMed]

5. Mundt, F.; Heidari-Hamedani, G.; Nilsonne, G.; Metintas, M.; Hjerpe, A.; Dobra, K. Diagnostic and prognostic value of soluble syndecan-1 in pleural malignancies. *Biomed Res. Int.* **2014**, *2014*, 419853. [CrossRef] [PubMed]

6. Hjerpe, A.; Ascoli, V.; Bedrossian, C.W.; Boon, M.E.; Creaney, J.; Davidson, B.; Dejmek, A.; Dobra, K.; Fassina, A.; Field, A.; et al. Guidelines for the cytopathologic diagnosis of epithelioid and mixed-type malignant mesothelioma: Complementary Statement from the International Mesothelioma Interest Group, Also Endorsed by the International Academy of Cytology and the Papanicolaou Society of Cytopathology. *Diagn. Cytopathol.* **2015**, *43*, 563–576.

7. Halimi, M.; BeheshtiRouy, S.; Salehi, D.; Rasihashemi, S.Z. The Role of Immunohistochemistry Studies in Distinguishing Malignant Mesothelioma from Metastatic Lung Carcinoma in Malignant Pleural Effusion. *Iran. J. Pathol.* **2019**, *14*, 122–126. [CrossRef]

8. Watabe, S.; Kikuchi, Y.; Morita, S.; Komura, D.; Numakura, S.; Kumagai-Togashi, A.; Watanabe, M.; Matsutani, N.; Kawamura, M.; Yasuda, M.; et al. Clinicopathological significance of microRNA-21 in extracellular vesicles of pleural lavage fluid of lung adenocarcinoma and its functions inducing the mesothelial to mesenchymal transition. *Cancer Med.* **2020**, *9*, 2879–2890. [CrossRef]

9. Stok, U.; Blokar, E.; Lenassi, M.; Holcar, M.; Frank-Bertoncelj, M.; Erman, A.; Resnik, N.; Sodin-Semrl, S.; Cucnik, S.; Pirkmajer, K.P.; et al. Characterization of Plasma-Derived Small Extracellular Vesicles Indicates Ongoing Endothelial and Platelet Activation in Patients with Thrombotic Antiphospholipid Syndrome. *Cells* **2020**, *9*, 1211. [CrossRef] [PubMed]

10. Rilla, K.; Mustonen, A.M.; Arasu, U.T.; Harkonen, K.; Matilainen, J.; Nieminen, P. Extracellular vesicles are integral and functional components of the extracellular matrix. *Matrix Biol.* **2019**, *75*, 201–219. [CrossRef]

11. Lee, J.S.; Hur, J.Y.; Kim, I.A.; Kim, H.J.; Choi, C.M.; Lee, J.C.; Kim, W.S.; Lee, K.Y. Liquid biopsy using the supernatant of a pleural effusion for EGFR genotyping in pulmonary adenocarcinoma patients: A comparison between cell-free DNA and extracellular vesicle-derived DNA. *BMC Cancer* **2018**, *18*, 1236. [CrossRef] [PubMed]

12. Samir, E.L.A.; Mager, I.; Breakefield, X.O.; Wood, M.J. Extracellular vesicles: Biology and emerging therapeutic opportunities. *Nat. Rev. Drug Discov.* **2013**, *12*, 347–357.

13. Guo, M.; Wu, F.; Hu, G.; Chen, L.; Xu, J.; Xu, P.; Wang, X.; Li, Y.; Liu, S.; Zhang, S.; et al. Autologous tumor cell-derived microparticle-based targeted chemotherapy in lung cancer patients with malignant pleural effusion. *Sci. Transl. Med.* **2019**, *11*. [CrossRef]

14. Mathieu, M.; Martin-Jaular, L.; Lavieu, G.; Thery, C. Specificities of secretion and uptake of exosomes and other extracellular vesicles for cell-to-cell communication. *Nat. Cell Biol.* **2019**, *21*, 9–17. [CrossRef] [PubMed]

15. Raposo, G.; Stoorvogel, W. Extracellular vesicles: Exosomes, microvesicles, and friends. *J. Cell Biol.* **2013**, *200*, 373–383. [CrossRef] [PubMed]

16. Nguyen, M.A.; Karunakaran, D.; Geoffrion, M.; Cheng, H.S.; Tandoc, K.; Perisic Matic, L.; Hedin, U.; Maegdefessel, L.; Fish, J.E.; Rayner, K.J. Extracellular Vesicles Secreted by Atherogenic Macrophages Transfer MicroRNA to Inhibit Cell Migration. *Arterioscler. Thromb. Vasc. Biol.* **2018**, *38*, 49–63. [CrossRef] [PubMed]

17. Ryu, F.; Takahashi, T.; Nakamura, K.; Takahashi, Y.; Kobayashi, T.; Shida, S.; Kameyama, T.; Mekada, E. Domain analysis of the tetraspanins: Studies of CD9/CD63 chimeric molecules on subcellular localization and upregulation activity for diphtheria toxin binding. *Cell Struct. Funct.* **2000**, *25*, 317–327. [CrossRef] [PubMed]

18. Kowal, J.; Arras, G.; Colombo, M.; Jouve, M.; Morath, J.P.; Primdal-Bengtson, B.; Dingli, F.; Loew, D.; Tkach, M.; Thery, C. Proteomic comparison defines novel markers to characterize heterogeneous populations of extracellular vesicle subtypes. *Proc. Natl. Acad. Sci. USA* **2016**, *113*, E968–E977. [CrossRef] [PubMed]

19. Greening, D.W.; Xu, R.; Gopal, S.K.; Rai, A.; Simpson, R.J. Proteomic insights into extracellular vesicle biology—Defining exosomes and shed microvesicles. *Expert Rev. Proteom.* **2017**, *14*, 69–95. [CrossRef] [PubMed]

20. Menck, K.; Bleckmann, A.; Schulz, M.; Ries, L.; Binder, C. Isolation and Characterization of Microvesicles from Peripheral Blood. *J. Vis. Exp.* **2017**, *119*, e55057. [CrossRef]

21. Thakur, B.K.; Zhang, H.; Becker, A.; Matei, I.; Huang, Y.; Costa-Silva, B.; Zheng, Y.; Hoshino, A.; Brazier, H.; Xiang, J.; et al. Double-stranded DNA in exosomes: A novel biomarker in cancer detection. *Cell Res.* **2014**, *24*, 766–769. [CrossRef] [PubMed]

22. Freeman, D.W.; Noren Hooten, N.; Eitan, E.; Green, J.; Mode, N.A.; Bodogai, M.; Zhang, Y.; Lehrmann, E.; Zonderman, A.B.; Biragyn, A.; et al. Altered Extracellular Vesicle Concentration, Cargo, and Function in Diabetes. *Diabetes* **2018**, *67*, 2377–2388. [CrossRef] [PubMed]

23. Yuan, T.; Huang, X.; Woodcock, M.; Du, M.; Dittmar, R.; Wang, Y.; Tsai, S.; Kohli, M.; Boardman, L.; Patel, T.; et al. Plasma extracellular RNA profiles in healthy and cancer patients. *Sci. Rep.* **2016**, *6*, 19413. [CrossRef] [PubMed]

24. Noppen, M.; De Waele, M.; Li, R.; Gucht, K.V.; D'Haese, J.; Gerlo, E.; Vincken, W. Volume and cellular content of normal pleural fluid in humans examined by pleural lavage. *Am. J. Respir. Crit. Care Med.* **2000**, *162*, 1023–1026. [CrossRef] [PubMed]

25. Gjomarkaj, M.; Pace, E.; Melis, M.; Spatafora, M.; Toews, G.B. Mononuclear cells in exudative malignant pleural effusions: Characterization of pleural phagocytic cells. *Chest* **1994**, *106*, 1042–1049. [CrossRef] [PubMed]

26. Valadi, H.; Ekstrom, K.; Bossios, A.; Sjostrand, M.; Lee, J.J.; Lotvall, J.O. Exosome-mediated transfer of mRNAs and microRNAs is a novel mechanism of genetic exchange between cells. *Nat. Cell Biol.* **2007**, *9*, 654–659. [CrossRef] [PubMed]

27. Lotvall, J.; Valadi, H. Cell to cell signalling via exosomes through esRNA. *Cell Adh. Migr.* **2007**, *1*, 156–158. [CrossRef]

28. Melo, S.A.; Luecke, L.B.; Kahlert, C.; Fernandez, A.F.; Gammon, S.T.; Kaye, J.; LeBleu, V.S.; Mittendorf, E.A.; Weitz, J.; Rahbari, N.; et al. Glypican-1 identifies cancer exosomes and detects early pancreatic cancer. *Nature* **2015**, *523*, 177–182. [CrossRef] [PubMed]

29. Javadi, J.; Dobra, K.; Hjerpe, A. Multiplex Soluble Biomarker Analysis from Pleural Effusion. *Biomolecules* **2020**, *10*, 1113. [CrossRef]

30. Wiklander, O.P.B.; Bostancioglu, R.B.; Welsh, J.A.; Zickler, A.M.; Murke, F.; Corso, G.; Felldin, U.; Hagey, D.W.; Evertsson, B.; Liang, X.M.; et al. Systematic Methodological Evaluation of a Multiplex Bead-Based Flow Cytometry Assay for Detection of Extracellular Vesicle Surface Signatures. *Front. Immunol.* **2018**, *9*, 1326. [CrossRef] [PubMed]

31. Hagey, D.W.; Kordes, M.; Gorgens, A.; Mowoe, M.O.; Nordin, J.Z.; Moro, C.F.; Lohr, J.M.; El Andaloussi, S. Extracellular vesicles are the primary source of blood-borne tumour-derived mutant KRAS DNA early in pancreatic cancer. *J. Extracell. Vesicles* **2021**, *10*, e12142. [CrossRef] [PubMed]

32. Carnell-Morris, P.; Tannetta, D.; Siupa, A.; Hole, P.; Dragovic, R. Analysis of Extracellular Vesicles Using Fluorescence Nanoparticle Tracking Analysis. *Methods Mol. Biol.* **2017**, *1660*, 153–173.

33. Dragovic, R.A.; Gardiner, C.; Brooks, A.S.; Tannetta, D.S.; Ferguson, D.J.; Hole, P.; Carr, B.; Redman, C.W.; Harris, A.L.; Dobson, P.J.; et al. Sizing and phenotyping of cellular vesicles using Nanoparticle Tracking Analysis. *Nanomedicine* **2011**, *7*, 780–788. [CrossRef] [PubMed]

34. Tiwari, S.; Kumar, V.; Randhawa, S.; Verma, S.K. Preparation and characterization of extracellular vesicles. *Am. J. Reprod. Immunol.* **2021**, *85*, e13367. [CrossRef] [PubMed]

35. Al-Nedawi, K.; Meehan, B.; Micallef, J.; Lhotak, V.; May, L.; Guha, A.; Rak, J. Intercellular transfer of the oncogenic receptor EGFRvIII by microvesicles derived from tumour cells. *Nat. Cell Biol.* **2008**, *10*, 619–624. [CrossRef]

36. Zhao, H.; Chen, Q.; Alam, A.; Cui, J.; Suen, K.C.; Soo, A.P.; Eguchi, S.; Gu, J.; Ma, D. The role of osteopontin in the progression of solid organ tumour. *Cell Death Dis.* **2018**, *9*, 356. [CrossRef] [PubMed]

37. Liu, F.T.; Rabinovich, G.A. Galectins as modulators of tumour progression. *Nat. Rev. Cancer* **2005**, *5*, 29–41. [CrossRef] [PubMed]

38. Thijssen, V.L.; Rabinovich, G.A.; Griffioen, A.W. Vascular galectins: Regulators of tumor progression and targets for cancer therapy. *Cytokine Growth Factor Rev.* **2013**, *24*, 547–558. [CrossRef] [PubMed]

39. Tang, Z.; Qian, M.; Ho, M. The role of mesothelin in tumor progression and targeted therapy. *Anticancer Agents Med. Chem.* **2013**, *13*, 276–280. [CrossRef] [PubMed]

40. Einama, T.; Kamachi, H.; Nishihara, H.; Homma, S.; Kanno, H.; Takahashi, K.; Sasaki, A.; Tahara, M.; Okada, K.; Muraoka, S.; et al. Co-expression of mesothelin and CA125 correlates with unfavorable patient outcome in pancreatic ductal adenocarcinoma. *Pancreas* **2011**, *40*, 1276–1282. [CrossRef] [PubMed]

41. Thompson, C.A.; Purushothaman, A.; Ramani, V.C.; Vlodavsky, I.; Sanderson, R.D. Heparanase regulates secretion, composition, and function of tumor cell-derived exosomes. *J. Biol. Chem.* **2013**, *288*, 10093–10099. [CrossRef] [PubMed]

42. Husain, A.N.; Colby, T.V.; Ordóñez, N.G.; Allen, T.C.; Attanoos, R.L.; Beasley, M.B.; Butnor, K.J.; Chirieac, L.R.; Churg, A.M.; Dacic, S.; et al. Guidelines for Pathologic Diagnosis of Malignant Mesothelioma 2017 Update of the Consensus Statement From the International Mesothelioma Interest Group. *Arch. Pathol. Lab. Med.* **2018**, *142*, 89–108. [CrossRef] [PubMed]

MDPI

St. Alban-Anlage 66

4052 Basel

Switzerland

Tel. +41 61 683 77 34

Fax +41 61 302 89 18

www.mdpi.com

Biomolecules Editorial Office

E-mail: biomolecules@mdpi.com

www.mdpi.com/journal/biomolecules